Sound to Sense, Sense to Sound

A State of the Art in
Sound and Music Computing

Pietro Polotti and Davide Rocchesso, editors

Pietro Polotti and Davide Rocchesso (Editors)
Dipartimento delle Arti e del Disegno Industriale
Università IUAV di Venezia
Dorsoduro 2206
30123 Venezia, Italia

This book has been developed within the Coordination Action $S2S^2$ (Sound to Sense, Sense to Sound) of the 6th Framework Programme - IST Future and Emerging Technologies: http://www.soundandmusiccomputing.org/

The Deutsche Nationalbibliothek lists this publication in the Deutsche Nationalbibliografie; detailed bibliographic data are available in the Internet at http://dnb.d-nb.de.

©Copyright Logos Verlag Berlin GmbH 2008

All rights reserved.

ISBN 978-3-8325-1600-0
Logos Verlag Berlin GmbH
Comeniushof, Gubener Str. 47,
10243 Berlin
Tel.: +49 (0)30 42 85 10 90
Fax: +49 (0)30 42 85 10 92
INTERNET: http://www.logos-verlag.de

Contents

Introduction

The purpose of this book is to give a wide account of state-of-the-art research in the sound domain, with a proper combination of human sciences, computational sciences, and technological realizations. This was one of the objectives of the European Coordination Action "Sound to Sense, Sense to Sound" ($S2S^2$) [1], altogether with the definition and the publication of a roadmap[2], aiming at providing a reference point for future research in Sound and Music Computing. The overall goal of the Action was to contribute to the formation of a solid and organic community focused on tying together the investigation on the sonic aspects of the world with the study of the sense conveyed through sound. As suggested by the title of the Action, the perspective is two-folded: from sound to sense, and from sense to sound. What we expect from the future is the emergence of methodologies for a higher-level sound analysis and processing, or for more engaging sound synthesis and control in musical and non-musical contexts. Eventually, only an integrated yet open sound-music research field may fill the existing gap between sound and sense and try to answer to the urgent requirements, coming from the world of pervasive and mobile technologies. Nowadays, a wide variety of techniques is available to

---

[1]The $S2S^2$ FET-open Coordination Action IST-2004-03773 was supported by the European Commission during a period of three years, from June 2004 until May 2007 (`http://www.soundandmusiccomputing.org`) involving partners from eleven research institutions through Europe, with the goal of forming a new and strongly multidisciplinary community around the main themes of sound and music computing research.

[2]`http://www.soundandmusiccomputing.org/roadmap`

generate and analyze sounds. However many are the fundamental yet unanswered questions emerging from the use of sound in these new scenarios:

- How to analyze sound to extract information that is genuinely meaningful?
- How to investigate learning processes concerning sound listening and perception in a pedagogical perspective and in order to enhance the use of sound for communication purposes?
- How to model and communicate sound embedded in multimodal contents and in multisensory experiences?
- How to synthesize sound by means of direct "manipulation" or other forms of intuitive control?
- How to synthesize sounds that are perceptually adequate for some specific purposes?
- How to design sound for context-aware environments and interactive applications?

A crucial and general issue emerging from the current state of affairs is that sound and sense are two separate domains and there is lack of methods to bridge them with two-way paths. So far, a number of fast-moving sciences ranging from signal processing to experimental psychology, from acoustics to cognitive musicology, have tapped the sound and music computing arena here and there. What we are still missing is an integrated multidisciplinary and multidirectional approach. Only by coordinating the actions of the most active contributors in different subfields, as we tried to do while preparing this book, we can hope to elicit fresh ideas and new, more general and robust paradigms. The potential impact on society is terrific, as there is already a number of mass application technologies that are stagnating because of the existing gap between sound and sense. Just to name a few: sound/music information retrieval and data mining, virtual and augmented environments, expressive multimodal communication, intelligent navigation, and many others.

The book covers a wide range of disciplines involved in sound and music computing, from musicology to pedagogy, from psychology of perception and action to artificial intelligence, from signal processing to ecological acoustics and interaction design. Most of the members of the S2S^2 Action actively contributed to the book in their respective fields of expertise providing the widest possible perspective on the state-of-the-art in Sound and Music Computing research. We acknowledge their commitment during all phases of selection of materials, preparation, writing and revising. The person who deserves the loudest "thank you!" is Nicola Bernardini, S2S^2 Action coordinator and tireless source of inspiration, help, and enthusiasm.

Organization of the book

Chapter 1 – "Sound, Sense and Music Mediation" – provides an historical/philosophical overview of the relationships between sound and sense. Chapter 2 – "Learning Music" – deals with auditory cognition investigated through learning processes. Chapter 3 – "Content Processing of Music Audio Signals" – illustrates the state of the art in automatic low level description of musical-audio signals. Chapter 4 – "From Sound to Sense via Feature Extraction and Machine Learning" – outlines the subject of high level descriptors for characterizing music. Chapter 5 – "Sense in Expressive Music Performance" – provides an overview on computational approaches to performance analysis. Chapter 6 – "Controlling Sound with Senses" – presents recent research on multimodal-gesture analysis for sound control in music performance. Chapter 7 – "Real-Time Control of Music Performance" – depicts methodologies for performance modeling. Chapter 8 – "Physics-Based Sound Synthesis" – provides the state-of-the-art of sound synthesis by physical models. Chapter 9 – "Interactive Sound" – discusses the ecological approach to sound synthesis. Finally, Chapter 10 – "Sound Design and Auditory Displays" – introduces emerging disciplines, such as auditory display and sound design, in the perspective of sound as a carrier of information. An appendix concludes the book, presenting some recent applications concerning sound control, performance tools and interaction design: a virtual DJ scratching system, a virtual guitar, a tabletop interface and an interactive book.

Chapter 1

Sound, Sense and Music Mediation: a Historical-Philosophical Perspective

Marc Leman[1], Frederik Styns[1] and Nicola Bernardini[2]

[1]IPEM, Department of Musicology, University of Ghent
[2]Conservatory of Padua

About this chapter

In this chapter it is shown that the modern scientific approach to sound and music computing has historical roots in research traditions that aimed at understanding the relationship between sound and sense, physics and meaning. This chapter gives a historical-philosophical overview of the different approaches that led to the current computational and empirical approaches to music cognition. It is shown that music cognition research has evolved from Cartesian dualism with a rather strict separation between sound and sense, to an approach in which sense is seen as embodied and strongly connected to sound. Along this development, music research has always been a driver for new developments that aim at bridging the gap between sound and sense. This culminates in recent studies on gesture and artistic applications of music mediation technologies.

1.1 Introduction

In all human musical activities, musical sound and sense are tightly related with each other. Sound appeals to the physical environment and its moving objects, whereas sense is about private feelings and the meanings it generates. Sound can be described in an objective way, using instruments and machines. In contrast, sense is a subjective experience which may require a personal interpretation in order to explicit this experience. How are the two related with each other? And what is the aim of understanding their relationship? Ideas for this chapter heavily draw upon "Embodied music cognition and mediation technology" (Leman, 2007).

1.2 From music philosophy to music science

Ancient Greek philosophers such as Pythagoras, Aristoxenos, Plato and Aristotle had quite different opinions about the relationship between sound and sense. Pythagoras mainly addressed the physical aspects by considering a mathematical order underlying harmonic pitch relationships. In contrast, Aristoxenos addressed perception and musical experience (Barker, 1984). Plato comes into the picture mainly because he attributed strong powers to music, thus, a strong effect from sound to sense. However, for him, it was a reason to abandon certain types of music because of the weakening effect music could have on the virtue of young people. Aristotle understood the relationship between sound and sense in terms of a mimesis theory (e.g. Politics, Part V). In this theory, he stated that rhythms and melodies contain similarities with the true nature of qualities in human character, such as anger, gentleness, courage, temperance and the contrary qualities. By imitating the qualities that these characters exhibit in music, our souls[1] are moved in a similar way, so that we become in tune with the affects we experience when confronted with the original. For example, when we hear imitations of men in action in music, then our feelings tend to move in sympathy with the original. When listening to music, our soul thus undergoes changes in tune with the affective character being imitated.

Understood in modern terms, Aristotle thus observed a close connection between sound and sense in that the soul would tend to move along or resonate with sound features in music that mimic dynamic processes (gestures, expressions) in humans. Anyhow, with these views on acoustics (Pythagoras' approach to music as ratios of numbers), musical experience (Aristoxenos' approach to music as perceived structure), and musical expressiveness (Aristotle's approach to music as imitation of reality), there was sufficient material for a few centuries of philosophical discussion about sound to sense relationships.

All this started again from scratch in the 17th century with the introduction of the so-called Cartesian dualism, which states that sound and sense are two entirely different things. In his *Musicae Compendium* (1618), the young Descartes gave a good summary of the state-of-the-art concerning the musical sound to sense relationship. Basically, he divided music into three basic components, namely, (i) the mathematical-physical aspect (Pythagoras), (ii) the sensory perception (Aristoxenos), and (iii) the ultimate effect of music perception on the individual listener's soul (or mind) (Aristotle). To Descartes, it is sound, and to some degree also sensory perception, that can be the subject of a scientific study. The reason is that sound, as well as the human ear, deal with physical objects. In his *Méditations Métaphysiques* (1641), he explains that since objects have extension and can be put into motion, we can apply our mathematical methods to them. In contrast, the effect of perception and the meaning that ensues from it resides in the soul. There it can be a subject of introspection. In that respect, sense is less suitable to scientific study because sense has no extension. Like thinking, it is a property of each person's soul or ego. Yet there is a subtle difference between sense and thinking. Because it is

---

[1]The concept of soul is an old philosophical concept while modern philosophers tend to relate to the more modern concepts of "self", "ego", or even "mind" (Metzinger, 2003).

based on our senses, sense is prone to errors and therefore not very suitable for building up reliable knowledge. Pure thinking (devoid from external input) is less vulnerable to errors because it leads to clear and distinct ideas. One of the most clear and distinct ideas, according to Descartes, is that "I" (or my ego) exist. Indeed, when "I" put everything into doubt, even the existence of the world (e.g. it could be a dream), the only thing that remains is the "I" that doubts or dreams, which proves the existence of the ego or soul. Therefore, in Descartes' opinion, the soul and the human body are two entirely different things. The soul can think and does not need extension, whereas the body cannot think and needs extension. Knowledge of the soul requires introspection, whereas knowledge of the body requires scientific methods and descriptions that focus on moving objects. According to Descartes, the link between "I" and the world is due to an organ in the brain that connects the parallel worlds of the subjective mind and the objective body.

In the *Méditations Mëtaphysiques*, this link is called the *sensus communis*, or the *Gemeinsinn*, that is, the part of the psyche responsible for binding the inputs of the individual sense organs into a coherent and intelligible representation. In more recent times, this concept will reappear as "body image" and "body schema". So far, Descartes' approach thus clearly distinguished sound and sense. His focus on moving objects opened the way for scientific investigations in acoustics and psychoacoustics, and it pushed matters related to sense and meaning a bit further away towards a disembodied mental phenomenon.

Like Descartes, many scientists working on music (Cohen, 1984) often stressed the mathematical and physical aspects, whereas the link with musical meaning was more a practical consequence. For example, the calculation of pitch tunings for clavichord instruments was often (and sometimes still is) considered to be a purely mathematical problem, yet it had consequences for the development of the harmonic and tonal system and the way it touches our sense of tone relationships and, ultimately, our mood and emotional involvement with music. Structural aspects of music perception, such as pitch scales and consonance, were clearly at the borderline of mathematical and physical studies. Pitch scales could be related to logic and reasoning (which, according to Descartes, belongs to the capacity of the soul), while consonance could be based on resonance and movement of objects, which belongs to physics. In that respect, emotions and expressive gestures were not considered to be a genuine topic of scientific study. Emotions and expressive gestures were too much influenced by sound, and therefore, since they were not based on pure thinking, they were prone to error and not reliable as a basis for scientific study and knowledge. Thus, while Plato and Aristotle saw a connection between sound and sense through mimesis, Descartes claimed that sound and sense had a different ontology.

Parallel with this approach, the traditions of Aristoxenos and Aristotle also culminated in rule-based accounts of musical practices such as Zarlino's, and later Rameau's and Mattheson's. In *Der Volkommene Kapelmeister* (1739), for example, Mattheson offers a manual of how to compose music in a convincing way that is expressive of certain affects. This work of Mattheson focuses on the way people deal with music and on the way they experience the musical sounds as something that tangles their most intimate feelings. These compositional formulae can be seen as handbooks for creating music that makes sense. Obviously, these approaches were based on musical intuition, harmonious combinations of notes, and musical intentions, in short on aspects that colour musical experience and meaning formation. However, they also contained pointers to acoustics. Somehow, there was the feeling that aspects of perception which closely adhere to the perception of syntax and structure had a foundation in acoustics. Yet not all aspects could be explained by it. The real experience of its existence, the associated feeling, mood and pleasure were believed to belong to the subject's private life, which was inaccessible to scientific investigation.

Descartes' dualism had a tremendous impact on scientific thinking and in particular also on music research. The science of sound and the practice of musical experience and sense were no longer connected by a common concept. Sound was the subject of a scientific theory, while sense was still considered to be the by-product of something subjective that is done with sound. Apart from sensory perception (e.g. roughness in view of tuning systems), there was no real scientific theory of sense, and so, the gap between mind and matter, sense and sound, remained large.

1.3 The cognitive approach

By jumping from Descartes to the late 19th century, we are neglecting important contributions from Spinoza, Kant, Leibniz, Hume, Locke and many others on matters that concern the relationship between sound and sense. This is justified by our focus on the moment where empirical studies of subjective involvement with music started to take place. This can indeed be situated in the 19th century, when Wundt started the first psychological laboratory in Leipzig in 1874. Through the disciplines of psychophysics and psychology, the idea was launched that between sound and sense there is the human brain, whose principles could also be understood in terms of psychic principles and later on, as principles of information processing. With this development, the processes that underlay musical sense come into the picture.

1.3.1 Psychoacoustics

With the introduction of psychoacoustics by von Helmholtz (1863), the foundations were laid for an information processing approach to the sound/sense relationship. Helmholtz assumed that musical sensing, and ultimately, its experience and sense, was based on physiological mechanisms in the human ear. This idea became very influential in music research because it provided an explanation of why some very fundamental structural aspects of musical sense, such as consonance and dissonance, harmony and tonality, had an impact on our sense. This impact was no longer purely a matter of acoustics, but also of the working of our sensing system. Through scientific experiments, the causality of the mechanisms (still seen as a moving object) could be understood and mathematical functions could capture the main input/output relationships. This approach provided the foundation for experimental psychology (Wundt) and later for Gestalt psychology in the first half of the 20th century, and the cognitive sciences approach of the second half of the 20th century.

1.3.2 Gestalt psychology

The Gestalt movement, which dates back to the late 19th century (Stumpf, Brentano), gained prominence by about 1920 thanks to the work of scholars such as Wertheimer, Kohler and Koffka. It had a major focus on sense as the perception and representation of musical structure, including the perception of tone distances and intervals, melodies, timbre, as well as rhythmic structures. Sense was less a matter of the ego's experience and pure thinking, but more a matter of the ego's representational system for which laws could be formulated and a psychoneural parallelism, that is a parallelism between brain and psyche, could be assumed.

After 1945, the Gestalt theory lost much of its attractiveness and internationally acclaimed innovative position (Leman and Schneider, 1997). Instead, it met severe criticisms especially from behavioristic and operationalistic quarters. There had been too many Gestalt laws, and perhaps not enough hardcore explanations to account for these, notwithstanding the great amount of experimental work that had been done over decades. However, after 1950, Gestalt thinking gradually gained a new impetus, and was found to be of particular importance in combination with then up-to-date trends in cybernetics and information science. The Gestalt approach influenced music research in that it promoted a thorough structural and cognitive account of music perception based on the idea that sense emerges as a global pattern from the information processing of patterns contained in musical sound.

1.3.3 Information theory

It also gradually became clear that technology would become an important methodological pillar of music research, next to experimentation. Soon after 1945, with the introduction of electronics and the collaboration between engineers and composers, electronic equipment was used for music production activities, and there was a need for tools that would connect musical thinking with sound energies. This was a major step in the development of technologies which extend the human mind to the electronic domain in which music is stored and processed. Notions such as entropy and channel capacity provided objective measures of the amount of information contained in music and the amount of information that could possibly be captured by the devices that process music (Moles, 1952, 1958; Winckel, 1960). The link from information to sense was easily made. Music, after all, was traditionally conceived of in terms of structural parameters such as pitch and duration. Information theory thus provided a measurement, and thus a higher-level description, for the formal aspects of musical sense. Owing to the fact that media technology allowed the realisation of these parameters into sonic forms, information theory could be seen as an approach to an objective and relevant description of musical sense.

1.3.4 Phenomenology

Schaeffer (1966) was the first to notice that an objective description of music does not always correspond with our subjective perception. For example, three frequencies in combination with each other do not always produce the sense of three frequencies. If they are multiples, then they produce the sense of only one single tone. In line with phenomenology and Gestalt theory, he felt that the description of musical structure, based on information theory, does not always tell us how music is actually perceived by subjects. Measurements of structures are certainly useful and necessary, but these measurements don't always reveal relationships with subjective understanding. Schaeffer therefore related perception of sounds to the manipulation of the analogue electronic sound-generating equipment of that time. He conceived musical sense in accordance with the new media technology of his time. Schaeffer therefore drew attention to the role of new media as mediators between sound and sense.

From that moment on, music research had an enormous impact on technology development. Music became a driver for the development of new and innovative technologies which prolonged the human mind into the electronic domain, thus offering a complete and unlimited control of sound. This unlimited control was a big challenge because it asked for a more thorough investigation of the relationship between sound and sense. Indeed, from now on, it was easy to produce sound, but when are sounds musically meaningful? In other words, when do sounds make any sense? And is sense, as such, something that we can model, consider as something autonomous, and perhaps automate?

1.3.5 Symbol-based modelling of cognition

The advent of computers marked a new area in music research. Computers could replace analogue equipment, but, apart from that, it also became possible to model the mind according to the Cartesian distinction between mind and matter. Based on information processing psychology and formal linguistics (Lindsay and Norman, 1977) it was believed that the *res cogitans* (related to that aspect of the mind which also Descartes had separated from matter) could be captured in terms of symbolic reasoning. Computers now made it possible to mimic human "intelligence" and develop an "artificial intelligence". Cognitive science, as the new trend was called, conceived the human mind in terms of a machine that manipulates representations of content on a formal basis (Fodor, 1981). The application of the symbol-based paradigm to music (Longuet Higgins, 1987; Laske, 1975; Baroni and Callegari, 1984; Balaban et al., 1992) was very appealing. However, the major feature of this approach is that it works with a conceptualisation of the world which is cast in symbols, while in general it is difficult to pre-define the algorithms that should extract the conceptualised features from the environment. The predefinition of knowledge atoms and the subsequent manipulation of those knowledge atoms in order to generate further knowledge is a main characteristic of a Cartesian or rationalist conception of the world. Symbol systems, when used in the context of rationalist modelling, should therefore be used with caution.

1.3.6 Subsymbol-based modelling of cognition

In the 1980s, based on the results of the so-called connectionist computation (Rumelhart et al., 1987; Kohonen, 1995) a shift of paradigm from symbol-based modelling to subsymbol-based modelling was initiated. Connectionism (re-

)introduced statistics as the main modelling technique for making connections between sound and sense. Given the limitations of the Cartesian approach that aimed at modelling the disembodied mind, this approach was rather appealing for music research because it could take into account the natural constraints of sound properties better than the symbol-based approach could (Todd et al., 1999). By including representations of sound properties (rather than focusing on symbolic descriptions which are devoid of these sound properties), the subsymbol-based approach was more in line with the naturalistic epistemology of traditional musicological thinking. It held the promise of an ecological theory of music in which sound and sense could be considered as a unity. The method promised an integrated approach to psychoacoustics, auditory physiology, Gestalt perception, self-organisation and cognition, but its major limitation, however, was that it still focused exclusively on perception. By doing this, it was vulnerable to the critique that it still adopted a disembodied approach to the notion of sense, in other words, that sense would be possible without body, that it could work by just picking up the structures that are already available in the environment.

1.4 Beyond cognition

The cognitive tradition was criticised for several reasons. One reason was the fact that it neglected the subjective component in the subject's involvement with the environment. Another reason was that it neglected action components in perception and therefore remained too much focused on structure and form. Criticism came from many different corners, first of all from inside cognitive science, in particular from scholars who stressed the phenomenological and embodied aspects of cognition (Maturana and Varela, 1987; Varela et al., 1992) and later also from the so-called postmodern musicology.

1.4.1 Subjectivism and postmodern musicology

Huron (1999) defines the so-called "New Musicology" as a methodological movement in music scholarship of the past two decades that is "loosely guided by a recognition of the limits of human understanding, an awareness of the social milieu in which scholarship is pursued, and a realisation of the political arena in which the fruits of scholarship are used and abused". DeNora (2003) argues that, in response to developments in other disciplines such as literary theory, philosophy, history, anthropology and sociology, new musicologists have called into question the separation of historical issues and musical form and that they have focused on the role of music as a social medium. New musicology, like postmodern thinking, assumes that there is no absolute truth to be known. More precisely, truth ought to be understood as a social construction that relates to a local or partial perspective on the world. So the focus of new musicology is on the socio-cultural contexts in which music is produced, perceived and studied and how such contexts guide the way people approach, experience and study music. Aspects of this school of thinking are certainly relevant to the sound/sense relationship (Hatten, 1994; Lidov, 2005; Cumming, 2000), although the methodology (called hermeneutic) is less involved with the development of an empirical and evidence-based approach to subjective matters related to musical sense. In addition, there is less attention to the problem of music mediation technologies. However, a main contribution is the awareness that music is functioning in a social and cultural context and that this context is also determinative for technology development.

1.4.2 Embodied music cognition

The action-based viewpoint put forward by Maturana, Varela and others (Varela et al., 1992; Maturana and Varela, 1987) has generated a lot of interest and a new perspective on how to approach the sound/sense relationship. In this approach, the link between sound and sense is based on the role of action as mediator between physical energy and meaning. In the cognitive approach the sound/sense relationship was mainly conceived from the point of view of mental processing. The approach was effective in acoustics and structural understanding of music, but it was less concerned with action, gestures and emotional involvement. In that respect, one could say that the Aristotelian component, with its focus on mimesis as binding component between sound and sense, was not part of the cognitive programme, nor was multi-modal information processing, or the issue of action-relevant perception (as reflected in the ecological psychology of Gibson).

Yet the idea that musical involvement is based on the embodied imitation of moving sonic forms has a long tradition. In fact, this tradition has been gradually rediscovered in the last decade. In systematic musicology, a school of researchers in the late 19th and early 20th Centuries had already a conception of musical involvement based on corporeal articulations (Lipps, 1903; Meirsmann, 1922/23; Heinitz, 1931; Becking, 1928; Truslit, 1938).

This approach differs from the Gestalt approach in that it puts more emphasis on action. Like Gestalt theory, this approach may be traced back to open problems in Kant's aesthetic theory, in particular the idea that beauty is in the formal structure. Unlike Gestalt theory, the emphasis was less on brain processes and the construction of good forms, but more on the phenomenology of the empathic relationship with these forms through movement and action. In this approach, Descartes' concept of ego and self is again fully connected with the body. The ego is no longer an entity that just thinks about its body (like a skipper who perceives the boat which he sails), but there is a strong component of feeling, or *gemeingefühl* (coenaesthesis), that is, sensing and awareness of body.

For example, Lipps (1903) argues that the understanding of an expressive movement (*Ausdrucksbewegung*) in music is based on empathy (*inneren Mitmachen, Einfülung*). While being involved with moving sonic forms, we imitate the movements as expressions. By doing this, we practice the motor muscles which are involved when genuine emotions are felt. As such, we have access to the intended emotional meaning of the music. According to Lipps, the act of (free or unbounded) imitation gives pleasure because it is an expression of the self (Lipps, 1903, p. 111). Similar ideas are found in the theory of optimal experience of Csikszentmihalyi (1990). Any expression of the self, or anything that contributes to its ordering, gives pleasure. As such, sad music may be a source of pleasure (*Lust*) because the moving sonic forms allow the subject to express an imitative movement (sadness). This imitation allows the subject to participate in the expressive movement without being emotionally involved, that is, without experiencing an emotional state of sadness.

Truslit (Truslit, 1938; Repp, 1993) also sees corporeal articulations as manifestations of the inner motion heard in music. He says that, "provided the sound has the dynamo-agogic development corresponding to a natural movement, it will evoke the impression of this movement in us" (Repp, 1993). Particularly striking is the example he gives of Beethoven who, while composing, would hum or growl up and down in pitch without singing specific notes. This is also a phenomenon often heard when jazz musicians are playing. Truslit used the technology of his time to extract information from acoustic patterns, as well as information from body movements with the idea of studying their correlations.

In *Gestaltung und Bewegung in der Musik*, Alexander Truslit argues that in order to fully experience music, it is essential to understand its most crucial characteristic. According to Truslit, this characteristic, the driving force of the music, is the expression of inner movement. The composer makes music that is full of inner movement. The musician gives shape to these inner movements by translating them into proper body gestures and the "good" music listener is able to trace and imitate these movements in order to experience and understand the music properly.

According to Truslit, not all music listeners are able to perceive the inner movements of the music. However, some music listeners have a special capacity to couple the auditive information to visual representations. Such visual representations are referred to as synoptic pictures. Listeners possessing this capability have a great advantage for understanding the musical inner movement. Central in Truslit's approach of musical movement are the notions of dynamics (intensity) and agogics (duration). If the music has the dynamo-agogic development corresponding to a natural movement, it will evoke the impression of this movement. Four basic movements are distinguished in order to identify and understand musical movement. These basic movements are: straight, open, closed and winding. Furthermore, it is stated that, based on this basic vocabulary of movements, it is possible to determine the shape of the inner movements of the music in an objective way. Once the shapes of the movements are determined, it is useful to make graphical representations of them. Such graphical representations can be used by musicians and music listeners as guidelines for understanding and examining music's inner movement. Truslit sees the inner movement of music first of all as something that is presented in the musical melody. The addition of rhythmic, metric or harmonic elements can only refine this inner movement. A distinction is made between rhythmic movement and the inner movement of the music that Truslit focuses on. In contrast to rhythmic movement, which is related to individual parts of the body, the inner movement forms the melody and is, via the labyrinth (which is situated in the vestibular system), related to the human body as a whole.

In accordance with Truslit, Becking (Becking, 1928; Nettheim, 1996) also makes a connection between music and movement, based on the idea of a dynamic rhythmic flow beyond the musical surface. This flow, a continuous up-down movement, connects points of metrical gravitude that vary in relative weight. Becking's most original idea was that these metrical weights vary from composer to composer. The analytical method Becking worked out in order to determine these weights was his method of accompanying movements, conducted with a light baton. Like Truslit, Becking determined some basic movements. These basic movements form the basic vocabulary that allowed him to classify the personal constants of different composers in different eras.

To sum up, the embodied cognition approach states that the sound/sense relationship is mediated by the human body, and this is put as an alternative to the disembodied cognition approach where the mind is considered to be functioning on its own. The embodied cognition approach of the early 20th century is largely in agreement with recent thinking about the connections between perception and action (Prinz and Hommel, 2002; Dautenhahn and Nehaniv, 2002).

1.4.3 Music and emotions

The study of subjective involvement with music draws upon a long tradition of experimental psychological research, initiated by Wundt in the late 19th century. Reference can be made to research in experimental psychology in which descriptions of emotion and affect are related to descriptions of musical structure (Hevner, 1936; Watson, 1942; Reinecke, 1964; Imberty, 1976; Wedin, 1972; Juslin and Sloboda, 2001; Gabrielsson and Juslin, 2003). These studies take into account a subjective experience with music. Few authors, however, have been able to relate descriptions of musical affect and emotions with descriptions of the physical structure that makes up the stimulus. Most studies, indeed, interpret the description of structure as a description of perceived structure, and not as a description of physical structure. In other words, description of musical sense proceeds in terms of perceptual categories related to pitch, duration, timbre, tempo, rhythms, and so on.

In that respect, Berlyne's work (Berlyne, 1971) on experimental aesthetics is important for having specified a relationship between subjective experience (e.g. arousal) and objective descriptions of complexity, uncertainty or redundancy. In Berlyne's concept, the latter provides an information-theoretic account of symbolic structures (e.g. melodies). They are not just based on perceived structures but are extracted directly from the stimulus (as symbolically represented). However, up to the present, most research has been based on a comparison between perceived musical structure and experienced musical affect. What is needed are comparisons of structure as perceived and structure which is directly extracted from the physical energy (Leman et al., 2003).

1.4.4 Gesture modelling

During the last decade, research has been strongly motivated by a demand for new tools in view of the interactive possibilities offered by digital media technology. This stimulated the interest in gestural foundations of musical involvement.[2] With the advent of powerful computing tools, and in particular real-time interactive music systems (Pressing, 1992; Rowe, 1992), gradually more attention has been devoted to the role of gesture in music (Wanderley and Battier, 2000; Camurri et al., 2001; Sundberg, 2000; Camurri et al., 2005). This gestural approach has been rather influential in that it puts more emphasis on sensorimotor feedback and integration, as well as on the cou-

[2]In 2004, the ConGAS COST-287 action, supported by the EU, established a European network of laboratories that focus on issues related to gesture and music.

pling of perception and action. With new sensor technology, gesture-based research has meanwhile become a vast domain of music research (Paradiso and O'Modhrain, 2003; Johannsen, 2004; Camurri and Rikakis, 2004; Camurri and Volpe, 2004), with consequences for the methodological and epistemological foundations of music cognition research. There is now convincing evidence that much of what happens in perception can be understood in terms of action (Jeannerod, 1994; Berthoz, 1997; Prinz and Hommel, 2002). Pioneering studies in music (Clynes, 1977; Todd et al., 1999; Friberg and Sundberg, 1999) addressed this coupling of perception and action in musical activity, yet the epistemological and methodological consequences of this approach have not been fully worked out in terms of a musicological paradigm (Leman, 1999). It is likely that more attention to the coupling of perception and action will result in more attention to the role of corporeal involvement in music, which in turn will require more attention to multi-sensory perception, perception of movement (kinaesthesia), affective involvement, and expressiveness of music (Leman and Camurri, 2005).

1.4.5 Physical modelling

Much of the recent interest in gesture modelling has been stimulated by advances in physical modelling. A physical model of a musical instrument generates sound on the basis of the movements of physical components that make up the musical instrument (for an overview, see (Karjalainen et al., 2001)). In contrast with spectral modelling, where the sound of a musical instrument is modelled using spectral characteristics of the signal that is produced by the instrument, physical modelling focuses on the parameters that describe the instrument physically, that is, in terms of moving material object components. Sound generation is then a matter of controlling the articulatory parameters of the moving components. Physical models, so far, are good at synthesising individual sounds of the modelled instrument. And although it is still far from evident how these models may synthesise a score in a musically interesting way – including phrasing and performance nuances – it is certain that a gesture-based account of physical modelling is the way to proceed (D'haes, 2004). Humans would typically add expressiveness to their interpretation, and this expressiveness would be based on the constraints of body movements that take particular forms and shapes, sometimes perhaps learned movement sequences and gestures depending on cultural traditions. One of the goals of gesture research related to music, therefore, aims at understanding the biomechanical and psychomotor laws that characterise human movement in the context of music production and perception (Camurri and Volpe, 2004).

1.4.6 Motor theory of perception

Physical models suggest a reconsideration of the nature of perception in view of stimulus-source relationships and gestural foundations of musical engagement. Purves and Lotto (2003), for example, argue that invariance in perception is based on statistics of proper relationships between the stimulus and the source that produces the stimulus. Their viewpoint is largely influenced by recent studies in visual perception. Instead of dealing with feature extraction and object reconstruction on the basis of properties of single stimuli, they argue that the brain is a statistical processor which constructs its perceptions by relating the stimulus to previous knowledge about stimulus-source relationships. Such a statistics, however, assumes that aspects related to human action should be taken into account because the source cannot be known unless through action. In that respect, this approach differs from previous studies in empirical modelling, which addressed perception irrespective of action related issues. Therefore, the emphasis of empirical modelling on properties of the stimulus should be extended with studies that focus on the relationship between stimulus and source, and between perception and action. Liberman and Mattingly (1989) had already assumed that the speech production-perception system is, in effect, an articulatory synthesiser. In the production mode, the synthesiser is activated by an abstract gestural pattern from which the synthesiser computes a series of articulatory movements that are needed to realise the gestures into muscle movements of the vocal tract. In the perception mode, then, the synthesiser computes the series of articulatory movements that could have produced the signal, and from this articulatory representation, the intended gestural pattern, contained in the stimulus, is obtained. Liberman and Mattingly assumed that a specialised module is responsible for both perception and production of phonetic structures. The perceptual side of this module converts automatically from acoustic signal to gesture. Perception of sound comes down to finding the proper parameters of the gesture that would allow the re-synthesis of what is heard. So, features related to sound are in fact picked up as parameters for the control of the articulatory system. Perception of a sound, in that view, is an inhibited re-synthesis of that sound, inhibited in the sense that the re-synthesis is not actually carried out but simulated. The things that need to be stored in memory, then, are not auditory images, but gestures, sequences of parameters that control the human articulatory (physical) system. The view also assumes that perception and action share a common representational system. Such models thus receive input from the sensors and produce appropriate actions as output and, by doing this, stimuli thus become meaningful in relation to their sources which are objects of action (Varela et al., 1992). Action, in other words, guarantees that the stimuli are connected to the object, the source of the physical energy that makes up the stimulus. The extension of empirical modelling with a motor theory of perception is currently a hot topic of research. It has some very important consequences for the way we conceive of music research, and in particular also for the way we look at music perception and empirical modelling.

1.5 Embodiment and mediation technology

The embodiment hypothesis entails that meaningful activities of humans proceed in terms of goals, values, intentions and interpretations, while the physical world in which these activities are embedded can be described from the point of view of physical energy, signal processing, features and descriptors. In normal life, where people use simple tools, this difference between the subject's experiences and the physical environment is bridged by the perceptive and active capabilities of the human body. In that perspective, the human body can be seen as the natural mediator between the subject and the physical world. The subject perceives the physical world on the basis of its subjective and action-oriented ontology, and acts accordingly using the body to realise its imagined goals. Tools are used to extend the limited capacities of natural body. This idea can be extended to the notion of mediation technology.

For example, to hit a nail into a piece of wood, I will use a hammer as an extension of my body. And by doing this, I'll focus on the nail rather than on the hammer. The hammer can easily become part of my own body image, that is, become part of the mental representation of my (extended) body. My extended body then allows my mental capacities to cross the borders of my natural human body, and by doing this, I can realise things that otherwise would not be possible. Apart from hitting nails, I can ride a bike to go to the library, I can make music by playing an instrument, or I can use my computer to access digital music. For that reason, technologies that bridge the gap between our mind and the surrounding physical environment are called mediation technologies. The hammer, the bike, the musical instrument and the computer are mediation technologies. They influence the way in which connections between human experience (sense) and the physical environment (e.g. sound) can take place.

Mediation concerns the intermediary processes that bridge the semantic gap between the human approach (subject-centered) and the physical approach (object or sound-centered), but which properties should be taken into account in order to make this translation effective? The hammer is just a straightforward case, but what about music that is digitally encoded in an mp3-player? How can we access it in a natural way, so that our mind can easily manipulate the digital environment in which music is encoded? What properties of the mediation technology would facilitate access to digitally encoded energy? What mediation tools are needed to make this access feasible and natural, and what are their properties? The answer to this question is highly dependent on our understanding of the sound/sense relationship as a natural relationship. This topic is at the core of current research in music and sound computing.

1.5.1 An object-centered approach to sound and sense

State-of-the-art engineering solutions are far from being sufficiently robust for use in practical sense/sound applications. For example, (Paivo, 2007) demonstrates that the classical bottom-up approach (he took the melody extraction from polyphonic audio as a case study, using state-of-the-art techniques in auditory modelling, pitch detection and frame-concatenation into music notes) has reached its performance platform. Similar observations have been made in rhythm and timbre recognition. The use of powerful stochastic and probabilistic modelling techniques (Hidden Markov Chains, Bayesian modelling, Support Vector Machines, Neural Networks) (see also `http://www.ismir.net/` for publications) do not really close this gap between sense and sound much further (De Mulder et al., 2006). The link between sound and sense turns out to be a hard problem. There is a growing awareness that the engineering techniques are excellent, but that the current approaches may be too narrow. The methodological problems relate to:

- Unimodality: the focus has been on musical audio exclusively, whereas humans process music in a multi-modal way, involving multiple senses (modalities) such as visual information and movement.

- Structuralism: the focus has been on the extraction of structure from musical audio files (such as pitch, melody, harmony, tonality, rhythm) whereas humans tend to access music using subjective experiences (movement, imitation, expression, mood, affect, emotion).

- Bottom-up: the focus has been on bottom-up (deterministic and learning) techniques whereas humans use a lot of top-down knowledge in signification practices.

- Perception oriented: the focus has been on the modelling of perception and cognition whereas human perception is based on action-relevant values.

- Object/Product-centered: research has focused on the features of the musical object (waveform), whereas the subjective factors and the social/cultural functional context in musical activities (e.g. gender, age, education, preferences, professional, amateur) have been largely ignored.

1.5.2 A subject-centered approach to sound and sense

Research on gesture and subjective factors such as affects and emotions show that more input should come from a better analysis of the subjective human being and its social/cultural context. That would imply:

- Multi-modality: the power of integrating and combining several senses that play a role in music such as auditory, visual, haptic and kinaesthetic sensing. Integration offers more than the sum of the contributing parts as it offers a reduction in variance of the final perceptual estimate.

- Context-based: the study of the broader social, cultural and professional context and its effect on information processing. Indeed, the context is of great value for the disambiguation of our perception. Similarly, the context may largely determine the goals and intended musical actions.

- Top-down: knowledge of the music idiom to better extract higher-level descriptors from music so that users can have easier access to these descriptors. Traditionally, top-down knowledge has been conceived as a language model. However, language models may be extended with gesture models as a way to handle stimulus disambiguation.

- Action: the action-oriented bias of humans, rather than the perception of structural form (or Gestalt). In other words, one could say that people do not move just in response to the music they perceive, rather they move to disambiguate their perception of music, and by doing this, they signify music.

- User-oriented: research should involve the user in every phase of the research. It is very important to better understand the subjective factors that determine the behavior of the user.

The subject-centered approach is complementary to the object-centered approach. Its grounding in an empirical and evidence-based methodology fits rather well with the more traditional engineering approaches. The main difference relates to its social and cultural orientation and the awareness that aspects of this orientation have a large impact on the development of mediation technology. After all, the relationship between sense and sound is not just a matter of one single individual person in relation to its musical environment. Rather, this single individual person lives in contact with other people, and in a cultural environment. Both the social and cultural environment will largely determine what music means and how it can be experienced.

1.6 Music as innovator

The above historical and partly philosophical overview gives but a brief account of the different approaches to the sound and sense relationship. This account is certainly incomplete and open to further refinement. Yet a striking fact in this overview is that music, in spanning a broad range of domains from sound to sense and social interaction, appears to be a major driver for innovation. This innovation appears both in the theoretical domain where the relationship between body, mind, and matter is a major issue, and in the practical domain, where music mediation technology is a major issue.

The historical overview shows that major philosophical ideas, as well as technical innovations, have come from inside music thinking and engagement. Descartes' very influential dualist philosophy of mind was first developed in a compendium on music. Gestalt theory was heavily based on music research. Later on, the embodied cognition approach was first explored by people having strong roots in music playing (e.g. Truslit was a music teacher). In a similar way, the first explorations in electronic music mediation technologies were driven by composers who wanted to have better access to the electronic tools for music creation. Many of these ideas come out of the fact that music is fully embedded in sound and that the human body tends to behave in resonance with sound, whereas the "mind's I" builds up experiences on top of this. Music nowadays challenges what is possible in terms of object-centered science and technology and it tends to push these approaches more in the direction of the human subject and its interaction with other subjects. The human way in which we deal with music is a major driver for innovation in science and technology, which often approaches music from the viewpoint of sound and derived sound-features. The innovative force coming from music is related to the subject-centered issues that are strongly associated with creativity and social-cultural factors.

The idea that music drives innovation rather than vice versa should not come as completely unexpected. Music is solidly anchored to scientific foundations and as such it is an epistemological domain which may be studied with the required scientific rigour. However, music is also an art and therefore certain ways of dealing with music do not require scientific justification per se because they justify themselves directly in signification practices. The requirements of musical expression can indeed provide a formidable thrust to scientific and technological innovation in a much more efficient way than the usual R&D cycles may ever dream of. In short, the musical research carried out in our time by a highly specialised category of professionals (the composers) may be thought as a sort of fundamental think tank from where science and technology have extracted (and indeed, may continue to extract in the future) essential, revolutionary ideas. In short, musical expression requirements depend, in general, on large scale societal changes whose essence is captured by the sensible and attuned composers. These requirements translate quickly into specific technical requirements and needs. Thus, music acts in fact as an opaque but direct knowledge transfer channel from the subliminal requirements of emerging societies to concrete developments in science and technology.

1.7 Conclusion

This chapter aims at tracing the historical and philosophical antecedents of sense/sound studies in view of a modern action-oriented and social-cultural oriented music epistemology. Indeed, recent developments seem to indicate that the current interest in embodied music cognition may be expanded to social aspects of music making. In order to cross the semantic gap between sense and sound, sound and music computing research tends to expand the object-centered approach engineering with a subject-centered approach from the human sciences. The subject-centered character of music, that is, its sense, has always been a major incentive for innovation in science and technology. The modern epistemology for sound and music computing is based on the idea that sound and sense are mediated by the human body, and that technology may form an extension of this natural mediator. The chapter aims at providing a perspective from which projections into the future can be made.

The chapter shows that the relationship between sound and sense is one of the main themes of the history and philosophy of music research. In this overview, attention has been drawn to the fact that three components of ancient Greek thinking already provided a basis for this discussion, namely, acoustics, perception, and feeling ("movement of the soul"). Scientific experiments and technological developments were first (17th – 18th century) based on an understanding of the physical principles and then (starting from the late 19th century) based on an understanding of the subjective principles, starting with principles of perception of structure, towards a better understanding of principles that underly emotional understanding.

During the course of history, the problem of music mediation was a main motivating factor for progress in scientific thinking about the sound/sense relationship. This problem was first explored as an extension of acoustic theory to the design of music instruments, in particular, the design of scale tuning. In modern times this problem is explored as an extension of the human body as mediator between sound and sense. In the 19th century, the main contribution was the introduction of an experimental methodology and the idea that the human brain is the actual mediator between sound and sense.

In the last decades, the scientific approach to the sound/sense relationship has been strongly driven by experiments and computer modelling. Technology has played an increasingly important role, first as measuring instrument, later as modelling tool, and more recently as music mediation tools which allow access to the digital domain. The approach started from a cognitive science (which adopted Cartesian dualism) and symbolic modelling, and evolved to sub-symbolic modelling and empirical modelling in the late 1980ies. In the recent decades, more attention has been drawn to the idea that the actual mediator between sound and sense is the human body.

With regards to new trends in embodied cognition, it turns out that the idea of the human body as a natural mediator between sound and sense is not entirely a recent phenomenon, because these ideas have been explored by researchers such as Lipps, Truslit, Becking, and many others. What it offers is a possible solution to the sound/sense dichotomy by saying that the mind is fully embodied, that is, connected to body. Scientific study of this relationship, based on novel insights of the close relationship between perception and action, is now possible thanks to modern technologies that former generations of thinkers did not have at their disposal.

A general conclusion to be drawn from this overview is that the scientific methodology has been expanding from purely physical issues (music as sound) to more subjective issues (music as sense). Scientists conceived these transition processes often in relation to philosophical issues such as the mind-body problem, the problem of intentionality and how perception relates to action. While the sound/sense relationship was first predominantly considered from a cognitive/structural point of view, this viewpoint has gradually been broadened and more attention has been devoted to the human body as the natural mediator between sound and sense. Perception is no longer conceived in terms of stimulus and extraction of structures. Instead, perception is conceived within the context of stimulus disambiguation and simulated action, with the possibility of having loops of action-driven perception. This change in approach has important consequences for the future research. Music has thereby been identified as an important driver for innovation in science and technology. The forces behind that achievement are rooted in the fact that music has a strong appeal to multi-modality, top-down knowledge, context-based influences and other subject-centered issues which strongly challenge the old disembodied Cartesian approaches to scientific thinking and technology development.

Bibliography

M. Balaban, K. Ebcioğlu, and O. E. Laske, editors. *Understanding music with AI: perspectives on music cognition*. AAAI Press, Cambridge (Mass.), 1992.

A. Barker. *Greek musical writings*. Cambridge readings in the literature of music. Cambridge University Press, Cambridge, 1984.

M. Baroni and L. Callegari. *Musical grammars and computer analysis*. Quaderni della Rivista italiana di musicologia 8. Olschki, Firenze, 1984.

G. Becking. *Der musikalische Rhythmus als Erkenntnisquelle*. B. Filser, Augsburg, 1928.

D. E. Berlyne. *Aesthetics and psychobiology*. Appleton-Century-Crofts, New York, 1971.

A. Berthoz. *Le sens du mouvement*. Editions O. Jacob, Paris, 1997.

A. Camurri and T. Rikakis. Multisensory communication and experience through multimedia. *Ieee Multimedia*, 11(3):17–19, 2004.

A. Camurri and G. Volpe, editors. *Gesture-based communication in human-computer interaction. Selected revised papers of the 5th Intl Gesture Workshop (GW2003)*. Lecture Notes in Artificial Intelligence, LNAI. Springer-Verlag, Berlin, 2004.

A. Camurri, G. De Poli, M. Leman, and G. Volpe. A multi-layered conceptual framework for expressive gesture applications. In X. Serra, editor, *Intl EU-TMR MOSART Workshop*, Barcelona, 2001. Univ Pompeu Fabra.

A. Camurri, G. Volpe, G. De Poli, and M. Leman. Communicating expressiveness and affect in multimodal interactive systems. *IEEE Multimedia*, 12(1): 43–53, 2005.

M. Clynes. *Sentics: the touch of emotions*. Anchor Press, New York, 1977.

H. F. Cohen. *Quantifying music: the science of music at the first stage of the scientific revolution, 1580 - 1650*. Reidel, Dordrecht, 1984.

M. Csikszentmihalyi. *Flow: the psychology of optimal experience*. Harper & Row, New York, 1990.

N. Cumming. *The sonic self: musical subjectivity and signification*. Advances in semiotics. Indiana University Press, Bloomington, 2000.

K. Dautenhahn and C. L. Nehaniv. *Imitation in animals and artifacts*. Complex adaptive systems. MIT Press, Cambridge (Mass.), 2002.

T. De Mulder, J. P. Martens, S. Pauws, F. Vignoli, M. Lesaffre, M. Leman, B. De Baets, and H. De Meyer. Factors affecting music retrieval in query-by-melody. *Ieee Transactions on Multimedia*, 8(4):728–739, 2006.

T. DeNora. *After Adorno: rethinking music sociology*. Cambridge University Press, Cambridge, 2003.

W. D'haes. *Automatic estimation of control parameters for musical synthesis algorithms*. PhD thesis, Universiteit Antwerpen, 2004.

J. A. Fodor. *Representations: philosophical essays on the foundations of cognitive science*. MIT Press, Cambridge (Mass.), 1st mit press edition, 1981.

A. Friberg and J. Sundberg. Does music performance allude to locomotion? A model of final ritardandi derived from measurements of stopping runners. *Journal of the Acoustical Society of America*, 105(3):1469–1484, 1999.

A. Gabrielsson and P. N. Juslin. Emotional expression in music. In H. H. Goldsmith, R. J. Davidson, and K. R. Scherer, editors, *Handbook of affective sciences*, pages 503–534. Oxford University Press, New York, 2003.

R. S. Hatten. *Musical meaning in Beethoven: markerdess, correlation, and interpretation*. Advances in semiotics. Indiana university press, Bloomington (Ind.), 1994.

W. Heinitz. *Strukturprobleme in Primitiver Musik.* Friederichsen, De Gruyter & Co. M. B. H., Hamburg, 1931.

K. Hevner. Experimental studies of the elements of expression in music. *American Journal of Psychology*, 48:246–248, 1936.

D. Huron. The new empiricism: systematic musicology in a postmodern age, 1999. URL `http://musiccog.ohio-state.edu/Music220/Bloch.lectures/3.Methodology.html`.

M. Imberty. Signification and meaning in music, 1976. Groupe de Recherches en Sémiologie Musicale, Faculté de Musique, Université de Montréal.

M. Jeannerod. The representing brain–neural correlates of motor intention and imagery. *Behavioral and Brain Sciences*, 17(2):187–202, 1994.

G. Johannsen, editor. *Engineering and music-supervisory, control and auditory communication (special issue)*, volume 29 of *Proceedings of the IEEE*. IEEE, 2004.

P. N. Juslin and J. A. Sloboda. *Music and emotion: theory and research.* Series in affective science. Oxford University Press, Oxford, 2001.

M. Karjalainen, T. Tolonen, V. Valimaki, C. Erkut, M. Laurson, and J. Hiipakka. An overview of new techniques and effects in model-based sound synthesis. *Journal of New Music Research*, 30(3):203–212, 2001.

T. Kohonen. *Self organizing maps*. Springer series in information sciences 30. Springer, Berlin, 1995.

O. E. Laske. *Introduction to a generative theory of music*. Sonological reports 1B. Utrecht State University, Institute of Sonology, Utrecht, 1975.

M. Leman. Naturalistic approaches to musical semiotics and the study of causal musical signification. In I. Zannos, editor, *Music and Signs, Semiotic and Cognitive Studies in Music*, pages 11–38. ASKO Art & Science, Bratislava, 1999.

M. Leman. *Embodied music cognition and mediation technology*. MIT Press, Cambridge (Mass.), 2007.

M. Leman and A. Camurri. Understanding musical expressiveness using interactive multimedia platforms. *Musicae Scientiae*, pages 209–233, 2005.

M. Leman and A. Schneider. Origin and nature of cognitive and systematic musicology: An introduction. In M. Leman, editor, *Music, Gestalt, and Computing: Studies in Cognitive and Systematic Musicology*, pages 13–29. Springer-Verlag, Berlin, Heidelberg, 1997.

M. Leman, V. Vermeulen, L. De Voogdt, J. Taelman, D. Moelants, and M. Lesaffre. Correlation of gestural musical audio cues and perceived expressive qualities. *Gesture-Based Communication in Human-Computer Interaction*, 2915: 40–54, 2003.

A. M. Liberman and I. G. Mattingly. A specialization for speech-perception. *Science*, 243:489–494, 1989.

D. Lidov. *Is language a Music? Writings on Musical Form and Signification*. Indiana University Press, Bloomington, 2005.

P. H. Lindsay and D. A. Norman. *Human information processing: An introduction to psychology*. Academic Press, New York, 2nd edition, 1977.

T. Lipps. *Ästhetik Psychologie des Schönen und der Kunst*. L. Voss, Hamburg und Leipzig, 1903.

H. C. Longuet Higgins. *Mental processes studies in cognitive science*. Explorations in cognitive science 1. MIT Press, Cambridge (Mass.), 1987.

H. R. Maturana and F. J. Varela. *The tree of knowledge: the biological roots of human understanding*. New Science Library, Boston, 1987.

H. Meirsmann. Versuch einer phänomologie der musik. *Zeitschrift für Musikwissenschaft*, 25:226–269, 1922/23.

T. Metzinger. *Being no one: the self-model theory of subjectivity*. MIT Press, Cambridge, Mass., 2003.

A. Moles. *Physique et technique du bruit*. Dunod, Paris, 1952.

A. Moles. *Théorie de l'information et perception esthétique*. Etudes de radio télévision. Flammarion, Paris, 1958.

N. Nettheim. How musical rhythm reveals human attitudes: Gustav Becking's theory. *International Review of the Aesthetics and Sociology of Music*, 27(2):101–122, 1996.

R. Paivo. *Melody Detection in Polyphonic Audio*. PhD thesis, University of Coimbra, 2007.

J. A. Paradiso and S. O'Modhrain, editors. *New interfaces for musical performance and interaction (special issue)*, volume 32 of *Journal of New Music Research*. Swets Zeitlinger, Lisse, 2003.

J. Pressing. *Synthesizer performance and real-time techniques*. The Computer music and digital audio series; v. 8. A-R Editions, Madison, Wis, 1992.

W. Prinz and B. Hommel, editors. *Common mechanisms in perception and action*. Attention and performance 19. Oxford University Press, Oxford, 2002.

D. Purves and R. B. Lotto. *Why we see what we do: an empirical theory of vision*. Sinauer Associates, Sunderland (Mass.), 2003.

H. P. Reinecke. *Experimentelle Beiträge zur Psychologie des musikalischen Hörens*. Universität Hamburg, Hamburg, 1964.

B. H. Repp. Music as motion: a synopsis of Alexander Truslit's (1938) Gestaltung und Bewegung in der Music. *Psychology of Music*, 12(1):48–72, 1993.

R. Rowe. *Interactive music systems: machine listening and composing*. MIT Press,, Cambridge, MA, 1992.

D. E. Rumelhart, J. L. McLelland, C. Asanuma, and PDP research group. *Parallel distributed processing explorations in the microstructure of cognition*. Computational models of cognition and perception. MIT press, Cambridge (Mass.), 6th print. edition, 1987.

P. Schaeffer. *Traité des objets musicaux*. Pierres vives. Seuil, Paris, 1966.

J. Sundberg, editor. *Music and Motion (special issue)*, volume 29 of *Journal of New Music Research*. Swets and Zeitlinger, Lisse, 2000.

N. P. M. Todd, D. J. O'Boyle, and C. S. Lee. A sensory-motor theory of rhythm, time perception and beat induction. *Journal of New Music Research*, 28(1): 5–28, 1999.

A. Truslit. *Gestaltung und Bewegung in der Musik; ein tönendes Buch vom musikalischen Vortrag und seinem bewegungserlebten Gestalten und Hören*. C. F. Vieweg, Berlin-Lichterfelde, 1938.

F. J. Varela, E. Rosch, and E. Thompson. *The embodied mind, cognitive science and human experience*. MIT Press, Cambridge (Mass.), 2nd print. edition, 1992.

H. von Helmholtz. *Die Lehre von den Tonempfindungen als physiologische Grundlage für die Theorie der Musik*. Fr. Vieweg u. Sohn, Braunschweig, 1863.

M. Wanderley and M. Battier, editors. *Trends in Gestural Control of Music*. IRCAM, Paris, 2000.

K. B. Watson. The nature and measurement of musical meanings. *Psychological Monographs*, 54:1–43, 1942.

L. Wedin. Multidimensional study of perceptual-emotional expression in music. *Scandinavian Journal of Psychology*, 13:241–257, 1972.

F. Winckel. *Vues Nouvelles sur le Monde des Sons*. Dunod, Paris, 1960.

Learning Music: Prospects about Implicit Knowledge in Music, New Technologies and Music Education

Emmanuel Bigand[1], Philippe Lalitte[1] and Barbara Tillmann[2]

[1]LEAD CNRS, Pôle AAFE-Esplanade Erasme, Université de Bourgogne
[2]CNRS-UMR 5020 and IFR 19, Lyon

About this chapter

The chapter proposes an overview of musical learning by underlining the force of the cognitive system, which is able to learn and to treat complex information at an implicit level. The first part summarises recent research in cognitive sciences, that studies the processes of implicit learning in music perception. These studies show that the abilities of non-musicians in perceiving music are very often comparable to those of musicians. The second part illustrates by means of some examples the use of multimedia tools for learning tonal and atonal music; these tools take advantage of the interaction between visual and auditive modalities.

2.1 Introduction

Delineating the musical abilities that are specifically linked to an intensive and formal training from those that emerge through mere exposure to music is a key issue for music cognition, music education, and all of the disciplines involved in sound and music computing, particularly for disciplines which deal with content processing of audio signals (see Chapter 3), machine learning (see Chapter 4), and sound design (see Chapter 10). Non-musicians do not learn a formal system with which they can describe and think about musical structures. Nevertheless, they have a considerable amount of experience with music: they hear music every day of their lives. They have all sung as children and in school, they are moving and dancing to musical rhythms, and most of them have attended concerts. Nowadays the new wearable digital audio players make it easy to listen to a large amount of music in all circumstances. How sophisticated are the emerging abilities to process music that result from this exposure when compared to the abilities caused by an intensive formal musical training? Given the huge differences in training, finding disparities between musically trained and untrained listeners would not be really surprising. However, research in auditory cognition domain has shown that even non-musician listeners have knowledge about the Western tonal musical system. Acquired by mere exposure, this implicit knowledge guides and shapes music perception. This chapter presents recent research studying implicit learning in music, and some examples of multimedia tools for learning Western tonal music as well as contemporary music. These tools are based on advances in cognitive psychology concerning the acquisition and the representation of knowledge, and the role of memory and of attention processes.

2.2 Implicit processing of musical structures

2.2.1 How do non-musician listeners acquire implicit knowledge of music?

Implicit learning processes enable the acquisition of highly complex information without complete verbalisable knowledge of what has been learned (Seger, 1994). Two examples of highly structured systems in our environment are language and music. Listeners become sensitive to the underlying regularities just by mere exposure to linguistic and musical material in everyday life. The implicitly acquired knowledge influences perception and interaction with the environment. This capacity of the cognitive system has been studied in the laboratory with artificial material containing statistical structures, such as finite-state grammars or artificial languages (i.e. Altmann et al., 1995; Reber, 1967, 1989; Saffran et al., 1996). Tonal acculturation is one example of the cognitive capacity to become sensitive to regularities in the environment. Francès (1958) was one of the first underlining the importance of statistical regularities in music for tonal acculturation, suggesting that mere exposure to musical pieces is sufficient to acquire tonal knowledge, even if it remains at an implicit level. In the music cognition domain, numerous research has provided evidence for non-musicians' knowledge about the tonal system (see Bigand and Poulin-Charronnat, 2006, for a review).

2.2.2 Implicit learning of Western pitch regularities

Western tonal music constitutes a constrained system of regularities (i.e. frequency of occurrence and co-occurrence of musical events, and psychoacoustic regularities) based on a limited number of elements. This section presents the tonal system from the perspective of cognitive psychology: it underlines the basic regularities between musical events, which appear in most musical styles of everyday life (e.g. classical music, pop music, jazz music, Latin music, etc.) and which can be acquired by implicit learning processes. The Western tonal system is based on 12 pitches repeated cyclically over octaves. Strong regularities of co-occurrence and frequencies of occurrence exist among these 12 pitch classes (referred to as the tones C, C#/Db, D, D#/Eb, E, F, F#/Gb, G, G#/Ab, A, A#/Bb, B): tones are combined into chords and into keys, forming a three-level organisational system (Figure 2.1). Based on tones and chords, keys (tonalities) have more or less close harmonic relations to each other. Keys sharing numerous tones and chords are said to be harmonically related. The strength of harmonic relations depends on the number of shared events. In music theory, major keys are conceived spatially as a circle (i.e. the circle of fifths), with harmonic distance represented by the number of steps on the circle. Inter-key distances are also defined between major and minor keys. The three levels of musical units (i.e. tones, chords, keys) occur with strong regularities of co-occurrence. Tones and chords belonging to the same key are more likely to co-occur in a musical piece than tones and chords belonging to different keys. Changes between keys are more likely to occur between closely related keys (e.g. C and G major) than between less-related ones (e.g. C and E major). Within each key, tones and chords have different tonal functions creating tonal and harmonic hierarchies. These within-key hierarchies are strongly correlated with the frequency of occurrence of tones and chords in Western musical pieces. Tones and chords used with higher frequency (and longer duration) correspond to events that are defined by music theory as having more important functions in a given key (Budge, 1943; Francès, 1958; Krumhansl, 1990).

This short description reveals a fundamental characteristic of the Western tonal music: functions of tones and chords depend on the established key. The same event can define an in-key or an out-of-key event and can take different levels of functional importance. For listeners, understanding context dependency of musical events' functions is crucial for the understanding of musical structures. Music cognition research suggests that mere exposure to Western musical pieces suffices to develop implicit, but nevertheless sophisticated, knowledge of the tonal system. Just by listening to music in everyday life, listeners become sensitive to the regularities of the tonal system without being necessarily able to verbalise them (Dowling and Harwood, 1986; Francès, 1958; Krumhansl, 1990). The seminal work by Krumhansl, Bharucha and colleagues has investigated the perception of relations between tones and

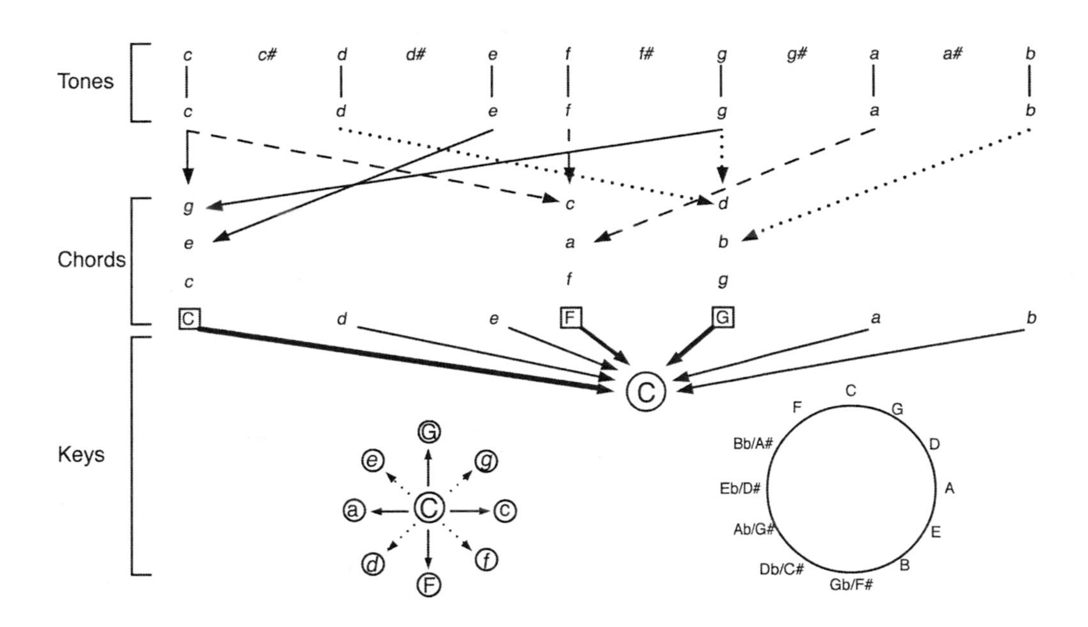

Figure 2.1: Schematic representations of the three organisational levels of the tonal system.
a) 12 pitch classes, followed by the diatonic scale in C Major. b) construction of three major chords, followed by the chord set in the key of C Major. c) relations of the C Major key with close major and minor keys (left) and with all major keys forming the circle of fifths (right). Tones are represented in italics, minor and major chords/keys in lower and upper case respectively (from Tillmann et al., 2001).

between chords as well as the influence of a changing tonal context on the perceived relations (see Krumhansl, 1990, for a review). The data showed the cognitive reality of tonal and harmonic hierarchies for listeners and the context dependency of musical tones and chords in perception and memorisation.

2.2.3 Connectionist model of musical knowledge representation and its acquisition

Bharucha (1987) proposed a connectionist account of tonal knowledge representation. In the MUSACT model (i.e. MUSical ACTivation), tonal knowledge is conceived as a network of interconnected units (Figure 2.2). The units are organised in three layers corresponding to tones, chords, and keys. Each tone unit is connected to the chords of which that tone is a component. Analogously, each chord unit is connected to the keys of which it is a member. Musical relations emerge from the activation that reverberates via connected links between tone, chord and key units. When a chord is played, the units representing the sounded component tones are activated and activation reverberates between the layers until equilibrium is reached (see Bharucha, 1987; Bigand et al., 1999, for more details). The emerging activation patterns reflect tonal and harmonic hierarchies of the established key: for example, units representing harmonically related chords are activated more strongly than units representing unrelated chords. The context dependency of musical events in the tonal system is thus not stored explicitly for each of the different keys, but emerges from activation spreading through the network.

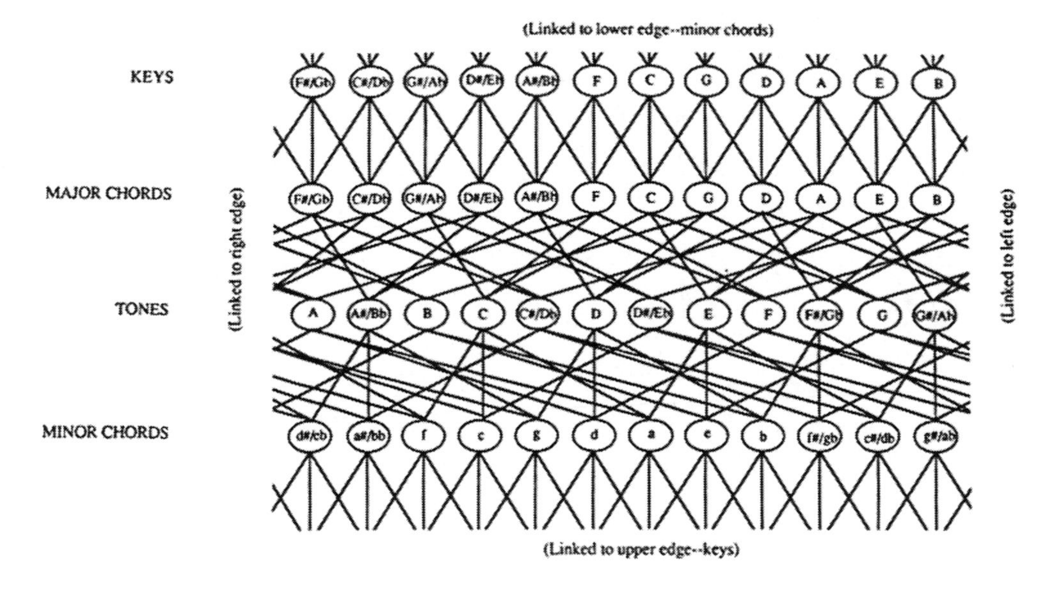

Figure 2.2: MUSACT model of tonal knowledge activation. The tone layer is the input layer, which is connected to the chord layer (consisting of major and minor chords). The chord layer is connected to the key layer (the third layer). Adapted from Bharucha (1987).

In Tillmann et al. (2000), we take advantage of the learning possibilities of artificial neural networks (e.g. connectionist models) to simulate tonal knowledge acquisition in non-musician listeners. For this purpose, unsupervised learning algorithms seem to be well suited: they extract statistical regularities via passive exposure and encode events that often occur together (Grossberg, 1970, 1976; Kohonen, 1995; Rumelhart and Zipser, 1985; von der Malsberg, 1973). Self-organizing maps (Kohonen, 1995) are one version of unsupervised learning algorithms that leads to a topological organisation of the learned information.

To simulate tonal acculturation, a hierarchical network composed of two self-organizing maps was exposed to short musical sequences (i.e., chord sequences). After learning, the connections in the network have changed and the units have specialised for the detection of chords and keys (the input layer coded the tones of the input material). The learned architecture is associated with a spreading activation process (as in MUSACT) to simulate top-down influences on the activation patterns. Interestingly, the learned connections and the activation patterns after reverberation mirror the outcome of the hard-wired network MUSACT, which has been conceived as an idealised end-state of implicit learning processes (see Tillmann et al., 2000). In collaboration with Michel Paindavoine (LE2I-CNRS, Dijon) and Charles Delbé (LEAD-CNRS, Dijon), the authors are currently working on several extensions of this connectionist approach. One of the projects concerns the construction of a set of "artificial musical ears" for this modelling approach. A step of auditory pre-processing will allow us to decode sound files and to work with musical stimuli having greater acoustic complexity. On the basis of this richer input, a network will be trained with a corpus of real recordings containing a variety of musical pieces.

2.2.4 Studying implicit learning processes with artificial materials

Implicit learning processes are supposed to be at the origin of listeners' tonal knowledge, acquired in everyday life. Implicit learning processes are studied more closely in the laboratory with artificial materials containing statistical regularities. In the seminal work by Reber (1967), participants were asked to memorise grammatical letter strings in a first phase of the experiment. They were unaware that rules existed. During the second phase of the experiment, they were informed that the previously seen sequences had been produced by a rule system (which was not described) and were asked to judge the grammaticality of new letter strings. Participants differentiated grammatical letter strings from new ungrammatical ones at better than chance level. Most of them were unable to explain the rules underlying the grammar in free verbal reports (e.g. Altmann et al., 1995; Dienes et al., 1991; Dienes and Longuet-Higgins, 2004; Reber, 1967, 1989).

Various findings are convergent in demonstrating the cognitive capacity to learn complex structures and regularities. The acquisition of regularities in the experimental material is not restricted to visual events (e.g. letters, lights, shapes), but has been extended to auditory events, such as sine waves (Altmann et al., 1995), musical timbres (e.g. gong, trumpet, piano, violin, voice in Bigand et al., 1998) or environmental sounds (e.g. drill, clap, steam in Howard and Ballas, 1980, 1982). Recent studies have started to consider the acoustical characteristics of the sound, such as prosodic cues (Johnson and Jusczyk, 2001; Thiessen and Saffran, 2003; Saffran et al., 1996) or acoustical similarities (Tillmann and McAdams, 2004). The aim of the studies was to test whether the relation between the statistical regularities and regularities inherent in the acoustical material could influence learning: conflicting information might hinder statistical learning, while converging information might facilitate learning. Tonal acculturation might represent a beneficial configuration: musical events appearing frequently together are also linked acoustically since they share (real and virtual) harmonics. To investigate whether convergence with acoustical features represent a facilitating or even necessary condition for statistical learning, Tillmann and McAdams (2004) systematically manipulated acoustical similarities between timbres so that they either underline the statistical regularities of the timbre units, contradict these regularities or are neutral to them. The outcome showed that listeners learned the statistical regularities of the complex auditory material and that the manipulated surface characteristics did not affect this statistical learning. The surface characteristics only affected grouping and overall preference bias for the different materials. This outcome suggests that tonal acculturation does not necessarily need the convergence between statistical and acoustical regularities. Supporting evidence can be found in acculturation to Arabic music, which is lacking the convergence between statistical and acoustic features (Ayari and McAdams, 2003). Together with the implicit learning study on twelve-tone music (see below), the data emits the rather encouraging hypothesis about the possibility to learn regularities of new musical styles.

2.2.5 Implicit learning of new musical systems

Music is an interesting medium to investigate implicit learning processes for several reasons. It is a highly complex structure of our environment that is too complex to be apprehended through explicit thoughts and deductive reasoning. Musical events *per se* are of no importance, yet musical pieces are more than a pleasing succession of coloured sounds. The psychological effects of musical sounds come from the complex multilevel relationships between musical events involved in a given piece (Meyer, 1956; Lerdahl and Jackendoff, 1983). The abstract associative and architectonic relationships that are not close in time define relevant structures in music. These relations cannot be easily articulated in an explicit way. Despite an eminent tradition in music history, as well as in contemporary music theory, to formalise the relevant structure of Western music (see Lerdahl and Jackendoff, 1983; Lerdahl, 2001; Narmour, 1990), none of these frameworks provides a complete and satisfactory account of the Western musical grammars. A further interesting feature of music for research on implicit learning is that musical structures are not always conceived for being explicitly processed. It is even of crucial importance for composers that listeners are sensitive to the structures that underlie a musical piece while still being unaware of them. And in fact, the most common impression among a general audience is that of being unable to verbally describe what they perceive. In some instances, people are even convinced that they do not perceive any underlying structure. The fact that musical events do not refer to any specific object in the external world probably contributes to the difficulty of apprehending musical structures in an explicit way.

A final interesting feature is that musical systems constantly evolve towards new musical grammars. Being faced with masterpieces that derive from an entirely new musical system is not an artificial situation for contemporary listeners and this raises a challenging issue for implicit learning theories. The considerable and persistent confusion reported by listeners to contemporary music suggests that some musical grammars may be too artificial to be internalised through passive exposure (Lerdahl, 1989). As a consequence, several cognitive constraints have been delineated, and musical grammars should obey these constraints in order to be learnable (Lerdahl, 1988, 2001). Contemporary music challenges the ability of the human brain to internalise all types of regularities. This raises a question with implications for cognitive science, music cognition, and contemporary music research.

To the best of our knowledge, very little research has directly addressed implicit learning with musical material (Bigand et al., 1998; Dienes et al., 1991). Numerous research in music cognition, however, deals indirectly with implicit learning processes by showing that explicit learning is not necessary for the development of a sensitivity to the underlying rules of Western music[1] (see section above). Only a few studies have addressed the implicit learning of new musical systems. Most of them have focused on the learning of serial music, a system that appeared in the West in the first half of the 20th century. During this period, the tonal system was overtaken by the serial system developed, in particular, by Schoenberg (Griffiths, 1978). Serial works of music obey compositional rules that differ from those that govern tonal music.

A serial musical piece is based on a specific ordering of the chromatic scale called the twelve-tone row. A twelve-tone row is an arrangement, into a certain order, of the twelve tones of the chromatic scale regardless of register (Figure 2.3). The tones of the row must be used in their chosen order (repetition of tones is allowed in certain circumstances and two or more successive tones of the row may appear as a chord), and once all twelve tones of the row have appeared, the row is repeated again and again until the end of the composition. The row may appear in any of its four basic forms: the original row, the

[1]The tonal system designates the most usual style of music in the West, including, Baroque (Bach), Classic (Mozart) and Romantic (Chopin) music, as well as to a certain extent folk music such as pop-music, jazz and Latin-music.

inverted form (in which ascending intervals of the original row are replaced by equivalent descending ones and vice versa), the retrograde form (in which the tones of the original row are read backwards), and the retrograde inversion (in which the tones of the inverted form are read backwards), and each of the four forms of the row may be transposed to any of the twelve tones of the chromatic scale, thus making available forty-eight permissible patterns of one row. In theory, each tone of the row should have roughly the same frequency of occurrence over the entire piece.

Each serial composition results from a complex combination of all of these transformations which are applied to one specific tone row. Schoenberg argued that these manipulations would produce an interesting balance between perceptual variety and unity. A critical point on which he insisted was that the initial row must remain unchanged throughout the entire piece. In other words, Schoenberg's cognitive intuition was that the perceptual coherence deriving from the serial grammar was unlikely to be immediately perceived but would result from a familiarisation with the row.

Several experimental studies have addressed the psychological reality of the organisation resulting from serial musical grammar. The oldest, by Francès (1958, exp. 6), consisted of presenting participants with 28 musical pieces based on a specific tone row and requiring participants to detect four pieces that violated the row. These odd pieces were actually derived from another row (the foil row). The analysis of accurate responses revealed that participants had considerable difficulty in detecting the four musical pieces that violated the initial row. Moreover, the fact that music theorists specialised in serial music did not respond differently from musically untrained participants suggests that extensive exposure to serial works is not sufficient for the internalisation of this new musical system. Although Francès' research is remarkable as pioneer work in this domain, the study contained several weaknesses relative to the experimental design as well as to the analysis of the data and this detracts from the impact of his conclusion. The most noticeable problem concerns the foil row, notably because it was strongly related to the tested row.

Empirical evidence supporting the perceptual reality of the rules of serial music was reported by Dowling (1972) with short melodies of 5 tones.

In Dowling's experiment, participants were trained to identify reversed, retrograde and retrograde-inversion of standard melodies of 5 tones with equal duration. The melodies were deliberately made with small pitch intervals in order to improve performance. Dowling observed that musically untrained participants managed to identify above chance the rules of the serial music, with highest accuracy for the reversed transformation and the lowest for the retrograde inversion. Given that Dowling's musical stimuli were extremely short and simple, it is difficult to conclude that the rules of serial music may be internalised from a passive hearing of serial music. Moreover, in a very similar experiment using 12 tones instead of 5, Delannoy (1972) reported that participants did not succeed above chance in distinguishing permitted transformations of a standard musical sequence from those that violated the serial rules.

More recently, Dienes and Longuet-Higgins (2004) have attempted to train participants in the grammar of serial music by presenting them with 50 musical sequences that illustrated one of the transformation rules of serial music. The second half of the row was a transformation of the first half (i.e., a reverse, a retrograde or a retrograde inversion transformation). After this familiarisation phase, participants were presented with a new set of 50 sequences, some of them violating the rules of serial music (i.e., the last 6 notes were not a permitted transformation of the first 6). Participants were required to differentiate grammatical pieces (according to serial rules) from non-grammatical ones. Accuracy rates generally did not differ from chance level, which is consistent with Francès (1958)' and Delannoy (1972)'s findings.

A critical feature of the experiment of Dienes and Longuet-Higgins (2004) is that participants had never been exposed to a single tone row. Participants were trained with the transformational rules of serial music, but these rules were always instantiated with a new set of tones. The temporal order of the first 6 notes was chosen at random. As a consequence, the referential row was constantly moving from one trial to the other. This procedure is very demanding since it consists in requiring participants to learn abstract rules which are illustrated by a constantly changing alphabet. To the best of our knowledge, there is no evidence in the implicit learning domain to show that learning can occur in this kind of situation. If participants do not have the opportunity to be exposed to an invariant tone row in the training phase, it is not surprising that they fail to exhibit sensitivity to the serial grammar in the test phase. It should be noticed that this situation violates the basic principle of serial music postulating that only one row should be used for one piece. Krumhansl and Sandell (1987) have provided the strongest support for the psychological relevance of serial rules. As illustrated in their study, experiments 1 and 2 were run with simple forms of two tone rows. That is to say, the tone rows were played with isochronous tones that never exceeded the pitch range of one octave. Experiments 3 and 4 were run with excerpts of *Wind Quintet* op. 26 and *String Quartet* op. 37 by Schoenberg. The results of the classification tasks used in Experiments 2 and 3 demonstrated that participants discriminated, above chance level, between inversion, retrograde, and retrograde inversion of the two tone rows with correct responses varying from 73% to 85% in Experiment 2, and from 60% to 80% in Experiment 3. At first glance, this high accuracy is surprising. However, it should be noticed that participants were exposed to very simple forms of the tone rows a great number of times during Experiment 1. It seems likely that this exposure helps to explain the good performance. In other words, the peculiar importance of this study lies in the suggestion that previous exposure to a tone row can be a critical feature for the perception of the rules of serial music. The question remains, however, about the type of learning that actually occurred during this prior exposure. Given that all participants had formal instruction in music education, we cannot rule out the possibility that they used their explicit knowledge of musical notation to mentally represent the structures of the two rows.

In order to define the nature (implicit/explicit) of the knowledge in learning serial music rules, Bigand, D'Adamo and Poulin (in revision) have tested the ability of musically untrained and trained listeners to internalise serial music rules with 80 two-voice pieces, especially designed by the composer D. A. D'Adamo. A set of 40 pieces defined various instantiations (transpositions) of one twelve-tone row (grammatical pieces). The other set of 40 pieces were derived from another twelve-tone row (ungrammatical pieces). As it is shown in Figure 2.3, each ungrammatical piece was matched to a grammatical piece according to their superficial features (rhythm, pitch ranges, overall form of melodic contour, duration, dynamics). The ungrammatical pieces differed from the grammatical pieces only in the twelve-tone row used. In the learning phase, 20 pieces were presented twice to the participants, who had simply to indicate whether a given piece was heard for the first time or for the second time. In the test phase, 20 pairs of pieces were presented to the participants. Each pair contained a grammatical piece which had not been heard in the training phase and a matched ungrammatical piece (Figure 2.3). Since the pieces of a pair shared the same musical surface (i.e. same pitch range, melodic contour and rhythm), even if they were derived from two different twelve-tone rows, they sounded very close. The participants were asked to indicate which piece of the pair was composed in the same way as the pieces of the learning phase had been. All participants reported extreme difficulties in performing the task. Numerous participants complained that it was difficult to differentiate the two pieces of the pairs. Both experimental groups nevertheless performed above chance with 61% correct responses for non-musicians and 62% of correct responses for musicians, and with no significant difference between the two groups. In a second experiment (run with musically untrained listeners only), the stimuli of the learning phase were identical to those of the previous experiment, whereas the stimuli of the test phase consisted of pairs in which one of the pieces was derived from a retrograde inversion of the tested row. The striking finding was that the participants continued to discriminate grammatical from ungrammatical pieces above chance (60% of correct responses), suggesting that even musically untrained listeners are able to internalise via passive exposure complex regularities derived from the twelve-tone technique. This conclusion is consistent with other findings showing that the structures of Western contemporary music are processed in a similar way by musically trained and untrained listeners. After a short exposition phase, listeners were sensitive to the structure of twelve-tone music. The perception of this music is assumed to be based on frequency distributions of tone intervals. These results shed some light on the implicit versus explicit nature of the acquired knowledge, and the content of the information internalised through listening to these pieces.

Most probably, the knowledge internalised during the listening to the serial pieces was inaccessible to the explicit thought of the participants. If knowledge internalised through exposure was represented at an explicit level, then experts should be more able than non-expert participants to explicitly use this knowledge. This difference should result in a clear advantage for musical experts over musically untrained listeners. If, however, the acquired knowledge is represented at an implicit level, no strong difference should be observed between musically expert and novice participants. The present study converges with conclusions drawn from several other studies run with Western tonal music and argues in favor of the implicit nature of acquired musical knowledge.

Figure 2.3: Higher panel: One of the two twelve-tone row used in the study (the "grammatical" one). The row is shown in these four basic forms: original (O), inverted (INV), retrograde (RET), and retrograde inversion (RET INV). Lower panels: Example of pairs of matched pieces composed using two different rows ("grammatical" and "ungrammatical"). Both pieces share the same superficial features (rhythm, pitch ranges, overall form of melodic contour, duration, dynamics).

2.3 Perspectives in musical learning: using multimedia technologies

2.3.1 How should the learning of Western tonal music be optimised with the help of multimedia technologies?

Explaining the theoretical core of the Western musical system is one of the most difficult tasks for music teachers, and it is generally assumed that this explanation should only occur at the end of the curriculum in both music conservatoire and university departments. Lerdahl's Tonal Pitch Space Theory (TPST, Lerdahl, 2001) is likely to contribute to the development of music tools that would help music lovers as well as those at an early stage of musical study to improve their understanding of Western tonal music. The TPST can be considered as an idealised knowledge representation of tonal hierarchy. The psychological representation of knowledge implies a certain number of questions, for which different solutions have been proposed (Krumhansl et al., 1982a,b; Krumhansl and Kessler, 1982; Longuet-Higgins, 1978). For all these approaches, tonal hierarchies are represented in the form of a multidimensional space, in which the distances of chords from the instantiated tonic correspond to their relative hierarchical importance. The more important the chord is, the smaller the distance. Lerdahl successfully explains the way in which the TPST synthesises various existing musicological and psychological models and suggests new solutions. In the opinion of the authors, the crucial contribution of the model is the description of a formal tool to quantify the tonal distances between any couple of events belonging to any key, a quantification that no other approach proposes.

The TPST model outlines several developments to the model initially described in an earlier series of articles (Lerdahl, 1988, 1991). We summarise here the basic ideas. According to the theory, tonal hierarchy is represented in three embedded levels. The first two (the pitch class level and chordal level) represent within-key hierarchies between tones and chords. The third level represents the distances between keys (region level). The pitch class level (basic space) represents the relation between the 12 pitch classes. It contains five sublevels (from level a to e), corresponding to the chromatic level (level e), diatonic level (level d), triadic level (level c), fifth level (level b) and the tonic level (level a). In a given context, a tonic tone, part of a tonic chord, will be represented at all five levels. The fifth and the third tones of a tonic chord will be represented at four levels (from b to e) and three levels (from c to e) respectively. A diatonic but non-chordal tone will be represented at two levels (from d to e). A non-diatonic chord will be represented at only one level (level e). The level at which a given pitch class is represented thus reflects its importance in the tonal context. For example, in the C major key, the tone C is represented at all levels (from a to e), the tone G, at four levels (from b to e), the tone E, at three levels (from c to e) and the diatonic tones of the C major scale are represented at two levels only (from d to e).

This representation has two implications. First, it allows an understanding as to why tones (e.g. of a C major chord), which are distant in interval (C-E-G-C), can nevertheless be perceived being as close as are adjacent notes (C-D-E-F-G). Though forming distant intervals, the notes of the chord are adjacent at the triadic level in the representational space (level c). Moreover, this explanation of musical tension bound to these forces of attraction constitutes a very promising development for psychology. The second implication concerns the computation of distances between chords. If the C major chord was played in the context of G major, the tone F# will be represented at two levels (from d to e), while the tone F would remain at only one level (level e). This would produce one change in pitch class. The central idea of the TPST is to consider the number of changes that occurs in this basic space when the musical context is changed (as in the present example) as a way to define the pitch-space distance between two musical events.

The second level of the model involves the chordal level, that is the distance between chords in a given key. The model computes the distances separating the seven diatonic chords taking into account the number of steps that separate the roots of the chords along the circle of fifths (C-G-D-A-E-B-F) and the number of changes in pitch-class levels created by the second chord. Let us consider the distance between the C and G major chords in the key of C major. The G major chord induces 4 changes in the pitch-class level. The dominant tone D is now represented at 2 additional levels (from b to e), the third tone B at one additional level (from c to e) and the tonic tone at one additional level (from a to e). The number of steps that separates the two chords on the circle of fifths equals 1. As a consequence the tonal pitch-space distance between these two chords in this key context equals 5. Following the same rationale, the distance in pitch-space between the tonic and the subdominant chords equals 5. The distance between the tonic and the submediant (sixth degree) chords equals 7, as does the distance between the tonic and the mediant chords (third degree). The distance between the tonic chord and the supertonic (second degree) equals 8 as does the distance between the tonic and the diminished seventh chords (seventh degree). This model quantifies the strength of relations in harmonic progressions. Accordingly, the succession tonic/submediant (I/vi) corresponds to a harmonic progression that creates stronger tension than the succession tonic/subdominant (I-IV).

The third level of the TPST model involves the regional level. It evaluates distances between chords of different regions by taking into account the distances between regions as well as the existence of a pivot region. The regional space of the TPST is created by combining the cycle of fifths and the parallel/relative major-minor cycle. That is to say, the shortest distance in regional space (i.e., 7) is found between a given major key (say C major) and its dominant (G), its subdominant (F), its parallel minor (C minor) and its relative minor key (A minor). The greatest distance (30) is found between a major key and the augmented fourth key (C and F#). The tonal distance between two chords of different keys depends on the musical interpretation of the second chord. For example, in the context of C major key the distance between a C major chord and a C# minor chord would equal 23 if the C# is interpreted as a sixth degree (vi) of the E major key. The distance equals 30 if the C# is understood as the tonic chord of the Db minor key. As a consequence, the distance in pitch-space between two events that belong to distant keys depends on the selected route between the two events. In most cases, the selected route is defined by the overall musical context. By default, the model computes this distance according to the principle of the shortest path: "the pitch-space distance between two events is preferably calculated for the smallest value" (Lerdahl, 2001, p.74). The shortest path principle is psychologically plausible.

It has the heuristic merit of being able to influence the analysis of time-span and prolongational reductions (Lerdahl and Jackendoff, 1983) by preferring an analysis that reduces the value of these distances. The implementation of this principle in an artificial system should fairly easily lead to "intelligent" systems capable of automatic harmonic analysis.

One of the main features of the TPST as an efficient learning tool is to bridge the intuitive mental representations of untrained listeners with the mental representations of experts. Current developments in multimedia offer considerable opportunities to evolve the naive representation of novices in a given domain. The basic strategy consists in combining different modes of knowledge representation (e.g. sounds, image, language, animation) to progressively transform the initial mental representation into a representation of the domain that fits as closely as possible with that of experts. In the present case, the use of a space to describe the inner structure of the Western tonal system considerably facilitates this transformation. The mental representation of a complex system in a two or three-dimensional space is a metaphor that is common in a large variety of domains and that is intuitively accessible even for a child. A musical learning tool may thus consist of a multimedia animation that illustrates how music progresses through pitch-space. As it is shown in Figure 2.4, the animation displays in real-time every distance travelled through pitch space. After having listened several times to the piece, the journey through the pitch-space of the piece would be stored in memory in both visual and auditory formats. After listening to several pieces of the same stylistic period, the journeys through the pitch-space specific to this style would be stored in memory. After listening to several pieces of the Western music repertoire, listeners would create a mental representation of the overall structure of the tonal pitch space that fits with that of the experts. From a teaching perspective, the interesting point is that this mental representation will emerge from mere exposure to musical pieces presented with this music tool. In other words, the tool allows a passive exploration of the tonal pitch space by visualizing in a comprehensible format the deep harmonic structure of the heard pieces.

The structure of the space can be adapted at will and should notably be adjusted to suit the age of the user. At this early stage of development of the multimedia tool, we chose a structure that mimics real space with planets and satellites. Given the circularity of the Western musical space, only one portion of the space can be seen at a given time point, but this portion will progressively change when the music is moving from one region to another. A planet metaphorically represents a key, while the satellites represent the seven diatonic chords. Satellites corresponding to played chords are lit up in yellow, thus representing the route of the harmonic progressions within each key. The colour of the planet representing the key intensifies when several chords from the key are played, thus imitating the fact that the feeling of the tonality increases with duration. When the music modulates to another key, chords from both the initial key and the new key light up, and the animation turns towards the new key, and then discovering another portion of the tonal pitch-space. When the piece of music progresses rapidly towards distant keys, as in the case of Chopin's *Prelude* in E major, the pivot keys are briefly highlighted and passed quickly. The journey depends upon the modulations that have occurred. With the present tool, the user can associate the visual journey through tonal pitch space with the auditory sensation created by the music. The animation contains sufficient music theoretic information to allow the user to describe this musical journey in terms that are close to those employed by musicologists. Of course, this animation may also bring other important elements for the comprehension of harmonic processes, such as the arrangement of chords and voice leading. Connected to a MIDI instrument, it may equally be transformed into a tool for tonal music composition. By chaining chords together, the user can follow his or her journey through tonal space, and explore the structure of the tonal space.

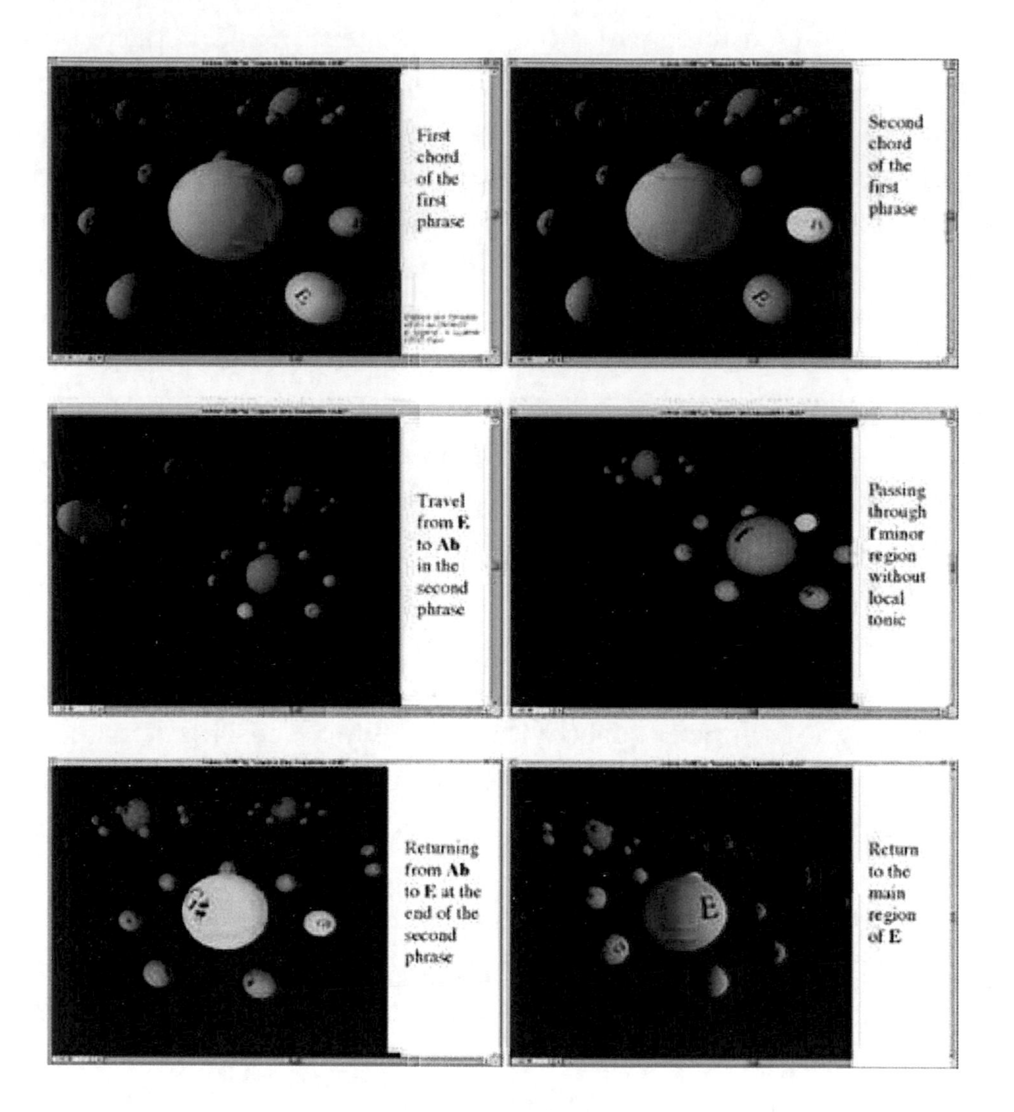

Figure 2.4: Musical tool derived from TPST and which is currently being developed by LEAD - This animation is available at the address `http://www.u-bourgogne.fr/LEAD/people/bigand_e.html`. This research is supported by a CNRS grant "Société de l'information". The animation was realised in collaboration with Aristide Quenel.

2.3.2 Creating learning multimedia tools for music with the contribution of cognitive sciences and ergonomics

The first multimedia works comprising music began to emerge at the beginning of the nineties. Since then, the number and diversity of musical multimedia products (CD-Rom, DVD-Rom, web sites) have been increasing considerably. However, multimedia products helping the user to integrate musical structures are rare. If we want to propose successful learning multimedia tools, the main question is how to use multimedia resources and why. We are going to present some assumptions of cognitive ergonomics that seem fundamental to multimedia dedicated to music education. These assumptions will be illustrated by two multimedia learning tools created at the LEAD. The perceptual supposed advantage of the tools will be evaluated during a next phase by several tests concerning attentional processing, memory and understanding.

The first principle is that the vantage point of music learning tools should be immediate representation formats of the non-experts. The tools have to combine in an advantageous way the various possibilities of multimodal representations to make these initial representations evolve towards those of experts (the principle of affordance; Gibson, 1977). Multimodality should be used as a powerful means to clarify the structure of complex systems, and to allow the user to easily develop a mental representation of the system. This mental representation should be compatible with the representation of experts. The aim of the project was to give listeners the access to a musical system that is often considered as complex (contemporary music) while being potentially of high educational value. It was an ideal opportunity to try out a multimedia approach to the learning of a complex system.

Reduction of information and optimisation of presentation forms

One of the main problems concerning multimedia is an overload of presentation forms. Multiplication of presentation forms (text, picture, animation, video, sound, etc.) often entails a cognitive cost that is high compared to the benefits in terms of training. This profusion of presentation forms often leads to an explosion of the quantity of information presented to the user, and to the lack of an objective analysis of the presentation forms used and of the combination of learning modes (visual & auditory, verbal & visual, etc.) available in the proposed tools. Information overload is often accompanied by an organisation of knowledge based on models that are not adapted to the initial knowledge of a user. The second fundamental principle in producing multimedia learning tools is thus reducing the quantity of information and optimizing the form in which it is presented. Properly used multimodality, particularly concerning the interaction between vision and audition, improves attentional processes, memorisation of musical material, and develops the capacity to represent the musical structures. There are several types of representation of the musical structures. The score is the best known of the notation. However, it requires specific knowledge and regular musical practice. There are other forms of music representation: tablatures of string instruments, sonograms (time-evolving spectrum), wave forms (amplitude), tracks or piano-rolls of sequencer softwares, etc. All these representation modes can certainly be employed in multimedia, but they often require expert knowledge.

The project carried out at the LEAD used graphical representations that can advantageously replace the representation forms of experts. These graphics consist of simple forms symbolizing one or more elements of musical structure (melodic contour, texture, harmonic density, rhythmic pattern, etc). The principal constraint is that these forms should not require additional coding, but induce the musical structure in an intuitive and direct way. Other presentation forms, which require expert knowledge, never intervene in the initial presentation of a musical excerpt. Figures 2.5 and 2.6 show two representation forms of a chord sequence in the piece *Couleurs de la Cité Céleste* by Oliver Messiaen. The constitution in terms of tones is identical for the 13 chords. It is the register, the duration and the change in instrumentation between the various instruments which give listeners an impression of a succession of sound colours (*Klangfarbenmelodie*). The excerpt is represented by blocks of colours whose width corresponds to the duration of the chords. The choice of the colours has been determined by the name of the colours written by the composer in the score. Their height symbolises the extent of chords (from the lowest to the highest). The position on the scale (on the left) represents the register. Blocks appear synchronously with the sound. With this type of representation, it is easy, for non-expert listeners, to perceive a degree of similarity between certain chords. Based on this type of representation, a user may intuitively become aware of the external structure of a sequence, even if it is not sufficient to form a precise representation of the musical structure. Figure 2.6 represents the same sequence of chords in form of a score. In order to focus listener's attention on the harmonic structure, the real duration of the chords was replaced by an equal duration for all chords. However, in contrast to a graphical representation of sound that was privileged here, this mode of representation is to give the users an opportunity to decompose musical structures. The users can choose what they want to listen to: the whole sequence, each chord separately, groups of instruments within a chord or each note of a chord.

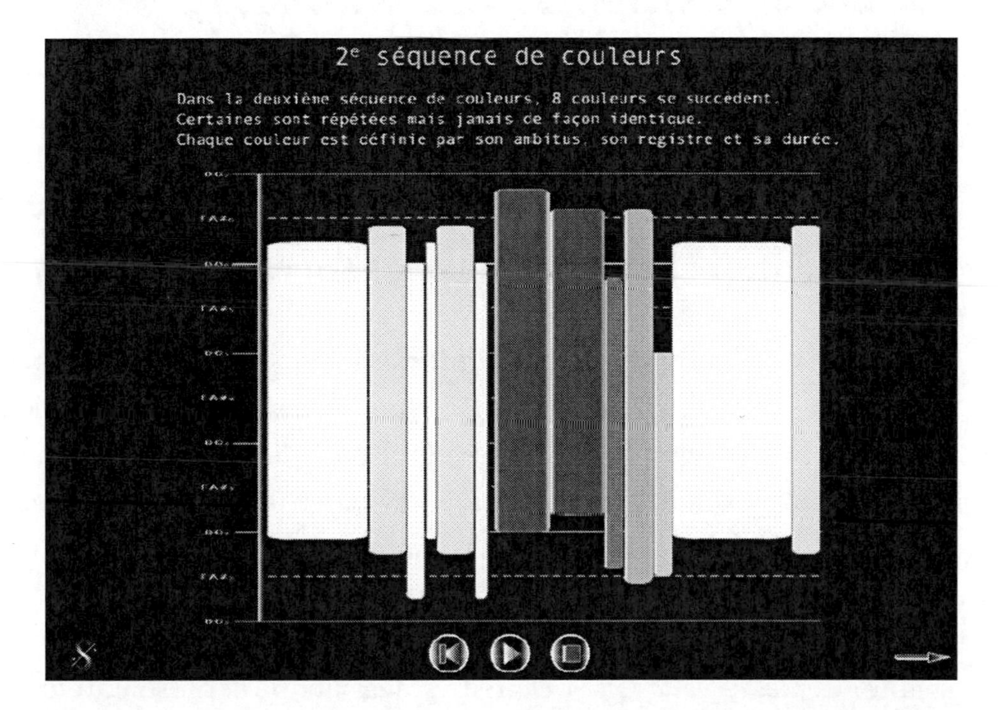

Figure 2.5: Multimedia tool for learning contemporary music. Graphic representation of a chord sequence of *Couleurs de la Cité céleste* by Olivier Messiaen.

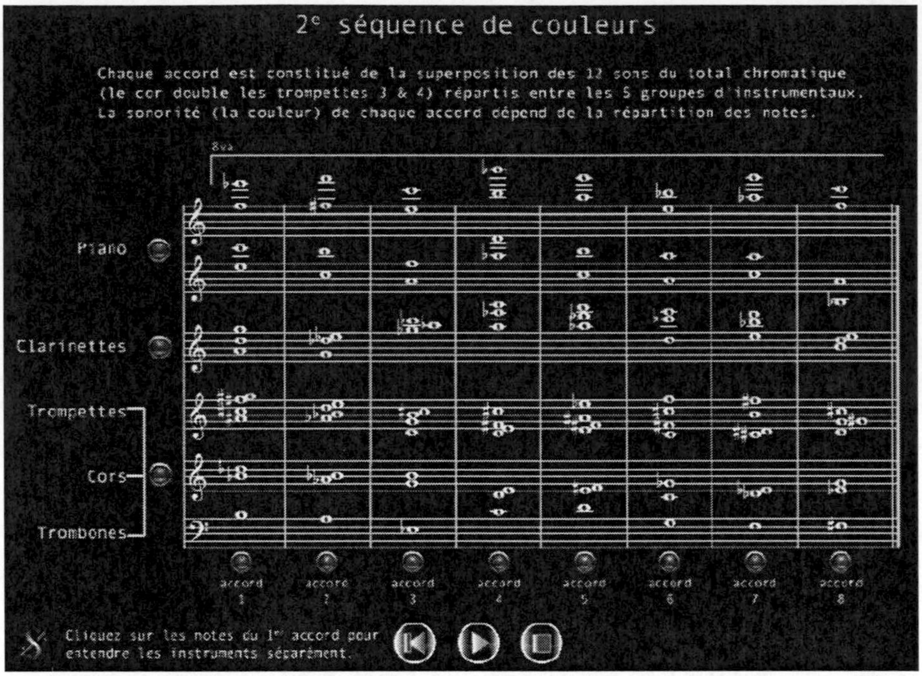

Figure 2.6: Multimedia tool for learning contemporary music. Score representation of the same chord sequence as Figure 2.5.

Synthesis of knowledge and implementation of continuity

Multimedia tools of learning should synthesise the knowledge of music in order to make it available to non-experts. Thus, it is necessary to implement this knowledge in a way adapted to the initial knowledge of the user. In the case of music (complex music in particular), it is important to raise the question of perceptibility of musical structures. It is a question of knowing exactly what should be emphasised. In our project, the pieces were selected according to the cognitive problems they represent (relating to their aesthetic differences). For example, *Couleurs de la Cité Céleste* by Messiaen is representative of the aesthetics where colour and timbre are important. This piece is composed of a great variety of musical elements that follow one another to form a sound mosaic. Globally, a multimedia learning tool must favour categorisation and memorisation of musical material, in order to allow the emergence of the mental representation of a temporal organisation of a piece. Figure 2.7 shows the main page of the multimedia learning tool of Messiaen's piece. One can see the representation of the formal structure of the excerpt in the center of the screen. Eight icons on the right and on the left of the screen give access to eight links (history of the piece, composer's biography, the orchestra, a large-scale structure of the excerpt, and four of the main materials of the piece: texts and metaphors of the Apocalypse, Gregorian chant, colours, and bird songs).

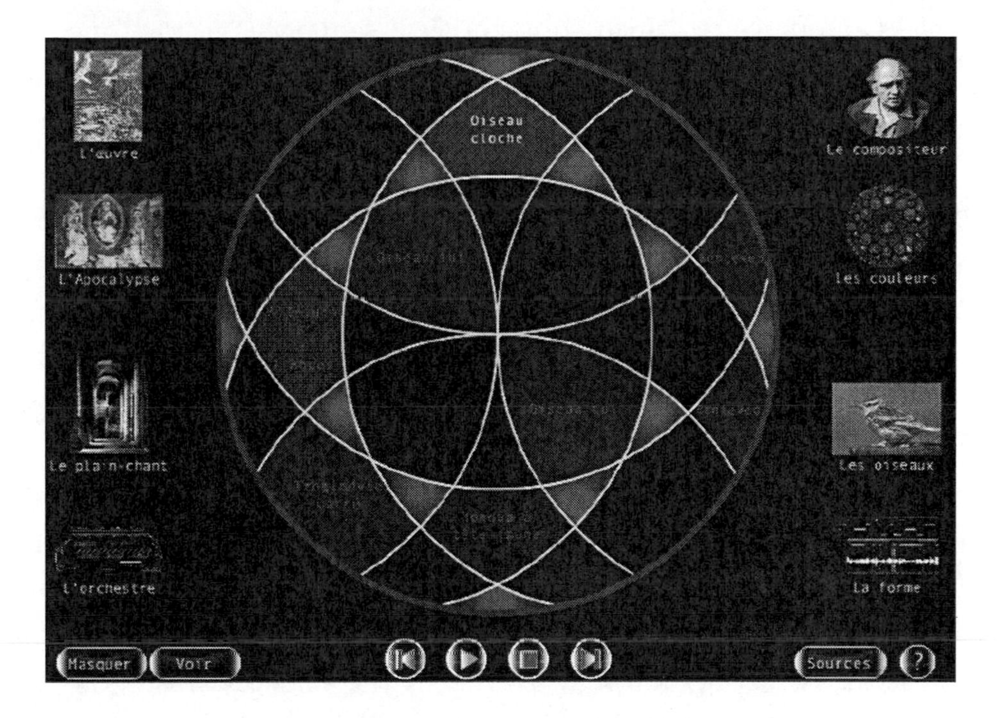

Figure 2.7: Multimedia tool for learning contemporary music. Main page of the multimedia learning tool of *Couleurs de la Cité céleste* de la Cité céleste by Olivier Messiaen.

One of the crucial problems in the pedagogy of listening is that of attention. Perception of musical structures is strongly dependent on attentional processes. In the simplest case, attentional processes are guided by the music itself (when, for example, a composer emphasises the principal melody by a discrete accompaniment) or by the performer (when he chooses to emphasise a specific structural element). However, most of the time, music has a complex and deliberately ambiguous structure. Contrary to the traditional methods in music education, multimedia tools make it possible to easily focus the listener's attention on internal or external elements of musical structure.

One part of our project consisted in seeking in the resources of multimedia the means of guiding attentional processes and, beyond, of favouring the memorisation and the comprehension of musical structures. The schematic representation of the formal structure of the beginning of *Couleurs de la Cité Céleste* (Figure 2.7) was conceived to facilitate mental representation of a complex formal structure in which short musical sequences follow one another in form of a mosaic that would emerge during perception. This multimedia animation does not contain any oral or textual explanation. Awareness of the structure emerges solely from the visual and auditory interaction. The choice of a circular representation corresponds to the form of a piece whose elements return in a recurrent and circular way. Each piece of the mosaic corresponds to a short musical sequence. Each colour represents a type of musical material (e.g. blue for bird songs). Nuances of colours differentiate the variations inside each category of sequence. The animation takes into account the cognitive processes of attention and memorisation. At the beginning of the animation, the stained-glass scheme is empty. Progressively, as the music unfolds, empty spaces are filled until the stained-glass is completed (all the sequences were played). When a sequence has ended, the corresponding stained-glass is gradually obscured (approximately 6 to 7 seconds, according to the maximum duration of the perceptual present; Fraisse, 1957). The luminosity of the piece of stained-glass solidifies at a very low rate. This process is very close to a trace of an event remaining in memory. When an identical sequence returns, the part that had been previously activated is briefly reactivated and then turns over in stand-by. This multimedia artifice supports the categorisation and the memorisation of materials. It also makes it possible to establish bonds of similarity and consequently gives direction to the formal structure which proceeds under the eyes of the user. This example illustrates how it is possible to guide listeners to focus their attention on sequential events. It is also useful to focus the attention of the user on simultaneous events. Figure 2.8 shows the animated representation of the formal structure of the beginning of *Eight Lines* by S. Reich. In contrast to the Messiaen's piece represented by stained-glass, the piece by Reich, whose unfolding follows a linear trajectory, is represented by 8 rectangular boxes. Coloured rectangular paving stones indicate the mo-

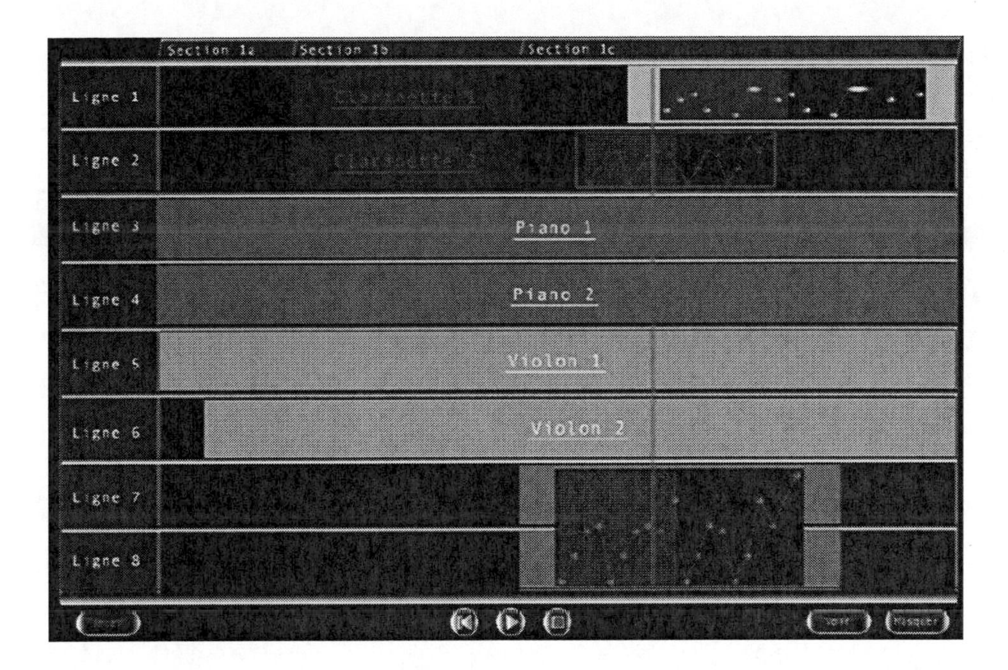

Figure 2.8: Multimedia tool for learning contemporary music. Formal structure represented in the multimedia learning tool of *Eight Lines* by S. Reich.

ment of appearance and the duration of intervention of each instrumental part. Synchronisation between the sound and the image is visualised by a vertical marker. When an instrument is active, its coloured paving stone is cleared up. In Figure 2.8, the active instrumental parts are the parts of viola and violoncello (lines 7 and 8, at the bottom), flute and bass clarinet (lines 1 and 2, at the top). Inside a paving stone, graphical animations, always synchronised with music, emerge to focus the attention on the melody of the instrumental part. The points indicate the onset of pitches, the lines indicate the melody contour.

New technologies of sound processing may provide other new possibilities for multimedia learning tools for music. The sound files obtained with these techniques or especially dedicated software can be integrated into the multimedia in order to improve the learning tools. The possibility of an interactive deconstruction or reconstruction of the musical structures combined to specific visual interfaces is certainly the most promising perspective in the nearest future.

Conclusion

The power of implicit learning is without doubt one of the major contributions of research into musical cognition. It reduces the distance between listeners (non-musicians and musicians) and leads to question the common practices in music education. Implicit learning supplies a solid scientific basis, together with the contributions of cognitive psychology (memory and attending processes), ergonomics and new technologies to create multimedia learning tools for music.

Bibliography

G. T. Altmann, Z. Dienes, and A. Goode. Modality independence of implicitly learned grammatical knowledge. *Journal of Experimental Psychology: Learning, Memory and Cognition*, 21:899–912, 1995.

M. Ayari and S. McAdams. Aural analysis of arabic improvised instrumental music (tagsim). *Music Perception*, 21:159–216, 2003.

J. J. Bharucha. Music cognition and perceptual facilitation: A connectionist framework. *Music Perception*, 5(1):1–30, 1987.

E. Bigand and B. Poulin-Charronnat. Are we "experienced listeners"? a review of the musical capacities that do not depend on formal musical training. *Cognition*, 100(1):100–130, 2006.

E. Bigand, P. Perruchet, and M. Boyer. Implicit learning of an artificial grammar of musical timbres. *Cahiers de Psychologie Cognitive/Current Psychology of Cognition*, 17(3):577–600, 1998.

E. Bigand, F. Madurell, B. Tillmann, and M. Pineau. Effect of global structure and temporal organization on chord processing. *Journal of Experimental Psychology: Human Perception and Performance*, 25(1):184–197, 1999.

H. Budge. *A study of chord frequencies*. Bureau of Publications, Teachers College, Columbia University, New York, 1943.

C. Delannoy. Detection and discrimination of dodecaphonic series. *Interface*, 1:13–27, 1972.

Z. Dienes and C. Longuet-Higgins. Can musical transformations be implicitly learned? *Cognitive Science*, 28:531–558, 2004.

Z. Dienes, D. Broadbent, and D. C. Berry. Implicit and explicit knowledge bases in artificial grammar learning. *Journal of Experimental Psychology: Learning, Memory and Cognition*, 17:875–887, 1991.

J. W. Dowling. Recognition of melodic transformations: inversion, retrograde, retrograde inversion. *Perception and Psychophysics*, 12:417–421, 1972.

J. W. Dowling and D. L. Harwood. *Music Cognition*. Academic Press, Orlando, Florida, 1986.

P. Fraisse. *Psychologie du temps*. Presses Universitaires de France, Paris, 1957.

R. Francès. *La perception de la musique*. Vrin, Paris, 1958.

J. J. Gibson. The theory of affordances. In R. E. Shaw and J. Bransford, editors, *Perceiving, Acting, and Knowing*. Lawrence Erlbaum Associates, Hillsdale, 1977.

P. Griffiths. *A Concise History of Modern Music*. Thames and Hudson, London, 1978.

S. Grossberg. Some networks that can learn, remember and reproduce any number of complicated space-time patterns. *Studies in Applied Mathematics*, 49:135–166, 1970.

S. Grossberg. Adaptive pattern classification and universal recoding: I. parallel development and coding of neural feature detectors. *Biological Cybernetics*, 23:121–134, 1976.

J. H. J. Howard and J. A. Ballas. Syntactic and semantic factors in the classification of nonspeech transient patterns. *Perception and Psychophysics*, 28(5): 431–439, 1980.

J. H. J. Howard and J. A. Ballas. Acquisition of acoustic pattern categories by exemplar observation. *Organization, Behavior and Human Performance*, 30: 157–173, 1982.

E. K. Johnson and P. W. Jusczyk. Word segmentation by 8-month-olds: When speech cues count more than statistics. *Journal of Memory and Language*, 44 (4):548–567, 2001.

T. Kohonen. *Self-Organizing Maps*. Springer, Berlin, 1995.

C. L. Krumhansl. *Cognitive foundations of musical pitch*. Oxford University Press, New York, 1990.

C. L. Krumhansl and E. J. Kessler. Tracing the dynamic changes in perceived tonal organization in a spatial representation of musical keys. *Psychol Rev*, 89(4):334–368, 1982.

C. L. Krumhansl and G. Sandell. The perception of tone hierarchies and mirror forms in twelve-tone serial music. *Music Perception*, 5(1):31–78, 1987.

C. L. Krumhansl, J. Bharucha, and M. A. Castellano. Key distance effects on perceived harmonic structure in music. *Percept Psychophys*, 32(2):96–108, 1982a.

C. L. Krumhansl, J. J. Bharucha, and E. J. Kessler. Perceived harmonic structures of chords in three related keys. *Journal of Experimental Psychology: Human Perception and Performance*, 8:24–36, 1982b.

F. Lerdahl. Tonal pitch space. *Music Perception*, 5(3):315–349, 1988.

F. Lerdahl. Structure de prolongation dans l'atonalite. In S. McAdams and I. Deliege, editors, *La musique et les sciences cognitives*, pages 171–179. Mardaga, Liège, 1989.

F. Lerdahl. Pitch-space journeys in two chopin preludes. In M.R. Jones and S. Holleran, editors, *Cognitive Bases of. Musical Communication*. Amer Psychological Assn, Washington, 1991.

F. Lerdahl. *Tonal Pitch Space*. Oxford University Press, Oxford, 2001.

F. Lerdahl and R. Jackendoff. *A Generative Theory of Tonal Music*. MIT Press, Cambridge (MA), 1983.

H. Longuet-Higgins. The perception of music. *Interdisciplinary Science Review*, 3:148–156, 1978.

F. Meyer. *Emotion and Meaning in Music*. University of Chicago Press, Chicago and London, 1956.

E. Narmour. *The Analysis and Cognition of Basic Melodic Structures*. University of Chicago Press, Chicago, 1990.

A. S. Reber. Implicit learning of artificial grammars. *Journal of Verbal Learning and Verbal Behavior*, 6:855–863, 1967.

A. S. Reber. Implicit learning and tacit knowledge. *Journal of Experimental Psychology: General*, 118:219–235, 1989.

D. E. Rumelhart and D. Zipser. Feature discovery by competitive learning. *Cognitive Science*, 9:75–112, 1985.

J. R. Saffran, E. L. Newport, and R. N. Aslin. Word segmentation : The role of distributional cues. *Journal of Memory and Language*, 35(4):606–621, 1996.

C. A. Seger. Implicit learning. *Psychological Bulletin*, 115:163–169, 1994.

E. D. Thiessen and J. R. Saffran. When cues collide: use of stress and statistical cues to word boundaries by 7- to 9-month-old infants. *Developmental Psychology*, 39(4):706–716, 2003.

B. Tillmann and S. McAdams. Implicit learning of musical timbre sequences : statistical regularities confronted with acoustical (dis)similarities. *Journal of Experimental Psychology: Learning, Memory and Cognition*, 30:1131–1142, 2004.

B. Tillmann, J. J. Bharucha, and E. Bigand. Implicit learning of tonality: a self-organizing approach. *Psychological Review*, 107(4):885–913, 2000.

B. Tillmann, J. J. Bharucha, and E. Bigand. Implicit learning of regularities in western tonal music by self-organization. In R. French and J. Sougné, editors, *Perspectives in Neural Computing series*, pages 175–184. Springer Verlag, London, 2001.

C. von der Malsberg. Self-organizing of orientation sensitive cells in the striate cortex. kybernetic. *Biological Cybernetics*, 14:85–100, 1973.

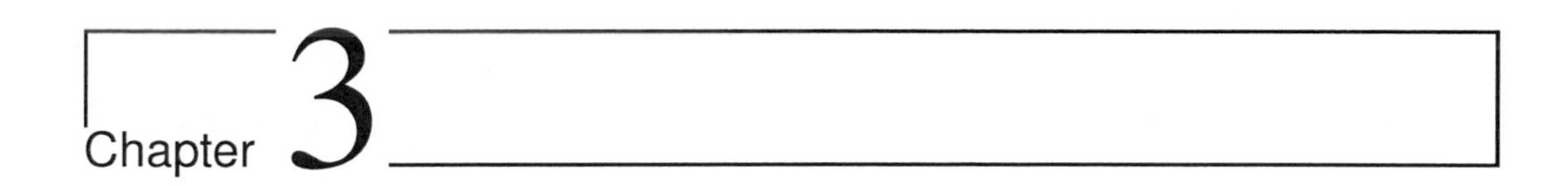

Content Processing of Music Audio Signals

Fabien Gouyon, Perfecto Herrera, Emilia Gómez, Pedro Cano, Jordi Bonada, Àlex Loscos, Xavier Amatriain, Xavier Serra

Music Technology Group, University Pompeu Fabra, Barcelona

About this chapter

In this chapter, we provide an overview of state-of-the-art algorithms for the automatic description of music audio signals, both from a low-level perspective (focusing on signal characteristics) and a more musical perspective (focusing on musically meaningful dimensions). We also provide examples of applications based on this description, such as music identification, music browsing and music signal transformations. Throughout the chapter, a special focus is put on promising research directions.

3.1 Introduction

Music Information Retrieval (MIR) is a young and very active research area. This is clearly shown in the constantly growing number and subjects of articles published in the Proceedings of the annual International Conference on Music Information Retrieval (ISMIR, the premier international scientific forum for researchers involved in MIR). MIR research is also increasingly published in high-standard scientific journals (e.g. the Communications of the ACM, or the IEEE Transactions on Audio, Speech and Language Processing)[1] and international conferences, as the ACM Multimedia, the International ACM SIGIR Conference, the IEEE International Conference on Acoustics, Speech, and Signal Processing, the IEEE International Conference on Multimedia and Expo, or the conference on Computer Music Modelling and Retrieval (CMMR), to name a few. In MIR, different long-established disciplines such as Musicology, Signal Processing, Psychoacoustics, Information Science, Computer Science, or Statistics converge by means of a multidisciplinary approach in order to address the wealth of scenarios for interacting with music posed by the digital technologies in the last decades (the standardisation of world-wide low-latency networks, the extensive use of efficient search engines in everyday life, the continuously growing amount of multimedia information on the web, in broadcast data streams or in personal and professional databases and the rapid development of on-line music stores). Applications are manifold; consider for instance automated music analysis, personalised music recommendation, online music access, query-based retrieval (e.g. "by-humming," "by-example") and automatic play-list generation. Among the vast number of disciplines and approaches to MIR (overviews of which can be found in Downie, 2003a and Orio, 2006), content processing of audio signals plays an important role. Music comes in many forms but content-based audio processing is only concerned with one of them: audio signals.[2] This chapter does not deal with the anal-

[1]both featuring recent special issues on MIR, see
`http://portal.acm.org/citation.cfm?id=1145287.1145308` and
`http://www.ewh.ieee.org/soc/sps/tap/sp_issue/cfp-mir.html`

[2]Hence the undifferentiated use in this chapter of the terms "music content processing" and "audio content processing."

ysis of symbolic music representations as e.g. digitised scores or structured representation of music events as MIDI. We also do not address the relatively new direction of research concerning the automated analysis of social, cultural and marketing dimensions of music networks (on these topics, we refer to Chapter 4, and the works by Cano et al., 2005b, 2006a and by Whitman, 2005). This section defines the notion of music content at diverse levels of abstraction and what we understand by *processing* music content: both its *description* and its *exploitation*. We also shortly mention representation issues in music content processing. Section 3.2 provides an overview of audio content description according to low-level features and diverse musically-meaningful dimensions as pitch, melody and harmony (see Section 3.2.4), rhythm (see Section 3.2.5), and music genre (see Section 3.2.6). The organisation follows increasing levels of abstraction. In Section 3.3, we address content exploitation and present different applications to content-based audio description. Finally, promising avenues for future work in the field are proposed in Section 3.4.

3.1.1 Music content: A functional view

A look at a dictionary reveals, at least, three senses for the word "content":

- Everything that is included in a collection;
- What a communication that is about something is about;
- The sum or range of what has been perceived, discovered or learned.

The disciplines of information science and linguistics offer interesting perspectives on the meaning of this term. However, we will rather focus on a more pragmatic view. The Society of Motion Picture and Television Engineers (SMPTE) and the European Broadcasting Union (EBU) have defined content as the combination of two entities termed *metadata* and *essence*. Essence is the raw program material itself, the data that directly encodes pictures, sounds, text, video, etc. Essence can also be referred to as media (although the former does not entail the physical carrier). In other words, essence is the encoded information that directly represents the actual message, and it is normally presented in a sequential, time-dependent manner. On the other hand, metadata (literally, "data about the data") is used to *describe* the essence and its different manifestations. Metadata can be classified, according to SMPTE/EBU, into several categories:

- Essential (meta-information that is necessary to reproduce the essence, like the number of audio channels, the Unique Material Identifier, the video format, etc.);
- Access (to provide control and access to the essence, i.e. copyright information);
- Parametric (to define parameters of the essence capture methods like camera set-up, microphone set-up, perspective, etc.);
- Relational (to achieve synchronisation between different content components, e.g. time-code);
- Descriptive (giving a description of the actual content or subject matter in order to facilitate the cataloging, search, retrieval and administration of content; i.e. title, cast, keywords, classifications of the images, sounds and texts, etc.).

In a quite similar way the National Information Standards Organisation considers three main types of metadata:

- Descriptive metadata, which describe a resource for purposes such as discovery and identification; they can include elements such as title, abstract, author, and keywords.
- Structural metadata, which indicate how compound objects are put together, for example, how visual or audio takes are ordered to form a seamless audiovisual excerpt.
- Administrative metadata, which provide information to help manage a resource, such as "when" and "how" it was created, file type and other

technical information, and who can access it. There are several subsets of administrative data; two of them that are sometimes listed as separate metadata types are:

- Right Management metadata, which deals with intellectual property rights;
- Preservation metadata, which contains information needed to archive and preserve a resource.

In accordance with these rather general definitions of the term "metadata," we propose to consider as content *all that can be predicated from a media essence*. Any piece of information related to a music piece that can be annotated, extracted, and that is in any way meaningful (i.e. it carries semantic information) to some user, can be technically denoted as metadata. Along this rationale, the MPEG-7 standard defines a content descriptor as "a distinctive characteristic of the data which signifies something to somebody" (Manjunath et al., 2002). This rather permissive view on the nature of music contents has a drawback: as they represent many different aspects of a music piece, metadata are not certain to be understandable by *any* user. This is part of the "user-modelling problem," whose lack of precision participates in the so-called *semantic gap*, that is, "the lack of coincidence between the information that one can extract from the (sensory) data and the interpretation that the same data has for a user in a given situation" (Smeulders et al., 2000). That has been signaled by several authors (Smeulders et al., 2000; Lew et al., 2002; Jermyn et al., 2003) as one of the recurrent open issues in systems dealing with audiovisual content. It is therefore important to consider metadata together with their functional values and address the question of which content means what to which users, and in which application. A way to address this issue is to consider content hierarchies with different levels of abstraction, any of them potentially useful for *some* users. In that sense, think of how different a content description of a music piece would be if the targeted user was a naive listener or an expert musicologist. Even a low-level descriptor such as the spectral envelope of a signal can be thought of as a particular level of content description targeted for the signal processing engineer. All these specifically targeted descriptions can be thought of as different instantiations of the same general content description scheme. Let us here propose the following distinction between descriptors of low, mid and high levels of abstraction (the latter being also sometimes referred to as "semantic" descriptors) (Lesaffre et al., 2003; Herrera, in print):

- A low-level descriptor can be computed from the essence data in a direct or derived way (i.e. after signal transformations like Fourier or Wavelet transforms, after statistical processing like averaging, after value quantisation like assignment of a discrete note name for a given series of pitch values, etc.). Most of low-level descriptors make little sense to the majority of users but, on the other hand, their exploitation by computing systems are usually easy. They can be also referred to as "signal-centered descriptors" (see Section 3.2.1).

- Mid-level descriptors require an induction operation that goes from available data towards an inferred generalisation about them. These descriptors usually pave the way for labelling contents, as for example a neural network model that makes decisions about music genre or about tonality, or a Hidden Markov Model that makes it possible to segment a song according to timbre similarities. Machine learning and statistical modelling make mid-level descriptors possible, but in order to take advantage of those techniques and grant the validity of the models, we need to gather large sets of observations. Mid-level descriptors are also sometimes referred to as "object-centered descriptors."

- The jump from low- or mid-level descriptors to high-level descriptors requires bridging the semantic gap. Semantic descriptors require an induction that has to be carried by means of a user-model (in order to yield the interpretation of the description), and not only a data-model as it was in the case of mid-level descriptors. As an example, let us imagine a simplistic "mood" descriptor consisting of labels "happy" and "sad." In order to compute such labels, one may[3] compute the tonality of the songs (i.e. "major" and "minor") and the tempo by means of knowledge-based analyzes of spectral and amplitude data. Using these mid-level descriptors, a model for computing the labels "happy" and "sad" would

[3]and it is only a speculation here

be elaborated by getting users' ratings of songs in terms of "happy" and "sad" and studying the relationships between these user-generated labels and values for tonality and tempo. High-level descriptors can also be referred to as "user-centered descriptors."

Standards In order to be properly exploited, music content (either low-, mid- or high-level content) has to be organised into knowledge structures such as taxonomies, description schemes, or ontologies. The Dublin Core and MPEG-7 are currently the most relevant standards for representing music content. The Dublin Core (DC) was specified by the Dublin Core Metadata Initiative, an institution that gathers organisations such as the Library of Congress, the National Science Foundation, or the Deutsche Bibliothek, to promote the widespread adoption of interoperable metadata standards. DC specifies a set of sixteen metadata elements, a core set of descriptive semantic definitions, which is deemed appropriate for the description of content in several industries, disciplines, and organisations. The elements are Title, Creator, Subject, Description, Publisher, Contributor, Date, Type, Format, Identifier, Source, Language, Relation, Coverage, Rights, and Audience. Description, for example, can be an abstract, a table of contents, a graphical representation or free text. DC also specifies a list of qualifiers that refine the meaning and use of the metadata elements, which open the door to refined descriptions and controlled-term descriptions. DC descriptions can be represented using different syntaxes, such as HTML or RDF/XML. On the other hand, MPEG-7 is a standardisation initiative of the ISO/IEC Moving Picture Expert Group that, contrasting with other MPEG standards, does not address the *encoding* of audiovisual essence. MPEG-7 aims at specifying an interface for the description of multimedia contents. MPEG-7 defines a series of elements that can be used to describe content, but it does not specify the algorithms required to compute values for those descriptions. The building blocks of MPEG-7 description are descriptors, description schemes (complex structures made of aggregations of descriptors), and the Description Definition Language (DDL), which defines the syntax that an MPEG-7 compliant description has to follow. The DDL makes hence possible the creation of non-standard, but compatible, additional descriptors and description schemes. This is an important feature because different needs will call for different kinds of structures, and for different instantiations of them. Depending on the theoretical and/or practical requirements of our problem, the required descriptors and description schemes will vary but, thanks to the DDL, we may build the proper structures to tailor our specific approach and required functionality. MPEG-7 descriptions are written in XML but a binary format has been defined to support their compression and streaming. The MPEG-7 standard definition covers eight different parts: Systems, DDL, Visual, Audio, Multimedia Description Schemes, Reference, Conformance and Extraction. In the audio section, we find music-specific descriptors for melody, rhythm or timbre, and in the Multimedia Description Schemes we find structures suitable to define classification schemes and a wealth of semantic information. As mentioned by Gómez et al. (2003a,b), the status of the original standard (see Manjunath et al., 2002, for an overview), as to representing music contents, is nevertheless a bit deceiving and it will probably require going beyond the current version for it to be adopted by the digital music community.

3.1.2 Processing music content: Description and exploitation

"Processing," beyond its straight meaning of "putting through a prescribed procedure," usually denotes a functional or computational approach to a wide range of scientific problems. "Signal processing" is the main term of reference here, but we could also mention "speech processing," "language processing," "visual processing" or "knowledge processing." A processing discipline focuses on the algorithmic level as defined by Marr (1982). The algorithmic level describes a system in terms of the steps that have to be carried out to solve a given problem. This type of description is, in principle, independent of the implementation level (as the algorithm can be effectively implemented in different ways). relationships and in terms of what is computed and why: for a given computational problem, several algorithms (each one implemented in several different ways) can be defined. The goal of a functional approach is that of developing systems that provide solutions to a given computational problem without considering the specific implementation of it. However, it is important to contrast the meaning of content processing with that of signal processing. The object of signal processing is the raw data captured by sensors, whereas content processing deals with an object that is within the signal, embedded in it like a second-order code, and to which we refer to using the word metadata. The processes of extraction and modelling these metadata require the synergy of, at least, four disciplines: Signal Processing, Artificial Intelligence, Information Retrieval, and Cognitive Science. Indeed, they require, among other things:

- Powerful signal analysis techniques that make it possible to address complex real-world problems, and to exploit context- and content-specific constraints in order to maximise their efficacy.
- Reliable automatic learning techniques that help building models about classes of objects that share specific properties, about processes that show e.g. temporal trends.
- Availability of large databases of describable objects, and the technologies required to manage (index, query, retrieve, visualise) them.
- Usable models of the human information processing involved in the processes of extracting and exploiting metadata (i.e. how humans perceive, associate, categorise, remember, recall, and integrate into their behavior plans the information that might be available to them by means of other content processing systems).

Looking for the origins of music content processing, we can spot different forerunners depending on the contributing discipline that we consider. When focusing on the discipline of Information Retrieval, Kassler (1966) and Lincoln (1967) are among the acknowledged pioneers. The former defines music information retrieval as "the task of extracting, from a large quantity of music data, the portions of that data with respect to which some particular musicological statement is true" (p. 66) and presents a computer language for addressing those issues. The latter discusses three criteria that should be met for automatic indexing of music material: eliminating the transcription by hand, effective input language for music, and an economic means for printing the music. This thread was later followed by Byrd (1984), Downie (1994), McNab et al. (1996) and Blackburn (2000) with works dealing with score processing, representation and matching of melodies as strings of symbols, or query by humming. Another batch of forerunners can be found when focusing on Digital Databases concepts and problems. Even though the oldest one dates back to the late eighties (Eaglestone, 1988), the trend towards databases for "content processing" emerges more clearly in the early nineties (de Koning and Oates, 1991; Eaglestone and Verschoor, 1991; Feiten et al., 1991; Keislar et al., 1995). These authors address the problems related to extracting and managing the acoustic information derived from a large amount of sound files. In this group of papers, we find questions about computing descriptors at different levels of abstraction, ways to query a content-based database using voice, text, and even external devices, and exploiting knowledge domain to enhance the functionalities of retrieval systems. To conclude with the antecedents for music content processing, we must also mention the efforts made since the last 30 years in the field of *Music Transcription*, whose goal is the automatic recovering of symbolic scores from acoustic signals (see Klapuri and Davy, 2006 for an exhaustive overview of music transcription research and Scheirer, 2000 for a critical perspective on music transcription). Central to music transcription is the segregation of the different music streams that coexist in a complex music rendition. Blind Source Separation (BSS) and Computational Auditory Scene Analysis (CASA) are two paradigms that address music stream segregation. An important conceptual difference between them is that, unlike the latter, the former intends to actually separate apart the different streams that summed together make up the multi-instrumental music signal. BSS is the agnostic approach to segregate music streams, as it usually does not assume any knowledge about the signals that have been mixed together. The strength of BSS models (but at the same time its main problem in music applications) is that only mutual statistical independence between the source signals is assumed, and no a priori information about the characteristics of the source signals (Casey and Westner, 2000; Smaragdis, 2001). CASA, on the other hand, is partially guided by the groundbreaking work of Bregman (1990) – who originally coined the term "Auditory Scene Analysis" (ASA) – on the perceptual mechanisms that enables a human listener to fuse or fission concurrent auditory events. CASA addresses the computational counterparts of ASA. Computer systems embedding ASA theories assume, and implement, specific heuristics that are hypothesised to play a role in the way humans perceive the music, as for example Gestalt principles. Worth mentioning here are the works by Mellinger (1991), Brown (1992), Ellis (1996), Kashino and Murase (1997), and Wang and Brown (2006). A comprehensive characterisation of the field of music content processing was offered by Leman (2003): "the science of musical content processing aims at explaining and modelling the mechanisms that transform information streams into meaningful musical units (both cognitive and emotional)." Music content processing is, for Leman, the object of study of his particular view of musicology, much akin to the so-called systematic musicology than to historic musicology. He additionally provides a definition of music content processing by extending it along three dimensions:

- The intuitive-speculative dimension, which includes semiotics of music, musicology, sociology, and philosophy of music. These disciplines provide a series of concepts and questions from a culture-centric point of view; music content is, following this dimension, a culture-dependent phenomenon.

- The empirical-experimental dimension, which includes research in physiology, psychoacoustics, music psychology, and neuro-musicology. These disciplines provide most of the empirical data needed to test, develop or ground some elements from the intuitive dimension; music content is, following this dimension, a percept in our auditory system.

- The computation-modelling dimension, which includes sound analysis and also computational modelling and simulation of perception, cognition and action. Music content is, following this dimension, a series of processes implemented in a computer, intended to emulate a human knowledge structure.

One can argue that these three dimensions address only the descriptive aspect of music content processing. According to Aigrain (1999), "content processing is meant as a general term covering feature extraction and modelling techniques for enabling basic retrieval, interaction and creation functionality." He also argues that music content processing technologies will provide "new aspects of listening, interacting with music, finding and comparing music, performing it, editing it, exchanging music with others or selling it, teaching it, analyzing it and criticizing it." We see here that music content processing can be characterised by two different tasks: *describing* and *exploiting* content. Furthermore, as mentioned above, the very meaning of "music content" cannot be entirely grasped without considering its functional aspects and including specific applications, targeted to specific users (Gouyon and Meudic, 2003). Hence, in addition to describing music content (as reviewed in Section 3.2), music content processing is also concerned with the design of computer systems that open the way to a more pragmatic content exploitation according to constraints posed by Leman's intuitive, empirical and computational dimensions (this exploitation aspect is the subject of Section 3.3).

3.2 Audio content description

3.2.1 Low-level audio features

Many different low-level features can be computed from audio signals. Literature in signal processing and speech processing provides us with a dramatic amount of techniques for signal modelling and signal representations over which features can be computed. Parametric methods (e.g. AR modelling, Prony modelling) directly provide such features, while additional post-processing is necessary to derive features from non-parametric methods (e.g. peaks can be extracted from spectral or cepstral representations). A comprehensive overview of signal representation and modelling techniques and their associated features is clearly beyond the scope of this chapter. Thus, we will only mention some features commonly used in music audio signal description, with a special focus on work published in the music transcription and MIR literature. Commonly, the audio signal is first digitised (if necessary) and converted to a general format, e.g. mono PCM (16 bits) with a fixed sampling rate (ranging from 5 to 44.1 KHz). A key assumption is that the signal can be regarded as being stationary over intervals of a few milliseconds. Therefore, the signal is divided into frames (short chunks of signal) of for example 10 ms. The number of frames computed per second is called *frame rate.* A tapered window function (e.g. a Gaussian or Hanning window) is applied to each frame to minimise the discontinuities at the beginning and end. Consecutive frames are usually considered with some *overlap* for smoother analyzes. The analysis step, the *hop size*, equals the frame rate minus the overlap.

Temporal features Many audio features can be computed directly from the temporal representation of these frames, for instance, the *mean* (but also the *maximum* or the *range*) of the amplitude of the samples in a frame, the *energy*, the *zero-crossing rate*, the *temporal centroid* (Gómez et al., 2005) and *auto-correlation coefficients* (Peeters, 2004). Some low-level features have also shown to correlate with perceptual attributes, for instance, amplitude is loosely correlated with *loudness*.

Spectral features It is also very common to compute features on a different representation of the audio, as for instance the spectral representation. Hence, a spectrum is obtained from each signal frame by applying a Discrete Fourier Transform (DFT), usually with the help of the Fast Fourier Transform (FFT). This procedure is called Short-Time Fourier Transform (STFT). Sometimes, the time-frequency representation is further processed by taking into account perceptual processing that takes place in the human auditory system as for instance the filtering performed by the middle-ear, loudness perception, temporal integration or frequency masking (Moore, 1995). Many features can be computed on the obtained representation, e.g. the spectrum *energy*, energy values in several frequency sub-bands (e.g. the perceptually-motivated *Bark bands*, Moore, 1995), the *mean*, *geometric mean*, *spread*, *centroid*, *flatness*, *kurtosis*, *skewness*, *spectral slope*, *high-frequency content* and *roll-off* of the spectrum frequency distribution or the *kurtosis* and *skewness* of the spectrum magnitude distribution (see Peeters, 2004 and Gómez et al., 2005 for more details on these numerous features). Further modelling of the spectral representation can be achieved through sinusoidal modelling (McAulay and Quatieri, 1986) or sinusoidal plus residual modelling (Serra, 1989). Other features can be computed on the series of *spectral peaks* corresponding to each frame and on the spectrum of the *residual* component. Let us mention, for instance, the mean (and the accumulated) *amplitude* of sinusoidal and residual components, the *noisiness*, the *harmonic distortion*, the *harmonic spectral centroid*, the *harmonic spectral tilt* and different ratios of peak amplitudes as the first, second and third *tristimulus* or the *odd-to-even ratio* (Serra and Bonada, 1998; Gómez et al., 2005). Bear in mind that other transforms can be applied instead of the DFT such as the Wavelet (Kronland-Martinet et al., 1987) or the Wigner-Ville transforms (Cohen, 1989).

Cepstral features *Mel-Frequency Cepstrum Coefficients* (MFCCs) are widespread descriptors in speech research. The cepstral representation has been shown to be of prime importance in this field, partly because of its ability to nicely separate the representation of voice excitation (the higher coefficients) from the subsequent filtering performed by the vocal tract (the lower coefficients). Roughly, lower coefficients represent spectral envelope (i.e. the formants) while higher ones represent finer details of the spectrum, among them the pitch (Oppenheim and Schafer, 2004). One way of computing the Mel-Frequency Cepstrum from a magnitude spectrum is the following:

1. Projection of the frequency axis from linear scale to the Mel scale of lower dimensionality (i.e. 20, by summing magnitudes in each of the 20 frequency bands of a Mel critical-band filter-bank);

2. Magnitude logarithm computation;

3. Discrete Cosine Transform (DCT).

The number of output coefficients of the DCT is variable. It is often set to 13, as in the standard implementation of the MFCCs detailed in the widely-used speech processing software Hidden Markov Model Toolkit (HTK).[4]

Temporal evolution of frame features Apart from the instantaneous, or frame, feature values, many authors focus on the temporal evolution of features (see Meng, 2006, for an overview). The simplest way to address temporal

[4]http://htk.eng.cam.ac.uk/

evolution of features is to compute the derivative of feature values (which can be estimated by a first-order differentiator). The degree of change can also be measured as the feature differential normalised with its magnitude (Klapuri et al., 2006). This is supposed to provide a better emulation of human audition, indeed, according to Weber's law, for humans, the just-noticeable-difference in the increment of a physical attribute depends linearly on its magnitude before incrementing. That is, $\Delta x/x$ (where x is a specific feature and Δx is the smallest perceptual increment) would be constant.

3.2.2 Segmentation and region features

Frame features represent a significant reduction of dimensionality with respect to the audio signal itself, however, it is possible to further reduce the dimensionality by focusing on features computed on groups of consecutive frames (Meng, 2006), often called *regions*. An important issue here is the determination of relevant region boundaries: i.e. the *segmentation* process. Once a given sound has been segmented into regions, it is possible to compute features as statistics of all of the frame features over the whole region (Serra and Bonada, 1998).

Segmentation Segmentation comes in different flavors. For McAdams and Bigand (1993), it "refers to the process of dividing an event sequence into distinct groups of sounds. The factors that play a role in segmentation are similar to the grouping principles addressed by Gestalt psychology." This definition implies that the segmentation process represents a step forward in the level of abstraction of data description. However, it may not necessarily be the case. Indeed, consider an adaptation of a classic definition coming from the visual segmentation area (Pal and Pal, 1993): "[sound] segmentation is a process of partitioning [the sound file/stream] into non-intersecting regions such that each region is homogeneous and the union of no two adjacent regions is homogeneous." The notion of homogeneity in this definition implies a property of signal or feature stationarity that may equate to a perceptual grouping process, but not necessarily. In what is sometimes referred to as *model-free* segmentation, the main idea is using the amount of change of a feature vector as a boundary detector: when this amount is higher than a given threshold, a boundary change decision is taken. Threshold adjustment requires a certain amount of trial-and-error, or fine-tuned adjustments regarding different segmentation classes. Usually, a smoothing window is considered in order to weight contributions from closer observations (Vidal and Marzal, 1990, p. 45). It is also possible to generalise the previous segmentation process to multi-dimensional feature vectors. There, the distance between consecutive frames can be computed with the help of different measures as for example the Mahalanobis distance (Tzanetakis and Cook, 1999). In the same vein, Foote (2000) uses MFCCs and the cosine distance measure between pairs of frames (not only consecutive frames), which yields a dissimilarity matrix that is further correlated with a specific kernel. Different kernels can be used for different types of segmentations (from short- to long-scale). The level of abstraction that can be attributed to the resulting regions may depend on the features used in the first place. For instance, if a set of low-level features is known to correlate strongly with a human percept (as the fundamental frequency correlates with the pitch and the energy in Bark bands correlates with the loudness) then the obtained regions may have some relevance as features of mid-level of abstraction (e.g. music notes in this case). *Model-based* segmentation on the other hand is more directly linked to the detection of mid-level feature boundaries. It corresponds to a focus on mid-level features that are thought, *a priori*, to make up the signal. A classical example can be found in speech processing where dynamical models of phonemes, or words, are built from observations of labelled data. The most popular models are Hidden Markov Models (HMM) (Rabiner, 1989). Applications of HMMs to the segmentation of music comprise segmentation of fundamental frequency envelopes in music notes (Raphael, 1999) and segmentation of MFCC-based temporal series in regions of globally-homogeneous timbres (Batlle and Cano, 2000). Rossignol (2000) proposes other examples of model-based segmentation and reports on the performance of different induction algorithms – Gaussian Mixture Models (GMM), k-Nearest Neigbours (k-NN) and Artificial Neural Networks (ANN) – in the tasks of speech/music segmentation and intra-note segmentation (see also Chapter 4). In the more general context of signal segmentation (not just music signals), Basseville and Nikiforov (1993) propose many segmentation techniques, some of which entail the use of signal models. For instance, they propose a time-domain technique in which two temporal windows are used: a sliding window with fixed size and a window with constant size increase. In this technique, a distance estimation is computed between two AR models built on each window (derived from the cross entropy between the conditional distributions of the two models). Here also, a threshold is used to determine whether the distance should be considered representative of a boundary or not. Application of this technique to music signals can be found in the works by Jehan (1997) and Thornburg and Gouyon (2000).

Note onset detection The detection of note onsets in music signals has attracted many computer music researchers since the early eighties (Gordon, 1984). Several methods have been designed, making use of diverse low-level features. The simplest focus on the temporal variation of a single feature, for instance the energy or the pitch. However, the combined use of multiple features (as energy *and* pitch) seems to provide better estimates, state-of-the-art algorithms often making use of band-wise energy processing (Klapuri, 1999; Bello, 2003). Model-based note onset segmentation has also been an active research field (Thornburg and Gouyon, 2000). The literature on onset detection is extensive and a review is beyond the scope of this chapter (for an exhaustive overview, see Bello, 2003; Bello et al., 2005).

Intra-note segmentation In addition to note onset detection, some research has also been dedicated to the segmentation of music signals in terms of Attack, Sustain and Release regions. This is especially relevant, from a feasibility point of view, when dealing with isolated instrument samples or musical phrases played by a monophonic instrument (Jenssen, 1999; Maestre and Gómez, 2005). Given starting and ending boundaries of these regions, it is possible to compute a number of features that relate to their durations as for example the *log-attack time* (Peeters, 2004). Some authors also focus on the variations of low-level frame features in these regions, such as the energy (Maestre and Gómez, 2005) or the fundamental frequency in sustain regions, characterizing therefore the *vibrato* (Herrera and Bonada, 1998; Rossignol et al., 1999; Collins, 2005).

Speech/Music segmentation A large body of work in automatic segmentation of audio signal also concerns the determination of boundaries of speech regions and music regions. This is usually achieved by model-based segmentation of multiple low-level features (Scheirer and Slaney, 1997; Harb and Chen, 2003; Pinquier et al., 2003; Kotti et al., 2006).

3.2.3 Audio fingerprints

Audio fingerprints have attracted a lot of attention for their usefulness in audio identification applications (see Section 3.3). Audio fingerprints are compact content-based signatures summarizing audio recordings (e.g. energies in specific frequency bands) that can be extracted from a music audio piece and stored in a database. Fingerprints of unlabelled pieces of audio can be calculated and matched against those stored in the database, providing a link to corresponding metadata (e.g. artist and song name). Section 3.3 provides more details on the main requirements of fingerprinting systems and application scenarios (for a general functional framework of audio fingerprinting systems and an overview of current technologies, see Cano et al., 2005a).[5] This section provides a short overview of audio features commonly used in the design of audio fingerprints.

Fingerprint extraction The fingerprint extraction derives a set of features from a recording in a concise and robust form. Fingerprint requirements include:

- Discrimination power over huge numbers of other fingerprints;
- Invariance to distortions;
- Compactness;

[5]Note that "fingerprinting" should not be mistaken for "watermarking," differences are explained in (Gomes et al., 2003).

- Computational simplicity.

The simplest approach one may think of – using directly the digitised waveform – is neither efficient nor effective. A more efficient implementation of this approach could use a hash method such as MD5 (Message Digest 5) or CRC (Cyclic Redundancy Checking) to obtain a compact representation of the binary file. However, hash values are fragile, a single bit flip is sufficient for the hash to completely change. They are also not robust to compression or distortions. Most fingerprint extraction systems consist of a front-end and a fingerprint modelling block (see Figure 3.1). The front-end computes low-level features from the signal and the fingerprint model defines the final fingerprint representation; we now briefly describe them in turn.

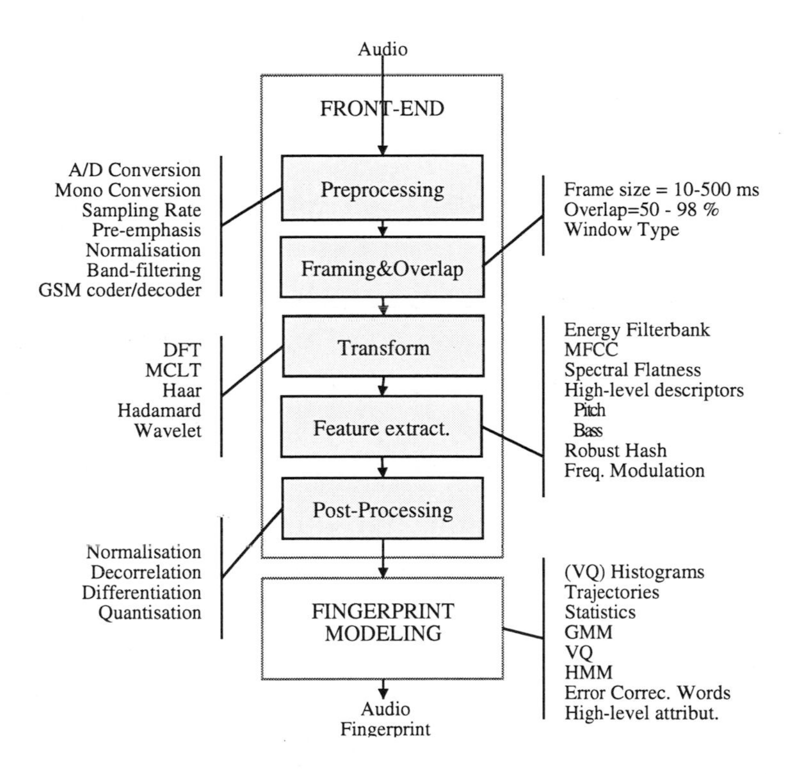

Figure 3.1: Fingerprint Extraction Framework: Front-end (top) and Fingerprint modelling (bottom).

Front-end Several driving forces co-exist in the design of the front-end: dimensionality reduction, perceptually meaningful parameters (similar to those used by the human auditory system), invariance/robustness (to channel distortions, background noise, etc.) and temporal correlation (systems that capture spectral dynamics). After the first step of audio digitisation, the audio is sometimes preprocessed to simulate the channel, for example band-pass filtered in a telephone identification task. Other types of processing are a GSM coder/decoder in a mobile phone identification system, pre-emphasis, amplitude normalisation (bounding the dynamic range to $[-1, 1]$). After framing the signal in small windows, overlap must be applied to assure robustness to shifting (i.e. when the input data is not perfectly aligned to the recording that was used for generating the fingerprint). There is a trade-off between the robustness to shifting and the computational complexity of the system: the higher the frame rate, the more robust to shifting the system is but at a cost of a higher computational load. Then, linear transforms are usually applied (see Figure 3.1). If the transform is suitably chosen, the redundancy is significantly reduced. There are optimal transforms in the sense of information packing and de-correlation properties, like Karhunen-Loeve (KL) or Singular Value Decomposition (SVD). These transforms, however, are computationally complex. For that reason, lower complexity transforms using fixed basis vectors are common (e.g. the DFT). Additional transformations are then applied in order to generate the final acoustic vectors. In this step, we find a great diversity of algorithms. The objective is again to reduce the dimensionality and, at the same time, to increase the invariance to distortions. It is very common to include knowledge of the transduction stages of the human auditory system to extract more perceptually meaningful parameters. Therefore, many systems extract several features performing a critical-band analysis of the spectrum. Resulting features are for example MFCCs, energies in Bark-scaled bands, geometric mean of the modulation frequency, estimation of the energy in Bark-spaced band-filters, etc., or many of the features presented in Section 3.2.1. Some examples are given in Figure 3.2. Most of the features described so far are absolute measurements. In order to better characterise temporal variations in the signal, higher order time derivatives are added to the signal model. Some systems compact the feature vector representation

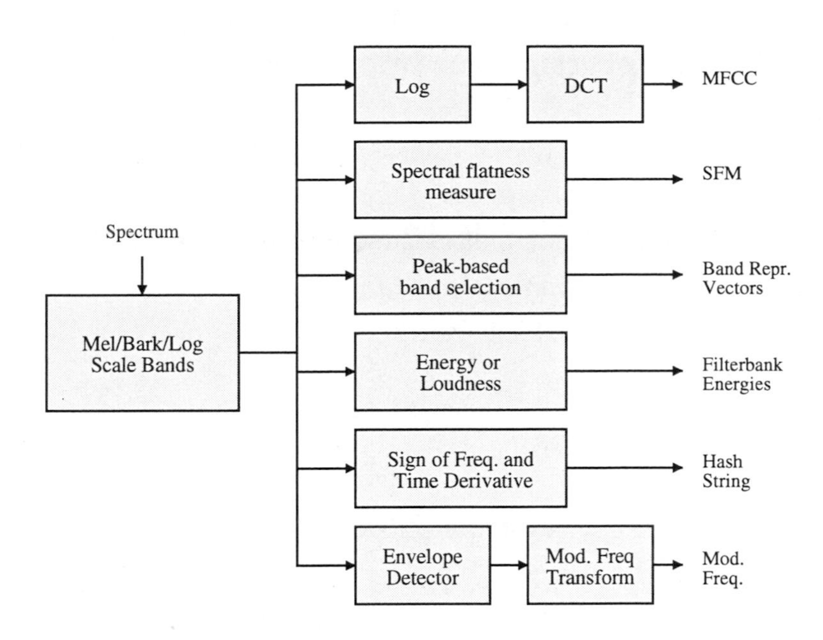

Figure 3.2: Feature Extraction Examples

using transforms such as Principal Component Analysis (PCA). It is also quite common to apply a very low resolution quantisation (ternary or binary) to the features, the purpose of which is to gain robustness against distortions and reduce the memory requirements.

Fingerprint models The sequence of features calculated on a frame-by-frame basis is then further reduced to a fingerprint model that usually implies statistics of frame values (mean and variance) and redundancies in frame vicinity. A compact representation can also be generated by clustering the feature vectors. The sequence of vectors is thus approximated by a much lower number of representative code vectors, a codebook. The temporal evolution of audio is lost with this approximation, but can be kept by collecting short-time statistics over regions of time or by HMM modelling (Batlle et al., 2002). At that point, some systems also derive musically-meaningful attributes from low-level features, as the *beats* (Kirovski and Attias, 2002) (see Section 3.2.5) or the *predominant pitch* (Blum et al., 1999) (see Section 3.2.4).

3.2.4 Tonal descriptors: From pitch to key

This section first reviews computational models of *pitch* description and then progressively addresses tonal aspects of higher levels of abstraction that imply different combinations of pitches: *melody* (sequence of single pitches combined over time), *pitch classes* and *chords* (simultaneous combinations of pitches), and *chord progressions*, *harmony* and *key* (temporal combinations of chords).

Pitch

The fundamental frequency is the main low-level descriptor to consider when describing melody and harmony. Due to the significance of pitch detection for speech and music analysis, a lot of research has been done in this field. We present here a brief review of the different approaches for pitch detection: fundamental frequency estimation for monophonic sounds, multi-pitch estimation and predominant pitch estimation. We refer to the paper by Gómez et al. (2003c) for an exhaustive review.

Fundamental frequency estimation for monophonic sounds As illustrated in Figure 3.3, the fundamental frequency detection process can be subdivided into three successive steps: the preprocessor, the basic extractor, and the post-processor (Hess, 1983). The basic extractor converts the input signal into a series of fundamental frequency estimates, one per analysis frame. Pitched/unpitched measures are often additionally computed to decide whether estimates are valid or should be discarded (Cano, 1998). The main task of the pre-processor is to facilitate the fundamental frequency extraction. Finally, the post-processor performs more diverse tasks, such as error detection and correction, or smoothing of an obtained contour. We now describe these three processing blocks in turn. Concerning the main extractor processing block, the first solution was to adapt the techniques proposed for speech (Hess, 1983). Later, other methods have been specifically designed for dealing with music signals. These methods can be classified according to their processing domain: time-domain algorithms vs frequency-domain algorithms. This distinction is not always so clear, as some of the algorithms can be expressed in both (time

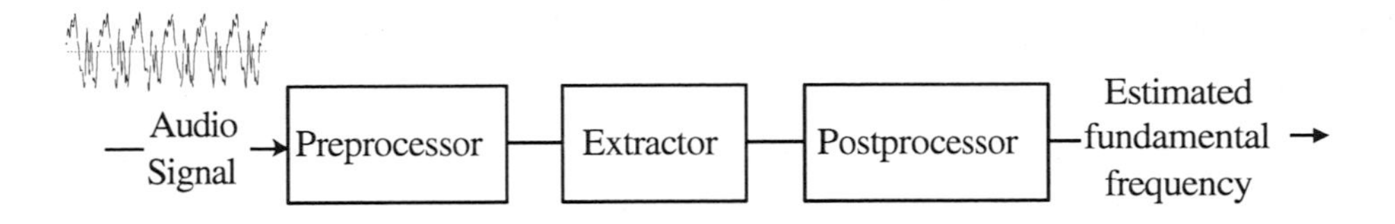

Figure 3.3: Steps of the fundamental frequency detection process

and frequency) domains, as the autocorrelation function (ACF) method. Another way of classifying the different methods, more adapted to the frequency domain, is to distinguish between spectral place algorithms and spectral interval algorithms (Klapuri, 2004). The spectral place algorithms weight spectral components according to their spectral location. Other systems use the information corresponding to spectral intervals between components. Then, the spectrum can be arbitrarily shifted without affecting the output value. These algorithms work relatively well for sounds that exhibit inharmonicity, because intervals between harmonics remain more stable than the places for the partials.

Time-domain algorithms The simplest time-domain technique is based on counting the number of times the signal crosses the 0-level reference, the zero-crossing rate (ZCR). This method is not very accurate when dealing with noisy signals or harmonic signals where the partials are stronger than the fundamental. Algorithms based on the time-domain autocorrelation function (ACF) have been among the most frequently used fundamental frequency estimators. ACF-based fundamental frequency detectors have been reported to be relatively noise immune but sensitive to formants and spectral peculiarities of the analyzed sound (Klapuri, 2004). Envelope periodicity algorithms find their roots in the observation that signals with more than one frequency component exhibit periodic fluctuations in their time domain amplitude envelope. The rate of these fluctuations depends on the frequency difference between the two frequency components. In the case of a harmonic sound, the fundamental frequency is clearly visible in the amplitude envelope of the signal. Recent models of human pitch perception tend to calculate envelope periodicity separately in distinct frequency bands and then combine the results across channels (Meddis and Hewitt, 1991; Terhardt et al., 1981). These methods attempt to estimate the perceived pitch, not only the physical periodicity, in acoustic signals of various kinds. A parallel processing approach (Gold and Rabiner, 1969; Rabiner and Schafer, 1978), designed to deal with speech signals, has been successfully used in a wide variety of applications. Instead of designing one very complex algorithm, the basic idea is to tackle the same problem with several, more simple processes in parallel and later combine their outputs. As Bregman (1998) points out, human perception appears to be redundant at many levels, several different processing principles seem to serve the same purpose, and when one of them fails, another is likely to succeed.

Frequency-domain algorithms another transformation. Noll (1967) introduced the idea of Cepstrum analysis for pitch determination of speech signals. The cepstrum is the inverse Fourier transform of the power spectrum logarithm of the signal. The Cepstrum computation (see Section 3.2.1) nicely separates the transfer function (spectral envelope) from the source, hence the pitch. Cepstrum fundamental frequency detection is closely similar to autocorrelation systems (Klapuri, 2004). Spectrum autocorrelation methods were inspired by the observation that a periodic but non-sinusoidal signal has a periodic magnitude spectrum, the period of which is the fundamental frequency. This period can be estimated by ACF (Klapuri, 2004). Harmonic matching methods extract a period from a set of spectral peaks of the magnitude spectrum of the signal. Once these peaks in the spectrum are identified, they are compared to the predicted harmonics for each of the possible candidate note frequencies, and a fitness measure can be developed. A particular fitness measure is described by Maher and Beauchamp (1993) as a "Two Way Mismatch" procedure. This method is used in the context of Spectral Modelling Synthesis (SMS), with some improvements, as pitch-dependent analysis window, enhanced peak selection, and optimisation of the search (Cano, 1998).

The idea behind Wavelet-based algorithms is to filter the signal using a wavelet with derivative properties. The output of this filter will have maxima where glottal-closure instants or zero crossings happen in the input signal. After detection of these maxima, the fundamental frequency can be estimated as the distance between consecutive maxima.

Klapuri (2004) proposes a band-wise processing algorithm that calculates independent fundamental frequency estimates in separate frequency bands. Then, these values are combined to yield a global estimate. This method presents several advantages: it solves the inharmonicity problem, it is robust with respect to heavy signal distortions, where only a fragment of the frequency range is reliable.

Preprocessing methods The main task of a preprocessor is to suppress noise prior to fundamental frequency estimation. Some preprocessing methods used in speech processing are detailed in the book by Hess (1983). Methods specifically defined for music signals are detailed in the already mentioned work by Klapuri (2004).

Post-processing methods The estimated series of pitches may be noisy and may present isolated errors, different methods have been proposed for correcting these. The first is low-pass filtering (linear smoothing) of the series. This may remove much of the local jitter and noise, but does not remove local gross measurement errors, and, in addition, it smears the intended discontinuities at the voiced-unvoiced transitions (Hess, 1983). Non-linear smoothing has been proposed to address these problems (Rabiner et al., 1975). Another procedure consists in storing several possible values for the fundamental frequency for each analysis frame (Laroche, 1995), assigning them a score (e.g. the value of the normalised autocorrelation). Several tracks are then considered and ranked (according to some continuity evaluation function) by for example dynamic programming. This approach minimises the abrupt fundamental frequency changes (e.g. octave errors) and gives good results in general. Its main disadvantage is its estimation delay and non-causal behavior. Usually, it is also useful to complement the forward estimation by a backward estimation (Cano, 1998).

Multi-pitch estimation Multi-pitch estimation is the simultaneous estimation of the pitches making up a polyphonic sound (a polyphonic instrument or several instruments playing together). Some algorithms used for monophonic pitch detection can be adapted to the simplest polyphonic situations (Maher and Beauchamp, 1993). However, they are usually not directly applicable to general cases, they require, among other differences, significantly longer time frames (around 100 ms) (Klapuri, 2004). Relatively successful algorithms implement principles of the perceptual mechanisms that enable a human listener to fuse or fission concurrent auditory streams (see references to "Auditory Scene Analysis" on page 92). For instance, Kashino et al. (1995) implement such principles in a Bayesian probability network, where bottom-up signal analysis can be integrated with temporal and musical predictions. An example following the same principles is detailed by Walmsley et al. (1999), where a comparable network estimates the parameters of a harmonic model jointly for a number of frames. Godsmark and Brown (1999) have developed a model that is able to resolve melodic lines from polyphonic music through the integration of diverse knowledge. Other methods are listed in the work by Klapuri (2004). The state-of-the-art multi-pitch estimators operate reasonably well for clean signals, frame-level error rates increasing progressively with the number of concurrent voices. Also, the number of concurrent voices is often underestimated and the performance usually decreases significantly in the presence of noise (Klapuri, 2004).

Predominant pitch estimation Predominant pitch estimation also aims at estimating pitches in polyphonic mixtures; however, contrarily to multi-pitch estimation, it assumes that a specific instrument is predominant and defines the melody. For instance, the system proposed by Goto (2000) detects melody and bass lines in polyphonic recordings using a multi-agent architecture by assuming that they occupy different frequency regions. Other relevant methods are reviewed by Gómez et al. (2003c) and Klapuri (2004).

Melody

Extracting melody from note sequences We have presented above several algorithms whose outputs are time sequences of pitches (or simultaneous combinations thereof). Now, we present some approaches that, building upon those, aim at identifying the notes that are likely to correspond to the main melody. We refer to the paper by Gómez et al. (2003c) for an exhaustive review of the state-of-the-art in melodic description and transformation from audio recordings. Melody extraction can be considered not only for polyphonic sounds, but also for monophonic sounds as they may contain notes that do not belong to the melody (for example, grace notes, passing notes or the case of several interleaved voices in a monophonic stream). As discussed by Nettheim (1992) and Selfridge-Field (1998, Section 1.1.3.), the derivation of a melody from a sequence of pitches faces the following issues:

- A single line played by a single instrument or voice may be formed by movement between two or more melodic or accompaniment strands.
- Two or more contrapuntal lines may have equal claim as "the melody."
- The melodic line may move from one voice to another, possibly with overlap.
- There may be passages of figuration not properly considered as melody.

Some approaches try to detect note groupings. Experiments have been done on the way listeners achieve melodic grouping (McAdams, 1994; Scheirer, 2000, p.131). These provide heuristics that can be taken as hypotheses in computational models. Other approaches make assumptions on the type of music to be analyzed. For instance, methods can be different according to the complexity of the music (monophonic or polyphonic music), the genre (classical with melodic ornamentations, jazz with singing voice, etc.) or the representation of the music (audio, MIDI, etc.). We refer to the works by Uitdenbogerd and Zobel (1998) and by Typke (2007) regarding melody extraction of MIDI data.

Melodic segmentation The goal of melodic segmentation is to establish a temporal structure on a sequence of notes. It may involve different levels of hierarchy, such as those defined by Lerdahl and Jackendoff (1983), and may include overlapping, as well as unclassified, segments. One relevant method proposed by Cambouropoulos (2001) is the Local Boundary Detection Model (LBDM). This model computes the transition strength of each interval of a melodic surface according to local discontinuities. function are considered as segment boundaries. This method is based on two rules: the *Change Rule* (measuring the degree of change between two consecutive intervals) and the *Proximity Rule* (each boundary is weighted according to the size of its absolute interval, so that segment boundaries are located at larger intervals). In his paper, Cambouropoulos (2001) uses not only pitch, but also temporal (inter-onset intervals, IOIs) and rest intervals. He compares this algorithm with the punctuation rules defined by Friberg and colleagues,[6] getting coherent results. The LBDM has been used by Melucci and Orio (1999) for content-based retrieval of melodies. Another approach can be found in the `Grouper`[7] module of the `Melisma` music analyzer, implemented by Temperley and Sleator. This module uses three criteria to select the note boundaries. The first one considers the gap score for each pair of notes that is the sum of the IOIs and the offset-to-onset interval (OOI). Phrases receive a weight proportional to the gap score between the notes at the boundary. The second one considers an optimal phrase length in number of notes. The third one is related to the metrical position of the phrase beginning, relative to the metrical position of the previous phrase beginning. Spevak et al. (2002) have compared several algorithms for melodic segmentation: LBDM, the Melisma Grouper, and a memory-based approach, the Data-Oriented Parsing (DOP) by Bod (2001). They also describe other approaches to melodic segmentation. To explore this issue, they have compared manual segmentation of different melodic excerpts. However, according to them, "it is typically not possible to determine one 'correct' segmentation, because the process is influenced by a rich and varied set of context."

Miscellaneous melodic descriptors Other descriptors can be derived from a numerical analysis of the pitches of a melody and used in diverse applications as comparative analysis (Toiviainen and Eerola, 2001), melody re-

[6]see `http://www.speech.kth.se/music/performance/performance_rules.html`, see also Chapter 7

[7]see `http://www.link.cs.cmu.edu/music-analysis/grouper.html`

trieval (Kostek, 1998; Tzanetakis, 2002), and algorithmic composition (Towsey et al., 2001). Some of these descriptors are computed using features related to structural, musical or perceptual aspects of sound. Some others are computed from note descriptors (therefore they require algorithms for note segmentation, see Section 3.2.2). Yet other descriptors can be computed as statistics of frame or sample features. One example is the pitch histogram features proposed by Tzanetakis (2002).

Pitch class distribution

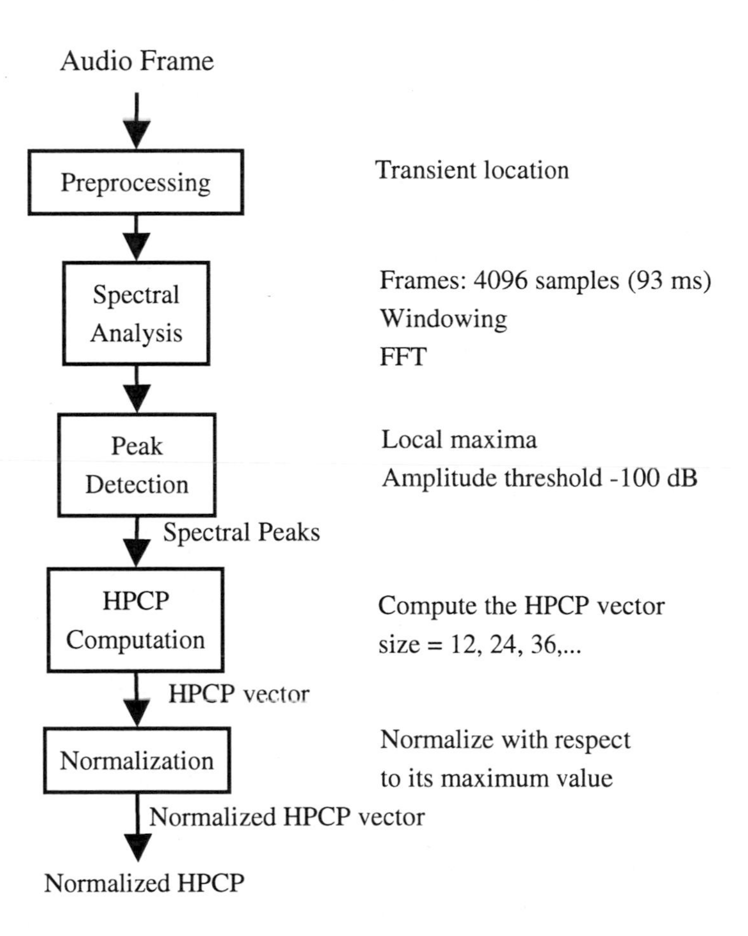

Figure 3.4: Block Diagram for HPCP Computation

Many efforts have been devoted to the analysis of chord sequences and key in MIDI representations of classical music, but little work has dealt directly with audio signals and other music genres. Adapting MIDI-oriented methods would require a previous step of automatic transcription of polyphonic audio, which, as mentioned by Scheirer (2000) and Klapuri (2004), is far from being solved. Some approaches extract information related to the pitch class distribution of music without performing automatic transcription. The pitch-class distribution is directly related to the chords and the tonality of a piece. Chords can be recognised from the pitch class distribution without requiring the detection of individual notes. Tonality can be also estimated from the pitch class distribution without a previous procedure of chord estimation. Fujishima (1999) proposes a system for chord recognition based on the pitch-class profile (PCP), a 12-dimensional low-level vector representing the intensities of the twelve semitone pitch classes. His chord recognition system compares this vector with a set of chord-type templates to estimate the played chord. In a paper by Sheh and Ellis (2003), chords are estimated from an audio recordings by modelling sequences of PCPs with an HMM. In the context of a key estimation system, Gómez (2004) proposes the Harmonic PCPs (HPCPs) as extension of the PCPs: only the spectral peaks in a certain frequency band are used (100 – 5000 Hz), a weight is introduced into the feature computation and a higher resolution is used in the HPCP bins (decreasing the quantisation level to less than a semitone). The procedure for HPCP computation is illustrated in Figure 3.4. A transient detection algorithm (Bonada, 2000) is used as preprocessing step in order to discard regions where the harmonic structure is noisy; the areas located 50 ms before and after the transients are not analyzed. As a post-processing step, HPCPs are normalised with respect to maximum values for each analysis frame, in order to store the relative relevance of each of the HPCP bins. In the context of beat estimation of drum-less audio signals, Goto and Muraoka (1999) also introduced the computation of a histogram of frequency components, used to detect chord changes. Note however that this method does not identify chord names. Constant Q profiles have also been used to characterise the tonal content of audio (Purwins et al., 2000). Constant Q profiles are twelve-dimensional vectors, each component referring to a pitch class, which are computed with the constant Q filter bank (Brown and Puckette, 1992). Purwins et al. (2003) present examples where constant Q profiles are used to track tonal centers. Later on, Purwins (2005) uses these features to

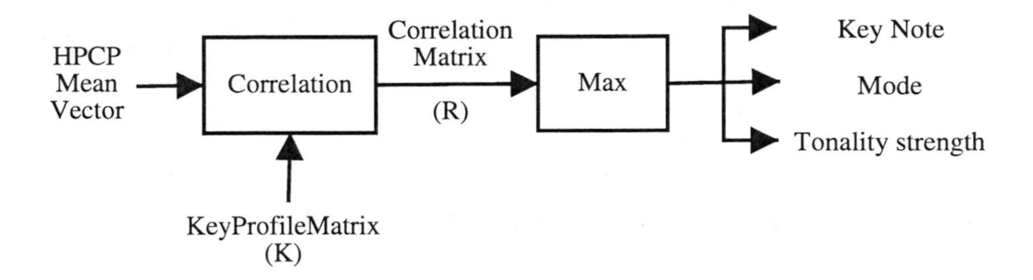

Figure 3.5: Block Diagram for Key Computation using HPCP

analyze the interdependence of pitch classes and key as well as key and composer. Tzanetakis (2002) proposes a set of features related to audio harmonic content in the context of music genre classification. These features are derived from a pitch histogram that can be computed from MIDI or audio data: the most common pitch class used in the piece, the frequency of occurrence of the main pitch class and the pitch range of a song.

Tonality: From chord to key

Pitch class distributions can be compared (correlated) with a tonal model to estimate the chords (when considering small time scales) or the key of a piece (when considering a larger time scale). This is the approach followed by Gómez (2004) to estimate the tonality of audio pieces at different temporal scales, as shown on Figure 3.5. To construct the key-profile matrix, Gómez (2004) follows the model for key estimation of MIDI files by Krumhansl (1990). This model considers that tonal hierarchies may be acquired through internalisation of the relative frequencies and durations of tones. The algorithm estimates the key from a set of note duration values, measuring how long each of the 12 pitch classes of an octave (C, C♯, etc.) have been played in a melodic line. In order to estimate the key of the melodic line, the vector of note durations is correlated to a set of key profiles or probe-tone profiles. These profiles represent the tonal hierarchies of the 24 major and minor keys, and each of them contains 12 values, which are the ratings of the degree to which each of the 12 chromatic scale tones fit a particular key. They were obtained by analyzing human judgments with regard to the relationship between pitch classes and keys (Krumhansl, 1990, pp. 78-81). Gómez (2004) adapts this model to deal with HPCPs (instead of note durations) and polyphonies (instead of melodic lines); details of evaluations can be found in (Gómez, 2006), together with an exhaustive review of computational models of tonality.

3.2.5 Rhythm

Representing rhythm

One way to represent the rhythm of a musical sequence is to specify an exhaustive and accurate list of onset times, maybe together with some other characterizing features such as durations, pitches or intensities (as it is done in MIDI). However, the problem with this representation is the lack of abstraction. There is more to rhythm than the absolute timings of successive music events, one must also consider *tempo*, *meter* and *timing* (Honing, 2001).

Tempo Cooper and Meyer (1960) define a pulse as "[...] one of a series of regularly recurring, precisely equivalent stimuli. [...] Pulses mark off equal units in the temporal continuum." Commonly, "pulse" and "beat" are often used indistinctly and refer both to one element in such a series and to the whole series itself. The tempo is defined as the number of beats in a time unit (usually the minute). There is usually a preferred pulse, which corresponds to the rate at which most people would tap or clap in time with the music. However, the perception of tempo exhibits a degree of variability. It is not always correct to assume that the pulse indicated in a score (Maelzel Metronome) corresponds to the "foot-tapping" rate, nor to the actual "physical tempo" that would be an inherent property of audio streams (Drake et al., 1999). Differences in human perception of tempo depend on age, musical training, music preferences and general listening context (Lapidaki, 1996). They are nevertheless far from random and most often correspond to a focus on a different metrical level and are quantifiable as simple ratios (e.g. 2, 3, $\frac{1}{2}$ or $\frac{1}{3}$).

Meter The metrical structure (or meter) of a music piece is based on the coexistence of several pulses (or "metrical levels"), from low levels (small time divisions) to high levels (longer time divisions). The segmentation of time by a given low-level pulse provides the basic time span to measure music event accentuation whose periodic recurrences define other, higher, metrical levels. The duration-less points in time, the beats, that define this discrete time grid obey a specific set of rules, formalised in the Generative Theory of Tonal Music (Lerdahl and Jackendoff, 1983, GTTM). Beats must be equally spaced. A beat at a high level must also be a beat at each lower level. At any metrical level, a beat which is also a beat at the next higher level is called a downbeat, and other beats are called upbeats. The notions of time signature, measure and bar lines reflect a focus solely on two (or occasionally three) metrical levels. Bar lines define the slower of the two levels (the measure) and the time signature defines the number of faster pulses that make up one measure. For instance, a $\frac{6}{8}$ time signature indicates that the basic temporal unit is an eighth-note and that between two bar lines there is room for six units. Two categories of meter are generally distinguished: duple and triple. This notion is contained in the numerator of the time signature: if the numerator is a multiple of two, then the meter is duple, if it is not a multiple of two but a multiple of three, the meter is triple. The GTTM specifies that there must be a beat of the metrical structure for every note in a music sequence. Accordingly, given a list of note onsets, the quantisation (or "rhythm-parsing") task aims at making it fit into Western music notation. Viable time points (metrical points) are those defined by the different coexisting metrical levels. Quantised durations are then rational numbers (e.g. 1, $\frac{1}{4}$, $\frac{1}{6}$) relative to a chosen time interval: the time signature denominator.

Timing A major weakness of the GTTM is that it does not deal with the deviations from strict metrical timing which occur in almost all styles of music. Thus it is only really suitable for representing the timing structures of music scores, where the expressive timing is not represented. There are conceptually two types of non-metrical timing: long-term tempo deviations (e.g. Rubato) and short-term timing deviations (e.g. "Swing"). One of the greatest difficulties in analyzing performance data is that the two dimensions of expressive timing are projected onto the single dimension of time. Mathematically, it is possible to represent any tempo change as a series of timing changes and vice-versa, but these descriptions are somewhat counterintuitive (Honing, 2001).

Challenges in automatic rhythm description

Automatic description of musical rhythm is not obvious. First of all because it seems to entail two dichotomic processes: a bottom-up process enabling very rapidly the percept of pulses from scratch, and a top-down process (a persistent mental framework) that lets this induced percept guide the organisation of incoming events (Desain and Honing, 1999). Implementing in a computer program both reactivity to the environment and persistence of internal representations is a challenge. Rhythm description does not solely call for the handling of timing features (onsets and offsets of tones). The definition and understanding of the relationships between rhythm perception and other music features such as intensity or pitches are still open research topics. Rhythm involves two dichotomic aspects that are readily perceived by humans: there are both a strong and complex structuring of phenomena occurring at different time scales and widespread departures from exact metrical timing. Indeed, inexact timings always occur because of expressive performances, sloppy performances and inaccurate collection of timing data (e.g. computational onset detection may have poor time precision and may suffer from false alarms). Furthermore, recent research indicates that even if perceived beats are strongly correlated to onsets of tones, they do not necessarily line up exactly with them, our perception rather favoring smooth tempo curves (Dixon et al., 2006).

Functional framework

The objective of automatic rhythm description is the parsing of acoustic events that occur in time into the more abstract notions of tempo, timing and meter. Algorithms described in the literature differ in their goals. Some of them derive the beats and the tempo of a single metrical level, others try to derive complete rhythmic transcriptions (i.e. musical scores), others aim at determining some timing features from musical performances (such as tempo changes, event

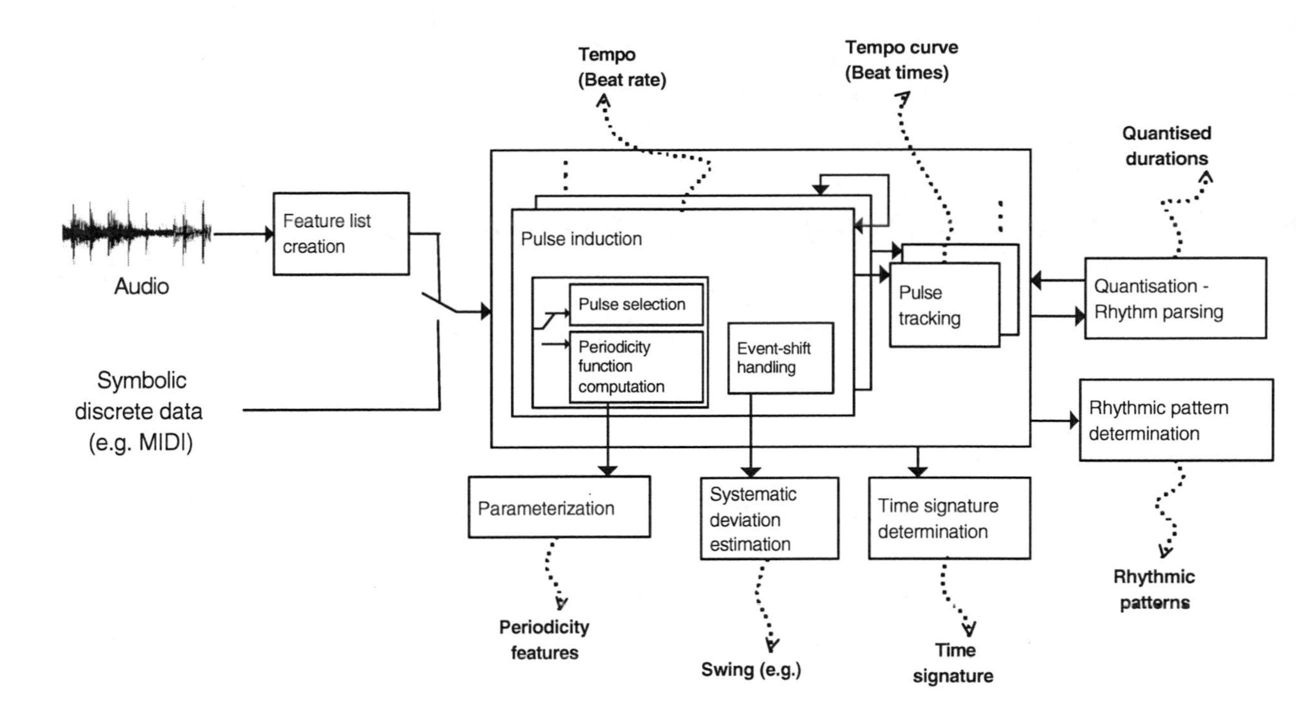

Figure 3.6: Functional units of rhythm description systems.

shifts or swing factors), others focus on the classification of music signals by their overall rhythmic similarities and others look for rhythm patterns. Nevertheless, these computer programs share some functional aspects that we represent as functional blocks of a general diagram in Figure 3.6. We briefly explain each of these functional blocks in the following paragraphs; we refer to the paper by Gouyon and Dixon (2005) for a more complete survey.

Feature list creation Either starting from MIDI, or taking into consideration other symbolic formats such as files containing solely onset times and durations (Brown, 1993), or even raw audio data, the first analysis step is the creation of a feature list, i.e. the parsing, or "filtering," of the data at hand into a sequence that is supposed to convey the predominant information relevant to a rhythmic analysis. These feature lists are defined here broadly, to include frame-based feature vectors as well as lists of symbolic events. The latter include onset times, durations (Brown, 1993), relative amplitude (Dixon, 2001), pitch (Dixon and Cambouropoulos, 2000), chords (Goto, 2001) and percussive instrument classes (Goto, 2001). Some systems refer to a data granularity of a lower level of abstraction, i.e. at the frame level. Section 3.2.1 describes usual low-level features that can be computed on signal frames. In rhythm analysis, common frame features are energy values and energy values in frequency sub-bands. Some systems also measure energy variations between consecutive frames (Scheirer, 2000; Klapuri et al., 2006). Low-level features other than energy (e.g. spectral flatness, temporal centroid) have also been recently advocated (Gouyon et al., 2006b).

Pulse induction A metrical level (a pulse) is defined by the periodic recurrence of some music event. Therefore, computer programs generally seek periodic behaviors in feature lists in order to select one (or some) pulse period(s) and also sometimes phase(s). This is the process of pulse induction. Concerning pulse induction, computer programs either proceed by *pulse selection*, i.e. evaluating the salience of a restricted number of possible periodicities (Parncutt, 1994), or by computing a periodicity function. In the latter case, a continuous function plots pulse salience versus pulse period (or frequency). Diverse transforms can be used: the Fourier transform, Wavelet transforms, the autocorrelation function, bank of comb filters, etc. In pulse induction, a fundamental assumption is made: the pulse period (and phase) is stable over the data used for its computation. That is, there is no speed variation in that part of the musical performance used for inducing a pulse. In that part of the data, remaining timing deviations (if any) are assumed to be short-time ones (considered as either errors or expressiveness features). They are either "smoothed out," by considering tolerance intervals or smoothing windows, or cautiously handled in order to derive patterns of systematic short-time timing deviations as e.g. the swing. Another step is necessary to output a discrete pulse period (and optionally its phase) rather than a continuous periodicity function. This is usually achieved by a peak-picking algorithm.

Pulse tracking Pulse tracking and pulse induction often occur as complementary processes. Pulse induction models consider short term timing deviations as noise, assuming a relatively stable tempo, whereas a pulse tracker handles the short term timing deviations and attempts to determine changes in the pulse period and phase, without assuming that the tempo remains constant. Another difference is that induction models work bottom-up, whereas tracking models tend to follow top-down approaches, driven by the pulse period and phase computed in a previous induction step. Pulse tracking is often a process of reconciliation between predictions (driven by previous period and phase computations) and the observed data. Diverse formalisms and techniques have been used in the design of pulse trackers: rule-based (Desain and Honing, 1999), problem-solving (Allen and Dannenberg, 1990), agents (Dixon, 2001), adaptive oscillators (Large and Kolen, 1994), dynamical systems (Cemgil et al., 2001), Bayesian statistics (Raphael, 2002) and particle filtering (Hainsworth and Macleod, 2004). A complete review can be found in the already mentioned paper by Gouyon and Dixon (2005). Some systems rather address pulse tracking by "repeated induction" (Scheirer, 2000; Laroche, 2003; Klapuri et al., 2006). A pulse is induced on a short analysis window (e.g. around 5 seconds of data), then the window is shifted in time and another induction takes place. Determining the tempo evolution then amounts to connecting the observations at each step. In addition to computational overload, one problem that arises with this approach to pulse tracking is the lack of continuity between successive observations and the difficulty of modelling sharp tempo changes.

Quantisation and time signature determination Few algorithms for time signature determination exist. The simplest approach is based on parsing the peaks of the periodicity function to find two significant peaks, which correspond respectively to a fast pulse, the time signature denominator, and a slower pulse, the numerator (Brown, 1993). The ratio between the pulse periods defines the time signature. Another approach is to consider all pairs of peaks as possible beat/measure combinations, and compute the fit of all periodicity peaks to each hypothesis (Dixon et al., 2003). Another strategy is to break the problem into several stages: determining the time signature denominator (e.g. by tempo induction and tracking), segmenting the music data with respect to this pulse and compute features at this temporal scope and finally detecting periodicities in the created feature lists (Gouyon and Herrera, 2003). Quantisation (or "rhythm parsing") can be seen as a by-product of the induction of several metrical levels, which together define a metrical grid. The rhythm of a given onset sequence can be parsed by assigning each onset (independently of its neighbors) to the closest element in this hierarchy. The weaknesses of such an approach are that it fails to account for musical context (e.g. a triplet note is usually followed by 2 more) and deviations from the metrical structure. Improvements to this first approach are considered by Desain and Honing (1989). Arguing that deviations from the metrical structure would be easier to determine if the quantised durations were known (Allen and Dannenberg, 1990), many researchers now consider rhythm parsing simultaneously with tempo tracking (Raphael, 2002; Cemgil and Kappen, 2003), rather than subsequent to it (hence the bi-directional arrow between these two modules in Figure 3.6).

Systematic deviation characterisation In the pulse induction process, short-term timing deviations can be "smoothed out" or cautiously handled so as to derive patterns of short-term timing deviations, such as swing: a "long-short" timing pattern of consecutive eight-notes. For instance, Laroche (2001) proposes to estimate the swing jointly with tempo and beats at the half-note level, assuming constant tempo: all pulse periods, phases and eight-note "long-short" patterns are enumerated and a search procedure determines which one best matches the onsets.

Rhythmic pattern determination Systematic short-term timing deviations are important music features. In addition, repetitive rhythmic patterns covering a longer temporal scope can also be characteristic of some music styles. For instance, many electronic musical devices feature templates of prototypical patterns such as Waltz, Cha Cha and the like. The length of such patterns is typically one bar, or a couple or bars. Few algorithms have been proposed for the automatic extraction of rhythmic patterns; they usually require the knowledge (or previous extraction) of part of the metrical structure, typically the beats and measure (Dixon et al., 2004).

Periodicity features Other rhythmic features, with a musical meaning less explicit than for example the tempo or the swing, have also been proposed, in particular in the context of designing rhythm similarity distances. Most of the time, these features are derived from a parametrisation of a periodicity function, e.g. the salience of several prominent peaks (Gouyon et al., 2004), their positions (Tzanetakis and Cook, 2002; Dixon et al., 2003), selected statistics (high-order moments, flatness, etc.) of the periodicity function considered as a probability density function (Gouyon et al., 2004) or simply the whole periodicity function itself (Foote et al., 2002).

Future research directions

Current research in rhythm description addresses all of these aspects, with varying degrees of success. For instance, determining the tempo of music with minor speed variations is feasible for almost all music styles if we do not expect that the system finds a specific metrical level (Gouyon et al., 2006a). Recent pulse tracking systems also reach high levels of accuracy. On the other hand, accurate quantisation, score transcription, determination of time signature and characterisation of intentional timing deviations are still open question. Particularly, it remains to be investigated how general recently proposed models are with respect to different music styles. New research directions include the determination of highly abstract rhythmic features required for music content processing and music information retrieval applications, the definition of the best rhythmic features and the most appropriate periodicity detection method (Gouyon, 2005).

3.2.6 Genre

Most music can be described in terms of dimensions such as melody, harmony, rhythm, etc. These high-level features characterise music and at least partially determine its genre, but, as mentioned in previous sections, they are difficult to compute automatically from raw audio signals. As a result, most audio-related music information retrieval research has focused on low-level features and induction algorithms to perform genre classification tasks. This approach has met with some success, but it is limited by the fact that the low level of representation may conceal many of the truly relevant aspects of a piece of music. See Chapter 4 and the works by Pampalk (2006), Ahrendt (2006), and Aucouturier (2006) for reviews of the current state-of-the-art in genre classification and more information on promising directions.

3.3 Audio content exploitation

We consider in this section a number of applications of content-based descriptions of audio signals. Although audio retrieval (see Section 3.3.1) is the one that has been addressed most often, others deserve a mention, e.g. content-based transformations (see Section 3.3.2).

3.3.1 Content-based search and retrieval

Searching a repository of music pieces can be greatly facilitated by automatic description of audio and music content (Cano, 2007), e.g. fingerprints, melodic features, tempo, etc. A content-based music retrieval system is a search engine at the interface of a repository, or organised database, of music pieces. Typically,

1. it receives a query, defined by means of musical strategies (e.g. humming, tapping, providing an audio excerpt or some measures of a score) or textual strategies (e.g. using "words" and/or "numbers" that describe some music feature like tempo, mood, etc.) referring to audio or music descriptors;
2. it has access to the set of music features extracted from the music files in the repository;
3. it returns a list of ranked files or excerpts that
    (a) are all relevant to the query (i.e. with high precision) or

   (b) constitute the set of all relevant files in the database (i.e. high recall);

4. (optionally) it processes some user-feedback information in order to improve its performance in the future.

Identification

With the help of fingerprinting systems it is possible to identify an unlabelled piece of audio and therefore provide a link to corresponding metadata (e.g. artist and song name). Depending on the application, different importance may be given to the following requirements:

Accuracy: The number of correct identifications, missed identifications, and wrong identifications (false positives).

Reliability: This is of major importance for copyright enforcement organisations.

Robustness: Ability to accurately identify an item, regardless of the level of compression and distortion or interference in the transmission channel. Other sources of degradation are pitching, equalisation, background noise, D/A-A/D conversion, audio coders (such as GSM and MP3), etc.

Granularity: Ability to identify whole titles from excerpts a few seconds long. It needs to deal with shifting, that is lack of synchronisation between the extracted fingerprint and those stored in the database and it adds complexity to the search (it needs to compare audio in all possible alignments).

Security: Vulnerability of the solution to cracking or tampering. In contrast with the robustness requirement, the manipulations to deal with are designed to fool the fingerprint identification algorithm.

Versatility: Ability to identify audio regardless of the audio format. Ability to use the same database for different applications.

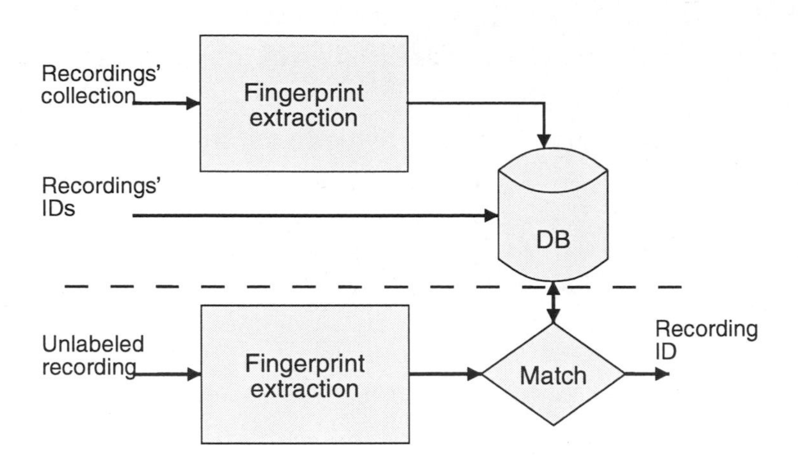

Figure 3.7: Content-based audio identification framework

Scalability: Performance with very large databases of titles or a large number of concurrent identifications. This affects the accuracy and the complexity of the system.

Complexity: It refers to the computational costs of the fingerprint extraction, the size of the fingerprint, the complexity of the search, the complexity of the fingerprint comparison, the cost of adding new items to the database, etc.

Fragility: Some applications, such as content-integrity verification systems, may require the detection of changes in the content. This is contrary to the robustness requirement, as the fingerprint should be robust to content-preserving transformations but not to other distortions.

The requirements of a complete fingerprinting system should be considered together with the fingerprint requirements listed in Section 3.2.3. Bear in mind that improving a certain requirement often implies loosing performance in some other. The overall identification process mimics the way humans perform the task. As seen in Figure 3.7, a memory of the recordings to be recognised is created off-line (top); in the identification mode (bottom), unlabelled audio is presented to the system to look for a match.

Audio Content Monitoring and Tracking One of the commercial usages of audio identification is that of remotely controlling the times a piece of music has been broadcasted, in order to ensure the broadcaster is doing the proper clearance for the rights involved (Cano et al., 2002).

Monitoring at the distributor end Content distributors may need to know whether they have the rights to broadcast certain content to consumers. Fingerprinting helps identify unlabelled audio in TV and radio channels' repositories. It can also identify unidentified audio content recovered from CD plants and distributors in anti-piracy investigations (e.g. screening of master recordings at CD manufacturing plants).

Monitoring at the transmission channel In many countries, radio stations must pay royalties for the music they air. Rights holders are eager to monitor radio transmissions in order to verify whether royalties are being properly paid. Even in countries where radio stations can freely air music, rights holders are interested in monitoring radio transmissions for statistical purposes. Advertisers are also willing to monitor radio and TV transmissions to verify whether commercials are being broadcast as agreed. The same is true for web broadcasts. Other uses include chart compilations for statistical analysis of program material or enforcement of cultural laws (e.g. in France, a certain percentage of the aired recordings must be in French). Fingerprinting-based monitoring systems can be used for this purpose. The system "listens" to the radio and continuously updates a play list of songs or commercials broadcast by each station. Of course, a database containing fingerprints of all songs and commercials to be identified must be available to the system, and this database must be updated as new songs come out. Examples of commercial providers of such services are: `http://www.musicreporter.net`, `http://www.audiblemagic.com`, `http://www.yacast.fr`, `http://www.musicip.com/` and `http://www.bmat.com/`. Additionally, audio content can be found in web pages and web-based peer-to-peer networks. Audio fingerprinting combined with a web crawler can identify their content and report it to the corresponding rights owners (e.g. `http://www.baytsp.com`).

Monitoring at the consumer end In usage-policy monitoring applications, the goal is to avoid misuse of audio signals by the consumer. We can conceive a system where a piece of music is identified by means of a fingerprint and a database is contacted to retrieve information about the rights. This information dictates the behavior of compliant devices (e.g. CD and DVD players and recorders, MP3 players or even computers) in accordance with the usage policy. Compliant devices are required to be connected to a network in order to access the database.

Added-value services Some systems store metadata related to audio files in databases accessible through the Internet. Such metadata can be relevant to a user for a given application and covers diverse types of information related to an audio file (e.g. how it was composed and how it was recorded, the composer, year of composition, the album cover image, album price, artist biography, information on the next concerts, etc.). Fingerprinting can then be used to identify a recording and retrieve the corresponding metadata. For example, MusicBrainz (`http://www.musicbrainz.org`), Id3man (`http://www.id3man.com`) or Moodlogic (`http://www.moodlogic.com`) automatically label collections of audio files. The user can download a compatible player that extracts fingerprints and submits them to a central server from which metadata associated to the recordings is downloaded. Gracenote (`http://www.gracenote.com`) recently enhanced their technology based on CDs' tables of contents with audio fingerprinting. Another application consists in finding or buying a song while it is being broadcast, by means of mobile-phone transmitting its GPS-quality received sound (e.g. `http://www.shazam.com`, `http://www.bmat.com`).

Summarisation

Summarisation, or thumbnailing, is essential for providing fast-browsing functionalities to content processing systems. An audiovisual summary that can be played, skipped upon, replayed or zoomed can save the user time and help him/her to get a glimpse of "what the music is about," especially when using personal media devices. Music summarisation consists in determining the key elements of a music sound file and rendering them in the most efficient way. There are two tasks here: first, extracting structure (Ong, 2007), and then, creating aural and visual representations of this structure (Peeters et al., 2002). Extracting a good summary from a sound file needs a comprehensive description of its content, plus some perceptual and cognitive constraints to be derived from users. An additional difficulty here is that different types of summaries can coexist, and that different users will probably require different summaries. Because of this amount of difficulty, the area of music summarisation is still under-developed. Reviews of recent promising approaches are presented by Ong (2007).

Play-list generation

This area concerns the design of lists of music pieces that satisfy some ordering criteria with respect to content descriptors previously computed, indicated (explicitly or implicitly) by the listener (Pampalk and Gasser, 2006). Play-list generation is usually constrained by time-evolving conditions (e.g. "start with slow-tempo pieces, then progressively increase tempo") (Pachet et al., 2000). Besides the play-list construction problem, we can also mention related problems such as achieving seamless transitions (in user-defined terms such as tempo, tonality, loudness) between the played pieces.

Music browsing and recommendation

Music browsing and recommendation are very much demanded, especially among youngsters. Recommendation consists in suggesting, providing guidance, or advising a potential consumer about interesting music files in on-line music stores, for instance. Nowadays, this is mainly possible by querying artist or song names (or other types of editorial data such as genre), or by browsing recommendations generated by collaborative filtering, i.e. using recommender systems that exploit information of the type "users that bought this album also bought this album." An obvious drawback of the first approach is that consumers need to know the name of the song or the artist beforehand. The second approach is only suitable when a considerable number of consumers has heard and rated the music. This situation makes it difficult for users to access and discover the vast amount of music composed and performed by unknown artists which is available in an increasing number of sites (e.g. `http://www.magnatune.com`) and which nobody has yet rated nor described. Content-based methods represent an alternative to these approaches. See the paper by Cano et al. (2005c) for the description of a large-scale music browsing and recommendation system based on automatic description of music content.[8] Other approaches to music recommendation are based on users' profiles: users' musical tastes and listening habits as well as complementary contextual (e.g. geographical) information (Celma, 2006a).[9] It is reasonable to assume that these different approaches will merge in the near future and result in improved music browsing and recommendation systems (Celma, 2006b).

Content visualisation

The last decade has witnessed great progress in the field of data visualisation. Massive amounts of data can be represented in multidimensional graphs in order to facilitate comparisons, grasp the patterns and relationships between data, and improve our understanding of them. Four purposes of information visualisation can be distinguished (Hearst, 1999):

Exploration, where visual interfaces can also be used as navigation and browsing interfaces.

Computation, where images are used as tools for supporting the analysis and reasoning about information. Data insight is usually facilitated by good data visualisations.

Communication, where images are used to summarise what otherwise would need many words and complex concepts to be understood. Music visualisation tools can be used to present concise information about relationships extracted from many interacting variables.

[8] See also `http://musicsurfer.iua.upf.edu`
[9] see `http://foafing-the-music.iua.upf.edu`

Decoration, where content data are used to create attractive pictures whose primary objective is not the presentation of information but aesthetic amusement.

It is likely that in the near future we will witness an increasing exploitation of data visualisation techniques in order to enhance song retrieval, collection navigation and music discovery (Pampalk and Goto, 2006).

3.3.2 Content-based audio transformations

Transformations of audio signals have a long tradition (Zölzer, 2002). A recent trend in this area of research is the editing and transformation of music audio signals triggered by explicit musically-meaningful representational elements, in contrast to low-level signal descriptors. These techniques are referred to as content-based audio transformations, or "adaptive digital audio effects" (Verfaille et al., 2006), and are based on the type of description of audio signals detailed above in Section 3.2. In this section, we give examples of such techniques, following increasing levels of abstraction in the corresponding content description.

Loudness modifications

The most commonly known effects related to loudness are the ones that modify the sound intensity level: volume change, tremolo, compressor, expander, noise gate and limiter (Verfaille et al., 2006). However, when combined with other low-level features, loudness is correlated to higher-level descriptions of sounds, such as the timbre or the musical intentions of a performer. It can therefore be used as a means to control musically-meaningful aspects of sounds. The mechanisms that relate the actions of a player to the sound level produced by a given instrument are usually so complex that this feature can seldom be decorrelated from others, such as timbre. Thus, differences between playing a soft and a loud note on an instrument do not reside only in loudness levels. Spectral modifications must also be accounted for. In the case of the singing voice, for instance, many studies have been carried out and are summarised by Sundberg (1987). Using Sundberg's nomenclature, it is possible, under certain conditions, to infer the source spectrum modifications from uttering the same vowel at different loudness of phonation. Building upon this assumption, Fabig and Janer (2004) propose a method for modifying the loudness of the singing voice by detecting the excitation slope automatically.

Time-scaling

In a musical context, time-scaling can be understood as changing the pace of a music signal, i.e. its tempo. If a musical performance is time-scaled to a different tempo, we should expect to listen to the same notes starting at a scaled time pattern, but with durations modified linearly according to the tempo change. The pitch of the notes should however remain unchanged, as well as the perceived expression. Thus, for example, vibratos should not change their depth, tremolo or rate characteristics. And of course, the audio quality should be preserved in such a way that if we had never listened to that music piece, we would not be able to know if we were listening to the original recording or to a transformed one. Time-scale modifications can be implemented in different ways. Generally, algorithms are grouped in three different categories: time domain techniques, phase-vocoder and variants, and signal models. In the remainder of this section we explain the basics of these approaches in turn.

Time domain techniques Time domain techniques are the simplest methods for performing time-scale modification. The simplest (and historically first) technique is the variable speed replay of analog audio tape recorders (McNally, 1984). A drawback of this technique is that during faster playback, the pitch of the sound is raised while the duration is shortened. On the other hand, during slower playback, the pitch of the sound is lowered while the duration is lengthened. Many papers show good results without scaling frequency by segmenting the input signal into several windowed sections and then placing these sections in new time locations and overlapping them to get the time-scaled version of the input signal. This set of algorithms is referred to as

Overlap-Add (OLA). To avoid phase discontinuities between segments, the synchronised OLA algorithm (SOLA) uses a cross-correlation approach to determine where to place the segment boundaries (Wayman et al., 1989). In TD-PSOLA (Moulines et al., 1989), the overlapping operation is performed pitch-synchronously to achieve high quality time-scale modification. This works well with signals having a prominent basic frequency and can be used with all kinds of signals consisting of a single signal source. When it comes to a mixture of signals, this method will produce satisfactory results only if the size of the overlapping segments is increased to include a multiple of cycles, thus averaging the phase error over a longer segment and making it less audible. WSOLA (Verhelst and Roelands, 1993) uses the concept of waveform similarity to ensure signal continuity at segment joints, providing high quality output with high algorithmic and computational efficiency and robustness. All the aforementioned techniques consider equally the transient and steady state parts of the input signal, and thus time-scale them both in the same way. To get better results, it is preferable to detect the transient regions and not time-scale them, just translate them into a new time position, while time-scaling the non-transient segments. The earliest mention of this technique can be found in the *Lexicon 2400* time compressor/expander from 1986. This system detects transients, and time-scales only the remaining audio using a TD-PSOLA-like algorithm. Lee et al. (1997) show that using time-scale modification on non-transient parts of speech alone improves the intelligibility and quality of the resulting time-scaled speech.

Phase vocoder and variants The phase vocoder is a relatively old technique that dates from the 70's (Portnoff, 1976). It is a frequency domain algorithm computationally quite more expensive than time domain algorithms. However it can achieve high-quality results even with high time-scale factors. Basically, the input signal is split into many frequency channels, uniformly spaced, usually using the FFT. Each frequency band (bin) is decomposed into magnitude and phase parameters, which are modified and re-synthesised by the IFFT or a bank of oscillators. With no transformations, the system allows a perfect reconstruction of the original signal. In the case of time-scale modification, the synthesis hop size is changed according to the desired time-scale factor.

Magnitudes are linearly interpolated and phases are modified in such a way that phase consistency are maintained across the new frame boundaries. The phase-vocoder introduces signal smearing for impulsive signals due to the loss of phase alignment of the partials. A typical drawback of the phase vocoder is the loss of vertical phase coherence that produces reverberation or loss of presence in the output. This effect is also referred to as "phasiness," which can be circumvented by phase-locking techniques (Laroche and Dolson, 1999) among bins around spectral peaks. Note that adding peak tracking to the spectral peaks, the phase-vocoder resembles the sinusoidal modelling algorithms, which is introduced in the next paragraph. Another traditional drawback of the phase vocoder is the bin resolution dilemma: the phase estimates are incorrect if more than one sinusoidal peak resides within a single spectral bin. Increasing the window may solve the phase estimation problem, but it implies a poor time resolution and smooths the fast frequency changes. And the situation gets worse in the case of polyphonic music sources because then the probability is higher that sinusoidal peaks from different sources will reside in the same spectrum bin. Different temporal resolutions for different frequencies can be obtained by convolution of the spectrum with a variable kernel function (Hoek, 1999). Thus, long windows are used to calculate low frequencies, while short windows are used to calculate high frequencies. Other approaches approximate a constant-Q phase-vocoder based on wavelet transforms or nonuniform sampling.

Techniques based on signal models Signal models have the ability to split the input signal into different components which can be parameterised and processed independently giving a lot of flexibility for transformations. Typically these components are sinusoids, transients and noise. In sinusoidal modelling (McAulay and Quatieri, 1986), the input signal is represented as a sum of sinusoids with time-varying amplitude, phase and frequency. Parameter estimation can be improved by using interpolation methods, signal derivatives and special windows. Time-scaling using sinusoidal modelling achieves good results with harmonic signals, especially when keeping the vertical phase coherence. However it fails to successfully represent and transform noise and transient signals. Attacks are smoothed and noise sounds artificial.

The idea of subtracting the estimated sinusoids from the original sound to obtain a residual signal was proposed by Smith and Serra (1987); this residual can then be modelled as a stochastic signal. This method allows the splitting of e.g. a flute sound into the air flow and the harmonics components, and to transform both parts independently. This technique successfully improves the quality of time-scale transformations but fails to handle transients, which are explicitly handled in (Verma et al., 1997). Then, all three components (sinusoidal, noise and transient) can be modified independently and re-synthesised. When time-scaling an input signal, transients can successfully be translated to a new onset location, preserving their perceptual characteristics.

Timbre modifications

Timbre is defined as all those characteristics that distinguish two sounds having the same pitch, duration and loudness. As a matter of fact, timbre perception depends on many characteristics of the signal such as its instantaneous spectral shape and its evolution, the relation of its harmonics, and some other features related to the attack, release and temporal structure. Timbre instrument modification can be achieved by many different techniques. One of them is to modify the input spectral shape by *timbre mapping*. Timbre mapping is a general transformation performed by warping the spectral shape of a sound by means of a mapping function $g(f)$ that maps frequencies of the transformed spectrum (f_y) to frequencies of the initial spectrum (f_x) via a simple equation $f_y = g(f_x)$. Linear *scaling* (compressing or expanding) is a particular case of timbre mapping in which the mapping function pertains to the family $f_y = k * f_x$, where k is the scale factor, usually between 0.5 and 2. The timbre scaling effect resembles modifications of the size and shape of the instrument. The *shifting* transformation is another particular case of the timbre mapping as well, in which $g(f)$ can be expressed as $f_y = f_x + c$, where c is an offset factor.

Morphing Another way of accomplishing timbre transformations is to modify the input spectral shape by means of a secondary spectral shape. This is usually referred to as *morphing* or *cross-synthesis*. In fact, morphing is a technique with which, out of two or more elements, we can generate new ones with hybrid properties. In the context of video processing, morphing has been widely developed and enjoys great popularity in commercials, video clips and films where faces of different people change one into another or chairs mutate into for example elephants. Analogously, in the context of audio processing, the goal of most of the developed morphing methods has been the smooth transformation from one sound to another. Along this transformation, the properties of both sounds combine and merge into a resulting hybrid sound. With different names, and using different signal processing techniques, the idea of audio morphing is well known in the computer music community (Serra, 1994; Slaney et al., 1996). In most algorithms, morphing is based on the interpolation of sound parameterisations resulting from analysis/synthesis techniques, such as the short-time Fourier transform, linear predictive coding or sinusoidal models.

Voice timbre Whenever the morphing is performed by means of modifying a reference voice signal in matching its individuality parameters to another, we can refer to it as voice conversion (Loscos, 2007). Some applications for the singing voice exist in the context of karaoke entertainment (Cano et al., 2000) and in the related topics of gender change (Cano et al., 2000) and unison choir generation (Bonada et al., 2006). We refer to the paper by Bonada and Serra (2007) regarding the general topic of singing voice synthesis. Still for the particular case of voice, other finer-grained transformations exist to modify the timbre character without resorting to a morphing between two spectral shapes: e.g. *rough*, *growl*, *breath* and *whisper* transformations. *Roughness* in voice can come from different pathologies such as biphonia, or diplophonia, and can combine with many other voice tags such as "hoarse" or "creaky." However, here we will refer to a rough voice as the one due to cycle to cycle variations of the fundamental frequency (jitter), and the period amplitude (shimmer). The most common techniques used to synthesise rough voices work with a source/filter model and reproduce the jitter and shimmer aperiodicities in the time domain (Childers, 1990). These aperiodicities can be applied to the voiced pulse-train excitation by taking real patterns that have been extracted from rough voice recordings or by using statistical models (Schoentgen, 2001).

Spectral domain techniques have also proved to be valid to emulate roughness (Loscos and Bonada, 2004). *Growl* phonation is often used when singing jazz, blues, pop and other music styles as an expressive accent. Perceptually, growl voices are close to other dysphonic voices such as "hoarse" or "creaky," however, unlike these others, growl is always a vocal effect and not a permanent vocal disorder. According to Sakakibara et al. (2004), growl comes from simultaneous vibrations of the vocal folds and supra-glottal structures of the larynx. The vocal folds vibrate half periodically to the aryepiglottic fold vibration generating sub-harmonics. Growl effect can be achieved by adding these sub-harmonics in frequency domain to the original input voice spectrum (Loscos and Bonada, 2004). These sub-harmonics follow certain magnitude and phase patterns that can be modelled from spectral analyses of real growl voice recordings. *Breath* can be achieved by different techniques. One is to increase the amount of the noisy residual component in those sound models in which there is a sinusoidal-noise decomposition. For sound models based on the phase-locked vocoder (see Section 3.3.2) a more breathy timbre can be achieved by filtering and distorting the harmonic peaks. The *whisper* effect can be obtained by equalizing a previously recorded and analyzed template of a whisper utterance. The time behavior of the template is preserved by adding to the equalisation the difference between the spectral shape of the frame of the template currently being used and an average spectral shape of the template. An "anti-proximity" filter may be applied to achieve a more natural and smoother effect (Fabig and Janer, 2004).

Rhythm transformations

In addition to tempo changes (see Section 3.3.2), some existing music editing softwares provide several rhythm transformation functionalities. For instance, any sequencer provides the means to adjust MIDI note timings to a metrical grid ("quantisation") or a predefined rhythmic template. By doing an appropriate mapping between MIDI notes and audio samples, it is therefore possible to apply similar timing changes to audio mixes. But when dealing with general polyphonic music excerpts, without corresponding MIDI scores, these techniques cannot be applied. Few commercial applications implement techniques to transform the rhythm of general polyphonic music excerpts. A review can be found in the paper by Gouyon et al. (2003a). A technique for swing transformation has also been proposed in the same paper by Gouyon et al. (2003a), which consists of a description module and a transformation module. The description module does onset detection and rhythmic analysis (see Section 3.2.5). Swing is relative to the length of consecutive eighth-notes, it is therefore necessary to determine the beat indexes of eighth-notes. It is also necessary to describe the excerpt at the next higher (slower) metrical level, the quarter-note, and determine the eighth-note "phase," that is, determine in a group of two eighth-notes which is the first one.[10] The existing ratio between consecutive eighth-notes is also estimated. This ratio can be changed by shortening or lengthening the first eighth-notes of each quarter-note, and lengthening or shortening the second eighth-notes accordingly. This is done by means of time-scaling techniques. In the papers by Gouyon et al. (2003a) and Janer et al. (2006), time-scaling is done in real-time and the user can continuously adjust the swing ratio while playing back the audio file. Having found evidence for the fact that deviations occurring within the scope of the smallest metrical pulse are very important for musical expressiveness, Bilmes (1993) proposes additional rhythmic transformations based on a high-level description of the rhythmic content of audio signals. Interesting recent applications of rhythm transformations can be found in the works by Wright and Berdahl (2006), Ramirez and Hazan (2006), Janer et al. (2006), Grachten et al. (2006), and Ravelli et al. (2007).

Melodic transformations

Melodic transformations such as *pitch discretisation to temperate scale* and *intonation* apply direct modifications to the fundamental frequency envelope. Arguably, these transformations may be considered low level transformations; however, they do change the way a high-level descriptor, namely the melody, is perceived by the listener. Intonation transformations are achieved by stretching or compressing the difference between the analysis pitch envelope and a

[10] Indeed, it is not at all the same to perform a "long-short" pattern as a "short-long" pattern.

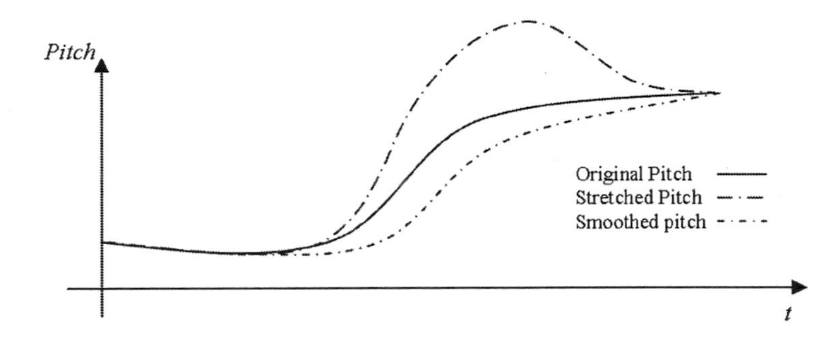

Figure 3.8: Intonation transformation.

low pass filtered version of it. The goal of the transformation is to increase or decrease the sharpness of the note attack, as illustrated in Figure 3.8. Pitch discretisation to temperate scale can be accomplished by forcing the pitch to take the nearest frequency value of the equal temperate scale. It is indeed a very particular case of pitch transposition where the pitch is quantified to one of the 12 semitones of an octave (Amatriain et al., 2003).[11] Other melodic transformations can be found in the software *Melodyne*[12] and in the paper by Gómez et al. (2003d) as *transposition* (global change of pitch), *horizontal symmetry*, in which the user can choose a pitch value (arbitrary or some global descriptor related to pitch distribution as minimum, maximum or mean pitch value of the melody) and perform a symmetric transformation of the note pitches with respect to this value on a horizontal axis, *contour direction changes* in which the user can change the interval direction without changing the interval depth (e.g. converting an ascending octave to a descending one), etc. Although these transformations are conceptually simple, they correspond to usual music composition procedures and can create dramatic changes that may enhance the original material (if used in the right creative context). Finally, melodies of monophonic instruments can also be transformed by applying changes on other high-level descriptors in addition to the pitch, such as tempo curves (Grachten et al., 2006) and note timing and loudness (Ramirez et al., 2004). See Chapter 5 for more information on analysis and generation of expressive musical performances.

[11]See also Antares' *Autotune*, `http://www.antarestech.com/`.
[12]`http://www.celemony.com/melodyne`

Harmony transformations

Harmonizing a sound can be defined as mixing a sound with several pitch-shifted versions of it (Amatriain et al., 2002; Verfaille et al., 2006). This requires two parameters: the number of voices and the pitch for each of these. Pitches of the voices to generate are typically specified by the key and chord of harmonisation. Where the key and chord are estimated from the analysis of the input pitch and the melodic context (Pachet and Roy, 1998), some refer to "intelligent harmonizing."[13] An application of harmonizing in real-time monophonic solo voices is detailed in (Bonada et al., 2006).

3.4 Perspectives

All areas of high level description of music audio signals (as for instance those addressed in this chapter – Tonality, Rhythm, etc. –) will, without doubt, witness rapid improvements in the near future. We believe however that the critical aspect to focus on for these improvements to happen is the systematic use of large-scale evaluations.

Evaluations Developing technologies related to content processing of music audio signals requires data (Cano et al., 2004). For instance, implementing algorithms for automatic instrument classification requires annotated samples of different instruments. Implementing a voice synthesis and transformation software calls for repositories of voice excerpts sung by professional singers. Testing a robust beat-tracking algorithm requires songs of different styles, instrumentation and tempi. Building models of music content with a Machine Learning rationale calls for large amounts of data. Besides, running an algorithm on big amounts of (diverse) data is a requirement to ensure the algorithm's quality and reliability. In other scientific disciplines long-term improvements have shown to be bounded to systematic evaluation of models. For instance, text retrieval techniques significantly improved over the year thanks

[13] see TC-Helicon's *Voice Pro*, `http://www.tc-helicon.com/VoicePro`.

to the TREC initiative (see `http://trec.nist.gov`). TREC evaluations proceed by giving research teams access to a standardised, large-scale test collection of text, a standardised set of test queries, and requesting a standardised way of generating and presenting the results. Different TREC tracks have been created over the years (text with moving images, web retrieval, speech retrieval, etc.) and each track has developed its own special test collections, queries and evaluation requirements. The standardisation of databases and evaluation metrics also greatly facilitated progress in the fields of Speech Recognition (Przybocki and Martin, 1989; Pearce and Hirsch, 2000), Machine Learning (Guyon et al., 2004) or Video Retrieval (see `http://www-nlpir.nist.gov/projects/trecvid/`). In 1992, the visionary Marvin Minsky declared: "the most critical thing, in both music research and general AI research, is to learn how to build a common music database" (Minsky and Laske, 1992). More than 10 years later, this is still an open issue. In the last few years, the music content processing community has recognised the necessity of conducting rigorous and comprehensive evaluations (Downie, 2002, 2003b). However, we are still far from having set a clear path to be followed for evaluating research progresses. Inspired by Downie (2003b), here follows a list of urgent methodological problems to be addressed by the research community:

1. There are no standard collections of music against which to test content description or exploitation techniques;

2. There are no standardised sets of performance tasks;

3. There are no standardised evaluation metrics.

As a first step, an audio description contest took place during the fifth edition of the ISMIR, in Barcelona, in October 2004. The goal of this contest was to compare state-of-the-art audio algorithms and systems relevant for some tasks of music content description, namely genre recognition, artist identification, tempo extraction, rhythm classification and melody extraction (Cano et al., 2006b). It was the first large-scale evaluation of audio description algorithms, and the first initiative to make data and legacy metadata publicly available (see `http://ismir2004.ismir.net` and Cano et al., 2006b, for more details). However, this competition addressed a small part of the bulk of research going on in music content processing. Following editions of the ISMIR have continued this effort: public evaluations now take place on an annual basis, in the Music Information Retrieval Evaluation eXchange (MIREX), organised during ISMIR conferences mainly by the International Music Information Retrieval Systems Evaluation Laboratory (IMIRSEL)[14] together with voluntary fellow researchers. MIREXes have widened the scope of the competitions and cover a broad range of tasks, including symbolic data description and retrieval. Future editions of MIREX are likely to make a further step, from evaluation of content description algorithms to evaluations of complete MIR systems.

Acknowledgments

This work has been partially funded by the European IST-507142 project SIMAC (Semantic Interaction with Music Audio Contents),[15] and the HARMOS E-Content project. The authors wish to thank their colleagues in the Music Technology Group for their help. Thanks also to Simon Dixon for participating in previous versions of Section 3.2.5.

[14] http://www.music-ir.org/evaluation/

[15] http://www.semanticaudio.org

Bibliography

P. Ahrendt. *Music Genre Classification Systems - A Computational Approach*. Unpublished PhD thesis, Technical University of Denmark, 2006.

P. Aigrain. New applications of content processing of music. *Journal of New Music Research*, 28(4):271–280, 1999.

P. Allen and R. Dannenberg. Tracking musical beats in real time. In Allen P. and Dannenberg R., editors, *Proceedings International Computer Music Conference*, pages 140–143, 1990.

X. Amatriain, J. Bonada, À. Loscos, and X. Serra. Spectral processing. In U. Zölzer, editor, *DAFX Digital Audio Effects*, pages 373–439. Wiley & Sons, 2002.

X. Amatriain, J. Bonada, À. Loscos, J. Arcos, and V. Verfaille. Content-based transformations. *Journal of New Music Research*, 32(1):95–114, 2003.

J-J. Aucouturier. *Ten Experiments on the Modelling of Polyphonic Timbre*. Unpublished PhD thesis, University of Paris 6, Paris, 2006.

M. Basseville and I. V. Nikiforov. *Detection of abrupt changes: Theory and application*. Prentice-Hall Inc., Englewood Cliffs, NJ, 1993.

E. Batlle and P. Cano. Automatic segmentation for music classification using competitive hidden markov models. In *Proceedings International Symposium on Music Information Retrieval*, 2000.

E. Batlle, J. Masip, and E. Guaus. Automatic song identification in noisy broadcast audio. In *Proceedings of IASTED International Conference on Signal and Image Processing*, 2002.

J. Bello. *Towards the Automated Analysis of Simple Polyphonic Music: A Knowledge-Based Approach*. Unpublished PhD thesis, Department of Electronic Engineering, Queen Mary University of London, 2003.

J. Bello, L. Daudet, S. Abdallah, C. Duxbury, M. Davies, and M. Sandler. A tutorial on onset detection in music signals. *IEEE Transactions on Speech and Audio Processing*, 13(5):1035–1047, 2005.

J. Bilmes. Techniques to foster drum machine expressivity. In *Proceedings International Computer Music Conference*, pages 276–283, 1993.

S. Blackburn. *Content based retrieval and navigation of music using melodic pitch contour*. Unpublished PhD thesis, University of Southampton, 2000.

T. Blum, D. Keislar, J. Wheaton, and E. Wold. Method and article of manufacture for content-based analysis, storage, retrieval and segmentation of audio information, U.S. patent 5,918,223, June 1999.

R. Bod. Memory-based models of melodic analysis: Challenging the Gestalt principles. *Journal of New Music Research*, 31(1):27–37, 2001.

J. Bonada. Automatic technique in frequency domain for near-lossless time-scale modification of audio. In *Proceedings International Computer Music Conference*, pages 396–399, 2000.

J. Bonada and X. Serra. Synthesis of the singing voice by performance sampling and spectral models. *IEEE Signal Processing Magazine*, 24(2):67–79, 2007.

J. Bonada, M. Blaauw, À. Loscos, and H. Kenmochi. Unisong: A choir singing synthesizer. In *Proceedings of 121st Convention of the AES*, 2006.

A. Bregman. Psychological data and computational auditory scene analysis. In D. Rosenthal and H. Okuno, editors, *Computational auditory scene analysis*. Lawrence Erlbaum Associates Inc., 1998.

A. Bregman. *Auditory scene analysis*. MIT Press, Harvard, MA, 1990.

G. Brown. *Computational auditory scene analysis: A representational approach*. Unpublished PhD thesis, University of Sheffield, 1992.

J. Brown. Determination of the meter of musical scores by autocorrelation. *Journal of the Acoustical Society of America*, 94(4):1953–1957, 1993.

J. Brown and M. Puckette. An efficient algorithm for the calculation of a constant Q transform. *Journal of the Acoustical Society of America*, 92(5):2698–2701, 1992.

D. Byrd. *Music notation by computer*. Unpublished PhD thesis, Indiana University, 1984.

E. Cambouropoulos. The local boundary detection model and its application in the study of expressive timing. In *Proceedings International Computer Music Conference*, 2001.

P. Cano. Fundamental frequency estimation in the SMS analysis. In *Proceedings Digital Audio Effects Conference*, 1998.

P. Cano. *Content-based audio search: From fingerprinting to semantic audio retrieval*. Unpublished PhD thesis, University Pompeu Fabra, Barcelona, 2007.

P. Cano, À. Loscos, J. Bonada, M. de Boer, and X. Serra. Voice morphing system for impersonating in karaoke applications. In *Proceedings International Computer Music Conference*, 2000.

P. Cano, E. Batlle, H. Mayer, and H. Neuschmied. Robust sound modeling for song detection in broadcast audio. In *Proceedings AES 112th International Convention*, 2002.

P. Cano, M. Koppenberger, S. Ferradans, A. Martinez, F. Gouyon, V. Sandvold, V. Tarasov, and N. Wack. MTG-DB: A repository for music audio processing. In *Proceedings International Conference on Web Delivering of Music*, 2004.

P. Cano, E. Batlle, T. Kalker, and J. Haitsma. A review of audio fingerprinting. *The Journal of VLSI Signal Processing*, 41(3):271–284, 2005a.

P. Cano, O. Celma, M. Koppenberger, and J. Martin-Buldú. The topology of music artists' graphs. In *Proceedings XII Congreso de Fisica Estadistica*, 2005b.

P. Cano, M. Koppenberger, N. Wack, J. Garcia, J. Masip, O. Celma, D. Garcia, E. Gómez, F. Gouyon, E. Guaus, P. Herrera, J. Massaguer, B. Ong, M. Ramirez, S. Streich, and X. Serra. An industrial-strength content-based music recommendation system. In *Proceedings International ACM SIGIR Conference*, page 673, 2005c.

P. Cano, O. Celma, M. Koppenberger, and J. Martin-Buldú. Topology of music recommendation networks. *Chaos: An Interdisciplinary Journal of Nonlinear Science*, 16, 2006a.

P. Cano, E. Gómez, F. Gouyon, P. Herrera, M. Koppenberger, B. Ong, X. Serra, S. Streich, and N. Wack. ISMIR 2004 audio description contest. *MTG Technical Report MTG-TR-2006-02*, 2006b.

M. Casey and A. Westner. Separation of mixed audio sources by independent subspace analysis. In *Proceedings International Computer Music Conference*, 2000.

O. Celma. Foafing the music: Bridging the semantic gap in music recommendation. In *Proceedings 5th International Semantic Web Conference*, 2006a.

O. Celma. *Music Recommendation: a multi-faceted approach*. Unpublished DEA thesis, University Pompeu Fabra, 2006b.

A. Cemgil and B. Kappen. Monte Carlo methods for tempo tracking and rhythm quantization. *Journal of Artificial Intelligence Research*, 18:45–81, 2003.

A. Cemgil, B. Kappen, P. Desain, and H. Honing. On tempo tracking: Tempogram representation and Kalman filtering. *Journal of New Music Research*, 28(4):259–273, 2001.

D. Childers. Speech processing and synthesis for assessing vocal disorders. *IEEE Magazine on Engineering in Medicine and Biology*, 9:69–71, 1990.

L. Cohen. Time-frequency distributions - A review. *Processings of the IEEE*, 77 (7):941–981, 1989.

N. Collins. Using a pitch detector for onset detection. In *International Conference on Music Information Retrieval*, pages 100–106, 2005.

G. Cooper and L. Meyer. *The rhythmic structure of music*. University of Chicago Press, 1960.

K. de Koning and S. Oates. Sound base: Phonetic searching in sound archives. In *Proceedings International Computer Music Conference*, pages 433–466, 1991.

P. Desain and H. Honing. The quantization of musical time: A connectionist approach. *Computer Music Journal*, 13(3):55–66, 1989.

P. Desain and H. Honing. Computational models of beat induction: The rule-based approach. *Journal of New Music Research*, 28(1):29–42, 1999.

S. Dixon. Automatic extraction of tempo and beat from expressive performances. *Journal of New Music Research*, 30(1):39–58, 2001.

S. Dixon and E. Cambouropoulos. Beat tracking with musical knowledge. In *Proceedings European Conference on Artificial Intelligence*, pages 626–630, 2000.

S. Dixon, E. Pampalk, and G. Widmer. Classification of dance music by periodicity patterns. In *Proceedings International Conference on Music Information Retrieval*, pages 159–165, 2003.

S. Dixon, F. Gouyon, and G. Widmer. Towards characterisation of music via rhythmic patterns. In *Proceedings International Conference on Music Information Retrieval*, pages 509–516, 2004.

S. Dixon, W. Goebl, and E. Cambouropoulos. Perceptual smoothness of tempo in expressively performed music. *Music Perception*, 23(3):195–214, 2006.

J. Downie, editor. *The MIR/MDL evaluation project white paper collection - Proceedings International Conference on Music Information Retrieval*. 2nd edition, 2002.

J. Downie. Music information retrieval. *Annual Review of Information Science and Technology*, 37:295–340, 2003a.

J. Downie. The scientific evaluation of music information retrieval systems: Foundations and the future. *Computer Music Journal*, 28(2):12–23, 2003b.

J. Downie. The MusiFind musical information retrieval project, phase II: User assessment survey. In *Proceedings 22nd Annual Conference of the Canadian Association for Information Science*, pages 149–166, 1994.

C. Drake, L. Gros, and A. Penel. How fast is that music? The relation between physical and perceived tempo. In S. Yi, editor, *Music, Mind and Science*, pages 190–203. Seoul National University Press, 1999.

B. Eaglestone. A database environment for musician-machine interaction experimentation. In *Proceedings International Computer Music Conference*, pages 20–27, 1988.

B. Eaglestone and A. Verschoor. An intelligent music repository. In *Proceedings International Computer Music Conference*, pages 437–440, 1991.

D. Ellis. *Prediction-driven computational auditory scene analysis*. Unpublished PhD thesis, Massachussetts Institute of Technology, 1996.

L. Fabig and J. Janer. Transforming singing voice expression - The sweetness effect. In *Proceedings Digital Audio Effects Conference*, pages 70–75, 2004.

B. Feiten, R. Frank, and T. Ungvary. Organizing sounds with neural nets. In *Proceedings International Computer Music Conference*, pages 441–444, 1991.

J. Foote. Automatic audio segmentation using a measure of audio novelty. In *Proceedings IEEE International Conference on Multimedia and Expo*, pages 452–455, 2000.

J. Foote, M. Cooper, and U. Nam. Audio retrieval by rhythmic similarity. In *Proceedings International Conference on Music Information Retrieval*, pages 265–266, 2002.

T. Fujishima. Real-time chord recognition of musical sound: A system using common lisp music. In *International Computer Music Conference*, pages 464–467, 1999.

D. Godsmark and G. J. Brown. A blackboard architecture for computational auditory scene analysis. *Speech Communication*, 27:351–366, 1999.

B. Gold and L. Rabiner. Parallel processing techniques for estimating pitch periods of speech in the time domain. *Journal of the Acoustical Society of America*, 46:442–448, 1969.

L. Gomes, P. Cano, E. Gómez, M. Bonnet, and E. Batlle. Audio watermarking and fingerprinting: For which applications? *Journal of New Music Research*, 32(1):65–82, 2003.

E. Gómez. Tonal description of polyphonic audio for music content processing. *INFORMS Journal on Computing, Special Issue on Computation in Music*, 18(3): 294–301, 2004.

E. Gómez. *Tonal Description of Music Audio Signals*. Unpublished PhD thesis, University Pompeu Fabra, Barcelona, 2006.

E. Gómez, F. Gouyon, P. Herrera, and X. Amatriain. MPEG-7 for content-based music processing. In *Proceedings 4th WIAMIS-Special session on Audio Segmentation and Digital Music*, 2003a.

E. Gómez, F. Gouyon, P. Herrera, and X. Amatriain. Using and enhancing the current MPEG-7 standard for a music content processing tool. In *Proceedings 114th AES Convention*, 2003b.

E. Gómez, A. Klapuri, and B. Meudic. Melody description and extraction in the context of music content processing. *Journal of New Music Research*, 32 (1):23–40, 2003c.

E. Gómez, G. Peterschmitt, X. Amatriain, and P. Herrera. Content-based melodic transformations of audio for a music processing application. In *Proceedings Digital Audio Effects Conference*, 2003d.

E. Gómez, J. P. Bello, M. Davies, D. Garcia, F. Gouyon, C. Harte, P. Herrera, C. Landone, K. Noland, B. Ong, V. Sandvold, S. Streich, and B. Wang. Front-end signal processing and low-level descriptors computation module. Technical Report D2.1.1, SIMAC IST Project, 2005.

J. Gordon. *Perception of attack transients in musical tones*. Unpublished PhD thesis, CCRMA, Stanford University, 1984.

M. Goto. A robust predominant-f0 estimation method for real-time detection of melody and bass lines in CD recordings. In *Proceedings IEEE International Conference on Acoustics Speech and Signal Processing*, pages 757–760, 2000.

M. Goto. An audio-based real-time beat tracking system for music with or without drums. *Journal of New Music Research*, 30(2):159–171, 2001.

M. Goto and Y. Muraoka. Real-time beat tracking for drumless audio signals: Chord change detection for musical decisions. *Speech Communication*, (27): 311–335, 1999.

F. Gouyon. *A computational approach to rhythm description*. Unpublished PhD thesis, Pompeu Fabra University, Barcelona, 2005.

F. Gouyon and S. Dixon. A review of automatic rhythm description systems. *Computer Music Journal*, 29(1):34–54, 2005.

F. Gouyon and P. Herrera. Determination of the meter of musical audio signals: Seeking recurrences in beat segment descriptors. In *Proceedings Audio Engineering Society, 114th Convention*, 2003.

F. Gouyon and B. Meudic. Towards rhythmic content processing of musical signals: Fostering complementary approaches. *Journal of New Music Research*, 32(1):41–64, 2003.

F. Gouyon, L. Fabig, and J. Bonada. Rhythmic expressiveness transformations of audio recordings: Swing modifications. In *Proceedings Digital Audio Effects Conference*, pages 94–99, 2003a.

F. Gouyon, S. Dixon, E. Pampalk, and G. Widmer. Evaluating rhythmic descriptors for musical genre classification. In *Proceedings 25th International AES Conference*, pages 196–204, 2004.

F. Gouyon, A. Klapuri, S. Dixon, M. Alonso, G. Tzanetakis, C. Uhle, and P. Cano. An experimental comparison of audio tempo induction algorithms. *IEEE Transactions on Speech and Audio Processing*, 14:1832–1844, 2006a.

F. Gouyon, G. Widmer, X. Serra, and A. Flexer. Acoustic cues to beat induction: A Machine Learning perspective. *Music Perception*, 24(2):181–194, 2006b.

M. Grachten, J. Arcos, and R. López de Mántaras. A case based approach to expressivity-aware tempo transformation. *Machine Learning Journal*, 65(2-3): 411–437, 2006.

I. Guyon, S. Gunn, A. Ben Hur, and G. Dror. Result analysis of the NIPS 2003 feature selection challenge. In *Proceedings Neural Information Processing Systems Conference*, pages 545–552, 2004.

S. Hainsworth and M. Macleod. Particle filtering applied to musical tempo tracking. *EURASIP Journal on Applied Signal Processing*, 15:2385–2395, 2004.

H. Harb and L. Chen. Robust speech music discrimination using spectrum's first order statistics and neural networks. In *Proceedings 7th International Symposium on Signal Processing and Its Applications*, pages 125–128, 2003.

M. Hearst. User interfaces and visualization. In R. Baeza-Yates and B. Ribeiro-Neto, editors, *Modern information retrieval*. Harlow, Essex: ACM Press, 1999.

P. Herrera. *Automatic classification of percussion sounds: From acoustic features to semantic descriptions*. Unpublished PhD thesis, University Pompeu Fabra, in print.

P. Herrera and J. Bonada. Vibrato extraction and parameterization in the spectral modeling synthesis framework. In *Proceedings Digital Audio Effects Conference*, 1998.

W. Hess. *Pitch Determination of Speech Signals. Algorithms and Devices.* Springer Series in Information Sciences. Springer-Verlag, Berlin, New York, Tokyo, 1983.

S. Hoek. Method and apparatus for signal processing for time-scale and/or pitch modification of audio signals, U.S. patent 6266003, 1999.

H. Honing. From time to time: The representation of timing and tempo. *Computer Music Journal*, 25(3):50–61, 2001.

J. Janer, J. Bonada, and S. Jordà. Groovator - An implementation of real-time rhythm transformations. In *Proceedings 121st Convention of the Audio Engineering Society*, 2006.

T. Jehan. *Musical signal parameter estimation*. Unpublished MSc thesis, Institut de Formation Supérieure en Informatique et Communication, Université Rennes I, 1997.

K. Jenssen. Envelope model of isolated musical sounds. In *Proceedings Digital Audio Effects Conference*, 1999.

I. Jermyn, C. Shaffrey, and N. Kingsbury. The methodology and practice of the evaluation of image retrieval systems and segmentation methods. Technical Report 4761, Institut National de la Recherche en Informatique et en Automatique, 2003.

K. Kashino and H. Murase. Sound source identification for ensemble music based on the music stream extraction. In *Proceedings International Joint Conference on Artificial Intelligence Workshop of Computational Auditory Scene Analysis*, pages 127–134, 1997.

K. Kashino, T. Kinoshita, and H. Tanaka. Organization of hierarchical perceptual sounds: Music scene analysis with autonomous processing modules and a quantitative information integration mechanism. In *Proceedings International Joint Conference On Artificial Intelligence*, 1995.

M. Kassler. Toward musical information retrieval. *Perspectives of New Music*, 4:59–67, 1966.

D. Keislar, T. Blum, J. Wheaton, and E. Wold. Audio analysis for content-based retrieval. In *Proceedings International Computer Music Conference*, pages 199–202, 1995.

D. Kirovski and H. Attias. Beat-ID: Identifying music via beat analysis. In *Proceedings IEEE International Workshop on Multimedia Signal Processing*, pages 190– 193, 2002.

A. Klapuri. Sound onset detection by applying psychoacoustic knowledge. In *Proceedings IEEE International Conference on Acoustics, Speech and Signal Processing*, volume 6, pages 3089 – 3092, 1999.

A. Klapuri. *Signal Processing Methods for the Automatic Transcription of Music*. Unpublished PhD thesis, Tampere University of Technology, Tampere, Finland, 2004.

A. Klapuri and M. Davy, editors. *Signal processing methods for music transcription*. Springer-Verlag, New York, 2006.

A. Klapuri, A. Eronen, and J. Astola. Analysis of the meter of acoustic musical signals. *IEEE Trans. Speech and Audio Processing*, 14(1):342–355, 2006.

B. Kostek. Computer-based recognition of musica phrases using the rough-set approach. *Information Sciences*, 104:15–30, 1998.

M. Kotti, L. Martins, E. Benetos, J. Santos Cardoso, and C. Kotropoulos. Automatic speaker segmentation using multiple features and distance measures: A comparison of three approaches. In *Proceedings IEEE International Conference on Multimedia and Expo*, pages 1101–1104, 2006.

R. Kronland-Martinet, J. Morlet, and Grossman. Analysis of sound patterns through wavelet transforms. *International Journal on Pattern Recognition and Artificial Intelligence*, 1(2):273–302, 1987.

C. Krumhansl. *Cognitive Foundations of Musical Pitch*. New York, 1990.

E. Lapidaki. *Consistency of tempo judgments as a measure of time experience in music listening*. Unpublished PhD thesis, Northwestern University, Evanston, IL, 1996.

E. Large and E. Kolen. Resonance and the perception of musical meter. *Connection Science*, 6:177–208, 1994.

J. Laroche. Traitement des signaux audio-fréquences. Technical report, Ecole National Supérieure de Télécommunications, 1995.

J. Laroche. Estimating tempo, swing and beat locations in audio recordings. In *Proceedings IEEE Workshop on Applications of Signal Processing to Audio and Acoustics*, pages 135–138, 2001.

J. Laroche. Efficient tempo and beat tracking in audio recordings. *Journal of the Audio Engineering Society*, 51(4):226–233, 2003.

J. Laroche and M. Dolson. Improved phase-vocoder. time-scale modification of audio. *IEEE Transactions on Speech and Audio Processing*, 7:323–332, 1999.

S. Lee, H. D. Kin, and H. S. Kim. Variable time-scale modification of speech using transient information. In *Proceedings International Conference of Acoustics, Speech, and Signal Processing*, volume 2, pages 1319 – 1322, 1997.

M. Leman. Foundations of musicology as content processing science. *Journal of Music and Meaning*, 1(3), 2003. URL `http://www.musicandmeaning.net/index.php`.

F. Lerdahl and R. Jackendoff. *A generative theory of tonal music*. MIT Press, Cambridge, Massachusetts, 1983.

M. Lesaffre, M. Leman, K. Tanghe, B. De Baets, H. De Meyer, and J. Martens. User-dependent taxonomy of musical features as a conceptual framework for musical audio-mining technology. In *Proceedings Stockholm Music Acoustics Conference*, 2003.

M. Lew, N. Sebe, and J. Eakins. Challenges of image and video retrieval. In *Proceedings International Conference on Image and Video Retrieval*, pages 1–6, 2002.

H. Lincoln. Some criteria and techniques for developing computerized thematic indices. In H. Heckman, editor, *Electronishe Datenverarbeitung in der Musikwissenschaft*. Regensburg: Gustave Bosse Verlag, 1967.

À. Loscos. *Spectral Processing of the Singing Voice*. Unpublished PhD thesis, University Pompeu Fabra, Barcelona, 2007.

À. Loscos and J. Bonada. Emulating rough and growl voice in spectral domain. In *Proceedings Digital Audio Effects Conference*, 2004.

E. Maestre and E. Gómez. Automatic characterization of dynamics and articulation of expressive monophonic recordings. In *Proceedings 118th Audio Engineering Society Convention*, 2005.

R. Maher and J. Beauchamp. Fundamental frequency estimation of musical signals using a two-way mismatch procedure. *Journal of the Acoustical Society of America*, 95:2254–2263, 1993.

B. Manjunath, P. Salembier, and T. Sikora. *Introduction to MPEG-7: Multimedia Content Description Language*. Wiley and Sons, New York, 2002.

D. Marr. *Vision*. W.H. Freeman and Co., San Fransisco, 1982.

S. McAdams. Audition: Physiologie, perception et cognition. In J. Requin, M. Robert, and M. Richelle, editors, *Traite de psychologie expérimentale*, pages 283–344. Presses Universitaires de France, 1994.

S. McAdams and E. Bigand. *Thinking in Sound: The Cognitive Psychology of Human Audition*. Clarendon, Oxford, 1993.

R. McAulay and T. Quatieri. Speech analysis/synthesis based on a sinusoidal representation. *IEEE Transactions on Acoustics, Speech and Signal Processing*, 34(4):744–754, 1986.

R. McNab, L. Smith, and I. Witten. Signal processing for melody transcription. In *Proceedings 19th Australasian Computer Science Conference*, 1996.

G. McNally. Variable speed replay of digital audio with constant output sampling rate. In *Proceedings 76th AES Convention*, 1984.

R. Meddis and M. Hewitt. Virtual pitch and phase sensitivity of a computer model of the auditory periphery. I: Pitch identification. *Journal of the Acoustical Society of America*, 89(6):2866–2882, 1991.

D. Mellinger. *Event formation and separation in musical sound*. Unpublished PhD thesis, Stanford University, 1991.

M. Melucci and N. Orio. Musical information retrieval using melodic surface. In *Proceedings ACM Conference on Digital Libraries*, pages 152–160, 1999.

A. Meng. *Temporal Feature Integration for Music Organisation*. Unpublished PhD thesis, Technical University of Denmark, 2006.

M. Minsky and O. Laske. A conversation with Marvin Minsky. *AI Magazine*, 13(3):31–45, 1992.

B. Moore. *Hearing – Handbook of perception and cognition*. Academic Press Inc., London, 2nd edition, 1995.

E. Moulines, F. Charpentier, and C. Hamon. A diphone synthesis system based on time-domain prosodic modifications of speech. In *Proceedings International Conference of Acoustics, Speech, and Signal Processing*, volume 1, pages 238–241, 1989.

N. Nettheim. On the spectral analysis of melody. *Journal of New Music Research*, 21:135–148, 1992.

A. Noll. Cepstrum pitch determination. *Journal of the Acoustical Society of America*, 41:293–309, 1967.

B. Ong. *Structural Analysis and Segmentation of Music Signals*. Unpublished PhD thesis, University Pompeu Fabra, Barcelona, 2007.

A. Oppenheim and R. Schafer. From frequency to quefrency: A history of the cepstrum. *IEEE Signal Processing Magazine*, 21(5):95–106, 2004.

N. Orio. Music retrieval: A tutorial and review. *Foundations and Trends in Information Retrieval*, 1(1):1–90, 2006.

F. Pachet and P. Roy. Reifying chords in automatic harmonization. In *Proceedings ECAI Workshop on Constraints for Artistic Applications*, 1998.

F. Pachet, P. Roy, and D. Cazaly. A combinatorial approach to content-based music selection. *IEEE Multimedia*, 7:44–51, 2000.

N. Pal and S. Pal. A review of image segmentation techniques. *Pattern Recognition*, 26:1277–1294, 1993.

E. Pampalk. *Computational Models of Music Similarity and their Application to Music Information Retrieval*. Unpublished PhD thesis, Vienna University of Technology, Vienna, 2006.

E. Pampalk and M. Gasser. An implementation of a simple playlist generator based on audio similarity measures and user feedback. In *Proceedings International Conference on Music Information Retrieval*, pages 389–390, 2006.

E. Pampalk and M. Goto. MusicRainbow: A new user interface to discover artists using audio-based similarity and web-based labeling. In *Proceedings International Conference on Music Information Retrieval*, pages 367–370, 2006.

R. Parncutt. A perceptual model of pulse salience and metrical accent in musical rhythms. *Music Perception*, 11(4):409–464, 1994.

D. Pearce and H. Hirsch. The Aurora experimental framework for the performance evaluation of speech recognition systems under noisy conditions. In *Proceedings International Conference on Spoken Language Processing*, 2000.

G. Peeters. A large set of audio features for sound description (similarity and classification) in the CUIDADO project. Technical report, CUIDADO IST Project, 2004.

G. Peeters, A. La Burthe, and X. Rodet. Toward automatic music audio summary generation from signal analysis. In *International Conference on Music Information Retrieval*, pages 94–100, 2002.

J. Pinquier, J.-L. Rouas, and R. André-Obrecht. A fusion study in speech/music classification. In *Proceedings International Conference on Multimedia and Expo*, pages 409–412, 2003.

M. Portnoff. Implementation of the digital phase vocoder using the fast fourier transform. *IEEE Transactions on Acoustics, Speech, and Signal Processing*, 24: 243–248, 1976.

M. Przybocki and A. Martin. NIST speaker recognition evaluations. In *Proceedings International Conference on Language Resources and Evaluations*, pages 331–335, 1989.

H. Purwins. *Profiles of Pitch Classes Circularity of Relative Pitch and Key – Experiments, Models, Computational Music Analysis, and Perspectives*. Unpublished PhD thesis, Technical University of Berlin, 2005.

H. Purwins, B. Blankertz, and K. Obermayer. A new method for tracking modulations in tonal music in audio data format. *Proceeding International Joint Conference on Neural Network*, pages 270–275, 2000.

H. Purwins, T. Graepel, B. Blankertz, and K. Obermayer. Correspondence analysis for visualizing interplay of pitch class, key, and composer. In E. Puebla, G. Mazzola, and T. Noll, editors, *Perspectives in Mathematical Music Theory*. Verlag, 2003.

L. Rabiner. A tutorial on hidden markov models and selected applications in speech recognition. *Proceedings IEEE*, 77(2):257–285, 1989.

L. Rabiner and R. Schafer. *Digital Processing of Speech Signals*. Prentice-Hall, 1978.

L. Rabiner, M. Sambur, and C. Schmidt. Applications of a nonlinear smoothing algorithm to speech processing. *IEEE Transactions on Acoustics, Speech and Signal Processing*, 23(6):552–557, 1975.

R. Ramirez and A. Hazan. A tool for generating and explaining expressive music performances of monophonic jazz melodies. *International Journal on Artificial Intelligence Tools*, 15(4):673–691, 2006.

R. Ramirez, A. Hazan, E. Gómez, and E. Maestre. A machine learning approach to expressive performance in jazz standards. In *Proceedings International Conference on Knowledge Discovery and Data Mining*, 2004.

C. Raphael. Automatic segmentation of acoustic musical signals using hidden markov models. *IEEE Transactions on Pattern Analysis and Machine Intelligence*, 21(4):360–370, 1999.

C. Raphael. A hybrid graphical model for rhythmic parsing. *Artificial Intelligence*, 137(1-2):217–238, 2002.

E. Ravelli, J. Bello, and M. Sandler. Automatic rhythm modification of drum loops. *IEEE Signal Processing Letters*, 14(4):228–231, 2007.

S. Rossignol. *Séparation, segmentation et identification d'objets sonores. Application á la représentation, á la manipulation des signaux sonores, et au codage dans les applications multimedias*. Unpublished PhD thesis, IRCAM, Paris, France, 2000.

S. Rossignol, X. Rodet, P. Depalle, J. Soumagne, and J.-L. Collette. Vibrato: Detection, estimation, extraction, modification. In *Proceedings Digital Audio Effects Conference*, 1999.

K. Sakakibara, L. Fuks, H. Imagawa, and N. Tayama. Growl voice in ethnic and pop styles. In *Proceedings International Symposium on Musical Acoustics*, 2004.

E. Scheirer. *Music listening systems*. Unpublished PhD thesis, Massachusets Institute of Technology, 2000.

E. Scheirer and M. Slaney. Construction and evaluation of a robust multifeature speech/music discriminator. In *Proceedings IEEE International Conference on Audio, Speech and Signal Processing*, pages 1331–1334, 1997.

J. Schoentgen. Stochastic models of jitter. *Journal of the Acoustical Society of America*, 109:1631–1650, 2001.

E. Selfridge-Field. Conceptual and representational issues in melodic comparison. In W. B. Hewlett and E. Selfridge-Field, editors, *Melodic Similarity: Concepts, Procedures, and Applications*. MIT Press, Cambridge, Massachusetts, 1998.

X. Serra. *A System for Sound Analysis/Transformation/Synthesis based on a Deterministic plus Stochastic Decomposition*. Unpublished PhD thesis, Stanford University, 1989.

X. Serra. Sound hybridization techniques based on a deterministic plus stochastic decomposition model. In *Proceedings International Computer Music Conference*, 1994.

X. Serra and J. Bonada. Sound transformations based on the SMS high level attributes. In *Proceedings Digital Audio Effects Conference*, Barcelona, 1998.

A. Sheh and D. Ellis. Chord segmentation and recognition using em-trained hidden markov models. In *Proceedings International Conference on Music Information Retrieval*, 2003.

M. Slaney, M. Covell, and B. Lassiter. Automatic audio morphing. In *Proceedings IEEE International Conference on Audio, Speech and Signal Processing*, pages 1001–1004, 1996.

P. Smaragdis. *Redundancy Reduction for Computational Audition, a Unifying Approach*. Unpublished PhD thesis, Massachusetts Institute of Technology, 2001.

A. Smeulders, M. Worring, S. Santini, A. Gupta, and R. Jain. Content-based image retrieval at the end of the early years. *IEEE Transactions on Pattern Analysis and Machine Intelligence*, 22:1349–1380, 2000.

J. Smith and X. Serra. PARSHL: An analysis/synthesis program for non-harmonic sounds based on a sinusoidal representation. In *Proceedings Internationtal Computer Music Conference*, pages 290–297, 1987.

C. Spevak, B. Thom, and K. Hothker. Evaluating melodic segmentation. In *Proceedings International Conference on Music and Artificial Intelligence*, 2002.

J. Sundberg. *The Science of the Singing Voice*. Northern Illinois University Press, 1987.

E. Terhardt, G. Stoll, and M. Seewann. Algorithm for extraction of pitch and pitch salience from complex tonal signals. *Journal of the Acoustical Society of America*, 71:679–688, 1981.

H. Thornburg and F. Gouyon. A flexible analysis/synthesis method for transients. In *Proceedings International Computer Music Conference*, 2000.

P. Toiviainen and T. Eerola. A method for comparative analysis of folk music based on musical feature extraction and neural networks. In *Proceedings*

International Symposium on Systematic and Comparative Musicology, and International Conference on Cognitive Musicology, 2001.

M. Towsey, A. Brown, S. Wright, and J. Diederich. Towards melodic extension using genetic algorithms. *Educational Technology & Society*, 4(2), 2001.

R. Typke. *Music Retrieval based on Melodic Similarity*. PhD thesis, Utrecht University, 2007.

G. Tzanetakis. *Manipulation, analysis and retrieval systems for audio signals*. Unpublished PhD thesis, Computer Science Department, Princeton University, June 2002.

G. Tzanetakis and P. Cook. Multifeature audio segmentation for browsing and annotation. In *Proceedings IEEE Workshop on Applications of Signal Processing to Audio and Acoustics*, pages 103–106, 1999.

G. Tzanetakis and P. Cook. Musical genre classification of audio signals. *IEEE Transactions on Speech and Audio Processing*, 10(5):293–302, 2002.

A. Uitdenbogerd and J. Zobel. Manipulation of music for melody matching. In *Proceedings ACM Conference on Multimedia*, pages 235–240, 1998.

V. Verfaille, U. Zölzer, and D. Arfib. Adaptive digital audio effects (A-DAFx): A new class of sound transformations. *IEEE Transactions on Audio, Speech and Language Processing*, 14(5):1817– 1831, 2006.

W. Verhelst and M. Roelands. An overlap-add technique based on waveform similarity (WSOLA) for high quality time-scale modification of speech. In *Proceedings IEEE International Conference of Acoustics, Speech, and Signal Processing*, volume 2, pages 554–557, 1993.

T. Verma, S. Levine, and T. Meng. Transient modeling synthesis: A flexible analysis/synthesis tool for transient signals. In *Proceedings Internationtal Computer Music Conference*, 1997.

E. Vidal and A. Marzal. A review and new approaches for automatic segmentation of speech signals. In L. Torres, E. Masgrau, and Lagunas M., editors, *Signal Processing V: Theories and Applications*. 1990.

P. Walmsley, S. Godsill, and P. Rayner. Bayesian graphical models for polyphonic pitch tracking. In *Proceedings Diderot Forum*, 1999.

D. Wang and G. Brown, editors. *Computational auditory scene analysis – Principles, Algorithms, and Applications*. IEEE Press - Wiley-Interscience, 2006.

J. Wayman, R. Reinke, and D. Wilson. High quality speech expansion, compression, and noise filtering using the SOLA method of time scale modification. In *Proceedings 23d Asilomar Conference on Signals, Systems and Computers*, volume 2, pages 714–717, 1989.

B. Whitman. *Learning the Meaning of Music*. Unpublished PhD thesis, Massachusetts Institute of Technology, MA, USA, 2005.

M. Wright and E. Berdahl. Towards machine learning of expressive microtiming in Brazilian drumming. In *Proceedings International Computer Music Conference*, 2006.

U. Zölzer, editor. *DAFX Digital Audio Effects*. Wiley & Sons, 2002.

Chapter 4

From Sound to Sense via Feature Extraction and Machine Learning: Deriving High-Level Descriptors for Characterising Music

Gerhard Widmer[1,2], Simon Dixon[1], Peter Knees[2], Elias Pampalk[1], Tim Pohle[2]

[1] Austrian Research Institute for Artificial Intelligence, Vienna
[2] Department of Computational Perception, Johannes Kepler University Linz

About this chapter

This chapter gives a broad overview of methods and approaches for automatically extracting musically meaningful (semantic) descriptors for the characterisation of music pieces. It is shown how high-level terms can be inferred via a combination of bottom-up audio descriptor extraction and the application of machine learning algorithms. Also, the chapter will briefly indicate that meaningful descriptors can be extracted not just from an analysis of the music (audio) itself, but also from extra-musical sources, such as the internet (via web mining).

4.1 Introduction

Research in intelligent music processing is experiencing an enormous boost these days due to the emergence of the new application and research field of Music Information Retrieval (MIR). The rapid growth of digital music collections and the concomitant shift of the music market towards digital music distribution urgently call for intelligent computational support in the automated handling of large amounts of digital music. Ideas for a large variety of content-based music services are currently being developed in the music industry and in the research community. They range from content-based music search engines to automatic music recommendation services, from intuitive interfaces on portable music players to methods for the automatic structuring and visualisation of large digital music collections, and from personalised radio stations to tools that permit the listener to actively modify and "play with" the music as it is being played.

What all of these content-based services have in common is that they require the computer to be able to "make sense of" and "understand" the actual content of the music, in the sense of being able to recognise and extract musically, perceptually and contextually mgful (semantic) patterns from recordings, and to asto associate descriptors with the music that make sense to human listeners.

There is a large variety of musical descriptors that are potentially of interest. They range from low-level features of the sound, such as its bass content or its harmonic richness, to high-concepts such as "hip hop" or "sad music". Also, semantic descriptors may come in the form of atomic, discrete labels like "rhythmic" or "waltz", or they may be complex, structured entities such as harmony and rhythmic structure. As it is impossible to cover all of these in one coherent chapter, we will have to limit ourselves to a particular class of semantic descriptors.

This chapter, then, focuses on methods for automatically extracting high-level atomic descriptors for the characterisation of music. It will be shown how high-level terms can be inferred via a combination of bottom-up audio descriptor extraction and the application of machine learning algorithms. Also, it will be shown that meaningful descriptors can be extracted not just from an analysis of the music (audio) itself, but also from extra-musical sources, such as the internet (via web mining).

Systems that learn to assign labels must be evaluated in systematic, controlled experiments. The most obvious and direct way is via classification experiments, where the labels to be assigned are interpreted as distinct classes. In particular, genre classification, i.e. the automatic assignment of an appropriate style label to a piece of music, has become a popular benchmark task in the MIR community (for many reasons, not the least of them being the fact that genre labels are generally much easier to obtain than other, more intuitive or personal descriptors). Accordingly, the current chapter will very much focus on genre classification as the kind of benchmark problem that measures the efficacy of machine learning (and the underlying descriptors) in assigning meaningful terms to music. However, in principle, one can try to predict any other high-level labels from low-level features, as long as there is a sufficient number of training examples with given labels. Some experiments regarding non-genre concepts will be briefly described in section 4.3.4, and in section 4.4.2 we will show how textual characterisations of music artists can be automatically derived from the Web.

The chapter is structured as follows. Section 4.2 deals with the extraction of music descriptors (both very basic ones like timbre and more abstract ones like melody or rhythm) from recordings via audio analysis. It focuses in particular on features that have been used in recent genre classification research. Section 4.3 shows how the gap between what can be extracted bottom-up and more abstract, human-centered concepts can be partly closed with the help of inductive machine learning. New approaches to inferring additional high-level knowledge about music from extra-musical sources (the Internet) are presented in section 4.4. Section 4.5, finally, discusses current research and application perspectives and identifies important questions that will have to be addressed in the future.

4.2 Bottom-up extraction of descriptors from audio

Extracting descriptors from audio recordings to characterise aspects of the audio contents is not a new area of research. Much effort has been spent on feature extraction in areas like speech processing or audio signal analysis. It is impossible to give a comprehensive overview of all the audio descriptors developed over the past decades. Instead, this chapter will focus solely on descriptors that are useful for, or have been evaluated in, music classification tasks, in the context of newer work in Music Information Retrieval. The real focus of this chapter is on extracting or predicting higher-level descriptors via machine learning. Besides, a more in-depth presentation of audio and music descriptors is offered in another chapter of this book (Chapter 3), so the following sections only briefly recapitulate those audio features that have played a major role in recent music classification work.

Connected to the concept of classification is the notion of music or generally sound similarity. Obviously, operational similarity metrics can be used directly for audio and music classification (e.g. via nearest-neighbour algorithms), but also for a wide variety of other tasks. In fact, some of the music description schemes presented in the following do not produce features or descriptors at all, but directly compute similarities; they will also be mentioned, where appropriate.

4.2.1 Simple audio descriptors for music classification

This section describes some common simple approaches to describe properties of audio (music) signals. For all algorithms discussed here, the continuous stream of audio information is cut into small, possibly overlapping fragments of equal length, called frames. The typical length of a frame is about 20 ms. Usually, for each frame one scalar value per descriptor is calculated, which can be done either on the time-domain or the frequency-domain representation of the signal. To obtain a (scalar) descriptor that pertains to an entire audio track, the values of all frames can be combined by, for example, applying simple statistics such as mean and standard deviation of all individual values.

Time-domain descriptors

On the time-domain representation of the audio signal, several descriptors can be calculated. An algorithm that mainly describes the power envelope of the audio signal is Root Mean Square (RMS): the individual values appearing in each frame are squared, and the root of the mean of these values is calculated. These values might be combined as described above, or by calculating which fraction of all RMS values is below, e.g. the average RMS value of a piece (Low Energy Rate). Comparable to the RMS values are the Amplitude Envelope values, which are the maximum absolute values of each frame. The amplitude envelope and RMS descriptors are commonly used as a first step in algorithms that detect rhythmic structure.

The time-domain representation might also be used to construct measures that model the concept of Loudness (i.e. the perceived volume). For example, a simple and effective way is to take the 0.23th power of the RMS values.

Another possibility is to approximately measure the perceived brightness with the phZero Crossing Rate. This descriptor simply counts how often the signal passes zero-level.

Also, the time-domain representation can be used to extract periodicity information from it. Common methods are autocorrelation and comb filterbanks. Autocorrelation gives for each given time lag the amount of self-similarity of the time domain samples by multiplying the signal with a time-lagged version of itself. In the comb filterbank approach, for each periodicity of interest, there is a comb filter with the appropriate resonance frequency.

Frequency-domain descriptors

A number of simple measures are commonly applied to describe properties of the frequency distribution of a frame:

- The Band Energy Ratio is the relation between the energy in the low frequency bands and the energy of the high frequency bands. This descriptor is vulnerable to producing unexpectedly high values when the energy in the low energy bands is close to zero.

- The Spectral Centroid is the center of gravity of the frequency distribution. Like the zero crossing rate, it can be regarded as a measure of perceived brightness or sharpness.
- The Spectral Rolloff frequency is the frequency below which a certain amount (e.g. 95%) of the frequency power distribution is concentrated.

These descriptors are calculated individually for each frame. The Spectral Flux is modelled to describe the temporal change of the spectrum. It is the Euclidean distance between the (normalised) frequency distributions of two consecutive frames, and can be regarded as a measure of the rate at which the spectrum changes locally.

The descriptors mentioned so far represent rather simple concepts. A more sophisticated approach is the Mel Frequency Cepstral Coefficients (MFCCs), which models the shape of the spectrum in a compressed form. They are calculated by representing the spectrum on the perceptually motivated Mel-Scale, and taking the logarithms of the amplitudes to simulate loudness perception. Afterwards, the discrete cosine transformation is applied, which results in a number of coefficients (MFCCs). Lower coefficients describe the coarse envelope of the frame's spectrum, and higher coefficients describe more detailed properties of the spectrum envelope. Usually, the higher-order MFCCs are discarded, and only the lower MFCCs are used to describe the music.

A popular way to compare two recorded pieces of music using MFCCs is to discard the temporal order of the frames, and to summarise them by clustering (e.g. Logan and Salomon, 2001; Aucouturier and Pachet, 2002b). In the case of the paper by Aucouturier and Pachet (2002b), for instance, the clustered MFCC representations of the frames are described by Gaussian Mixture Models (GMMs), which are the features for the piece of music. A way to compare GMMs is sampling: one GMM is used to produce random points with the distribution of this GMM, and the likelihood that the other GMM produces these points is checked.

It might seem that discarding the temporal order information altogether ignores highly important information. But recent research (Flexer et al., 2005) has shown that MFCC-based description models using Hidden Markov Models (which explicitly model the temporal structure of the data) do not improve classification accuracy (as already noted by Aucouturier and Pachet, 2004), though they do seem to better capture details of the sound of musical recordings (at least in terms of statistical likelihoods). Whether this really makes a difference in actual applications still remains to be shown.

The interested reader is referred to Chapter 3 of this book for a much more comprehensive review of audio descriptors and music description schemes.

4.2.2 Extracting higher-level musical patterns

The basic intuition behind research on classification by higher-level descriptors is that many musical categories can be defined in terms of high-level musical concepts. To some extent it is possible to define musical genre, for example, in terms of the melody, rhythm, harmony and instrumentation that are typical of each genre. Thus genre classification can be reduced to a set of sub-problems: recognising particular types of melodies, rhythms, harmonies and instruments. Each of these sub-problems is interesting in itself, and has attracted considerable research interest, which we review here.

Early work on music signal analysis is reviewed by Roads (1996). The problems that received the most attention were pitch detection, rhythm recognition and spectral analysis, corresponding respectively to the most important features of music: melody, rhythm and timbre (harmony and instrumentation).

Pitch detection is the estimation of the fundamental frequency of a signal, usually assuming it to be monophonic. Common methods include: time domain algorithms such as counting of zero-crossings and autocorrelation; frequency domain methods such as Fourier analysis and the phase vocoder; and auditory models that combine time and frequency domain information based on an understanding of human auditory processing. Recent work extends these methods to find the predominant pitch (usually the melody note) in polyphonic mixtures (Goto and Hayamizu, 1999; Gómez et al., 2003).

The problem of extracting rhythmic contents from a musical performance, and in particular finding the rate and temporal location of musical beats, has attracted considerable interest. A review of this work is found in the work of Gouyon and Dixon (2005). Initial attempts focussed on rhythmic parsing of musical scores, that is without the tempo and timing variations that characterise performed music, but recent tempo and beat tracking systems work quite successfully on a wide range of performed music. The use of rhythm for classification of dance music was explored by Dixon et al. (2003, 2004).

Spectral analysis examines the time-frequency contents of a signal, which is essential for extracting information about instruments and harmony. Short-time Fourier analysis is the most widely used technique, but many others are available for analysing specific types of signals, most of which are built upon the Fourier transform. MFCCs, already mentioned in section 4.2.1 above, model the spectral contour rather than examining spectral contents in detail, and thus can be seen as implicitly capturing the instruments playing (rather than the notes that were played). Specific work on instrument identification can be found in a paper by Herrera et al. (2003).

Regarding harmony, extensive research has been performed on the extraction of multiple simultaneous notes in the context of automatic transcription systems, which are reviewed by Klapuri (2004). Transcription typically involves the following steps: producing a time-frequency representation of the signal, finding peaks in the frequency dimension, tracking these peaks over the time dimension to produce a set of partials, and combining the partials to produce a set of notes. The differences between systems are usually related to the assumptions made about the input signal (for example the number of simultaneous notes, types of instruments, fastest notes, or musical style), and the means of decision making (for example using heuristics, neural nets or probabilistic reasoning).

Despite considerable successes, the research described above makes it increasingly clear that precise, correct, and general solutions to problems like automatic rhythm identification or harmonic structure analysis are not to be expected in the near future — the problems are simply too hard and would require the computer to possess the kind of broad musical experience and knowledge that human listeners seem to apply so effortlessly when listening to music. Recent work in the field of Music Information Retrieval has thus started to focus more on approximate solutions to problems like melody extraction (Eggink and Brown, 2004) or chord transcription (Yoshioka et al., 2004), or on more specialised problems, like the estimation of global tempo (Alonso et al., 2004) or tonality (Gómez and Herrera, 2004), or the identification of drum patterns (Yoshii et al., 2004).

Each of these areas provides a limited high-level musical description of an audio signal. Systems have yet to be defined which combine all of these aspects, but this is likely to be seen in the near future.

4.3 Closing the gap: Prediction of high-level descriptors via machine learning

While the bottom-up extraction of features and patterns from audio continues to be a very active research area, it is also clear that there are strict limits as to the kinds of music descriptions that can be directly extracted from the audio signal. When it comes to intuitive, human-centered characterisations such as "peaceful" or "aggressive music" or highly personal categorisations such as "music I like to listen to while working", there is little hope of analytically defining audio features that unequivocally and universally define these concepts. Yet such concepts play a central role in the way people organise, interact with and use their music.

That is where automatic learning comes in. The only way one can hope to build a machine that can associate such high-level concepts with music items is by having the machine learning the correct associations between low-level audio features and high-level concepts, from examples of music items that have been labelled with the appropriate concepts. In this section, we give a very brief introduction to the basic concepts of machine learning and pattern classification, and review some typical results with machine learning algorithms in musical classification tasks. In particular, the automatic labelling of music pieces with genres has received a lot of interest lately, and section 4.3.3 focuses specifically on genre classification. Section 4.3.4 then reports on recent experiments with more subjective concepts, which clearly show that a lot of improvement is still needed. One possible avenue towards achieving this improvement will then be discussed in section 4.4.

4.3.1 Classification via machine learning

Inductive learning as the automatic construction of classifiers from pre-classified training examples has a long tradition in several sub-fields of computer science. The field of statistical pattern classification (Duda et al., 2001; Hastie et al., 2001) has developed a multitude of methods for deriving classifiers from examples, where a classifier, for the purposes of this chapter, can be regarded as a black box that takes as input a new object to be classified (described via a set of features) and outputs a prediction regarding the most likely class the object belongs to. Classifiers are automatically constructed via learning algorithms that take as input a set of example objects labelled with the correct class, and construct a classifier from these that is (more or less) consistent with the given training examples, but also makes predictions on new, unseen objects – that is the classifier is a generalisation of the training examples.

In the context of this chapter, training examples would be music items (e.g. songs) characterised by a list of audio features and labelled with the appropriate high-level concept (e.g. "this is a piece I like to listen to while working"), and the task of the learning algorithm is to produce a classifier that can predict the appropriate high-level concept for new songs (again represented by their audio features).

Common training and classification algorithms in statistical pattern classification (Duda et al., 2001) include nearest neighbour classifiers (k-NN), Gaussian Mixture Models, neural networks (mostly multi-layer feed-forward perceptrons), and support vector machines (Cristianini and Shawe-Taylor, 2000).

The field of Machine Learning (Mitchell, 1997) is particularly concerned with algorithms that induce classifiers that are interpretable, i.e. that explicitly describe the criteria that are associated with or define a given class. Typical examples of machine learning algorithms that are also used in music classification are decision trees (Quinlan, 1986) and rule learning algorithms (Fürnkranz, 1999).

Learned classifiers must be evaluated empirically, in order to assess the kind of prediction accuracy that may be expected on new, unseen cases. This is essentially done by testing the classifier on new (labelled) examples which have not been used in any way in learning, and recording the rate of prediction errors made by the classifier. There is a multitude of procedures for doing this, and a lot of scientific literature on advantages and shortcomings of the various methods. The basic idea is to set aside a part of the available examples for testing (the test set), then inducing the classifier from the remaining data (the training set), and then testing the classifier on the test set. A systematic method most commonly used is known as n-fold cross-validation, where the available data set is randomly split into n subsets (folds), and the above procedure is carried out n times, each time using one of the n folds for testing, and the remaining $n - 1$ folds for training. The error (or, conversely, accuracy) rates reported in most learning papers are based on experiments of this type.

A central issue that deserves some discussion is the training data required for learning. Attractive as the machine learning approach may be, it does require (large) collections of representative labelled training examples, e.g. music recordings with the correct categorisation attached. Manually labelling music examples is a very laborious and time-consuming process, especially when it involves listening to the pieces before deciding on the category. Additionally, there is the copyright issue. Ideally, the research community would like to be able to share common training corpora. If a researcher wants to test her own features in a classification experiment, she/he needs access to the actual audio files.

There are some efforts currently being undertaken in the Music Information Retrieval community to compile large repositories of labelled music that can be made available to all interested researchers without copyright problems. Noteworthy examples of this are Masataka Goto's RWC Music Database [1], the

[1]http://staff.aist.go.jp/m.goto/RWC-MDB

IMIRSEL (International Music Information Retrieval System Evaluation Laboratory) project at the University of Illinois at Urbana-Champaign[2] (Downie et al., 2004), and the new FreeSound Initiative [3].

4.3.2 Learning algorithms commonly used in music classification

In this section, we briefly review some of the most common learning algorithms that are used in music classification and learning tasks.

Decision trees (Quinlan, 1986) are probably the most popular class of classification models in machine learning. Essentially, a decision tree corresponds to a set of classification rules (represented in the form of a tree) that predict the class of an object from a combination (conjunction) of specific characteristic feature values, which are determined via simple information-theoretic measures. Decision trees are widely used also in Music Information Retrieval. In a paper by West and Cox (2004), for instance, decision tree learning algorithms have been used to build a model of the distribution of frame values.

Because of its known merits, nearest-neighbour (NN) classification is widely used, also in MIR. Here, the idea is to compare a new (test) object, to be classified, with all the training instances and predict the class of the most similar training instance(s) for the new object. In order to obtain a measure of similarity, the feature values – possibly after feature selection – of each piece are regarded as a vector, and the euclidean distance from the test object to all training instances (e.g. Costa et al., 2004; Gouyon et al., 2004) or to representative reference vectors (e.g. Hellmuth et al., 2004; Kastner et al., 2004) is used as a (dis)similarity metric for classification.

Support Vector Machines (SVMs) are also applied to music classification. In essence, an SVM learns an optimal linear classification boundary between the classes in a high-dimensional space, which is implicitly computed from the

---

[2]`http://www.music-ir.org/evaluation`

[3]`http://freesound.iua.upf.edu`

original features via a so-called kernel function. Xu et al. (2003), for instance, use SVMs for genre classification, and Li and Ogihara (2003) train several SVMs to recognise mood labels, where each SVM decides if one specific label is present in the music.

Gaussian Mixture Models (GMMs) are useful for estimating the distribution of feature values. A GMM models a multivariate probability density as a weighted combination of Gaussians. GMMs can be used for classification by modelling each class as a GMM; an instance is then classified by calculating, for each class (GMM), the probability that the instance was produced by the respective GMM (i.e. the likelihood of the GMM, given the observed instance), and predicting the class with the maximum likelihood. In a paper by Liu et al. (2003), mood detection in classical music is performed based on this approach. GMM classifiers have also been used by Burred and Lerch (2003) and Tzanetakis and Cook (2002) for genre classification.

Neural Networks have also been applied to music classification — in particular, the so-called multi-layer perceptron or feed-forward network. Costa et al. (2004) use a multilayer perceptron to determine the class of a piece given its feature vector. Hellmuth et al. (2004) use a more elaborate approach by training a separate neural network for each class, and an additional one that combines the outputs of these networks.

4.3.3 Genre classification: Typical experimental results

The experimental results found in the literature on genre classification are not easy to compare, as researchers use many different music collections to evaluate their methods. Also, the ways of annotating the collections vary: some researchers label the pieces according to their own judgment, while others use online databases for the assignment of genre labels. Additionally, different authors often tackle slightly different problems (such as categorical vs. probabilistic classification), which makes a comparison of the results even more difficult. These facts should be kept in mind when assessing the examples given in this section.

Generally, when trying to separate the classes Pop and Classical, very high accuracies are reached, suggesting that this task is not too difficult. For example, Costa et al. (2004) achieve up to 90.3% classification accuracy, and Mierswa and Morik (2005) report even 100% on 200 pieces. In both cases, the baseline is one half. Although Xu et al. (2003) report a classification accuracy of 93% for four genres, in general the classification accuracy decreases when the number of genres grows.

For classification into dance music genres, Gouyon et al. (2004) obtain up to 78.9% accuracy (15.9% baseline) when classifying 698 pieces of music into eight classes. This classification is based on a number of rhythmic descriptors and a rule-based classifier whose rules were designed manually. For a wider range of musical contents, divided into eleven genres, Uhle and Dittmar (2004) report a classification accuracy of 67.6%, also based on rhythm features.

At the ISMIR 2004 conference, a comparison of different audio description algorithms was conducted in the form of a contest[4]. For the section of genre classification, the winning algorithm achieved a classification accuracy of 84.07% correct answers. The test collection consisted of 729 pieces, divided into six classes, with a baseline of 43.9%.

4.3.4 Trying to predict labels other than genre

Genre or style is a descriptor that is useful for many applications, especially in commercial settings. Even though the concept of "genre" is not well defined (see e.g. Aucouturier and Pachet, 2003), it is still much more objective than the kinds of personal characterisations human listeners attach to their music. But it is precisely these personal, subjective categorisations ("happy music", "aggressive music", "music I like when I am sad", "music that one can dance to") that, if learnable by computers, would open new possibilities for intelligent and rewarding musical interactions between humans and machines.

A small preliminary experiment on the learnability of subjective, non-genre categorisations is reported in this section. As will be seen, the results are rather poor, and a lot of improvement is still needed. Web-based learning

[4] http://ismir2004.ismir.net/ISMIR_Contest.html

about music is a promising alternative that might help to overcome the current limitations; that is the topic of the next section (Section 4.4).

The experiment presented here aimed to investigate the learnability of the categorisations mood (happy / neutral / sad), perceived tempo (very slow / slow / medium / fast / very fast / varying), complexity (low / medium / high), emotion (soft / neutral / aggressive), focus (vocal / both / instruments), and genre (blues / classical / electronica / folk / jazz / new age / noise / rock / world). To this end, each piece in a music collection of 729 pieces was labelled with the according value.

This data basis was used to examine the discriminative power of several descriptor sets in combination with a number of machine learning algorithms. The descriptor sets consisted mainly of descriptors that are widely used for music classification tasks (see section 4.2.1 above). Three different descriptor sets were tested: the set that was also used by Tzanetakis and Cook (2002), a set made from some Mpeg7 Low Level Descriptors, and a set that contained all descriptors of the above sets, together with some additional ones for rhythm and melody description.

To train the machine learning algorithms, mean and variance of the descriptors' values for a 30-second excerpt of the piece of music were taken as attributes. Table 4.1 shows the highest classification accuracies that were achieved with different learning algorithms; accuracy was estimated via stratified tenfold cross-validation. The evaluated learning algorithms were J48 (a decision tree learner, available — like all the other learning algorithms mentioned here — in the machine learning toolkit WEKA[5], SMO (a support vector machine), Naive Bayes, Naive Bayes with Kernel Estimation, Boosting, Boosting with J48, Regression with MP5, Linear Regression, and k-NN with k = 1, 3, 5, 10. The table also lists the results obtained when applying the algorithm from Aucouturier and Pachet (2004) to the same categorisations. For this algorithm, the best values obtained for k-NN classification with k = 1, 3, 5, 10 are shown. The other learning algorithms were not applicable to its feature data. Also, the baseline is given (i.e. the classification accuracy achieved when always guessing the most frequent class).

[5]Software freely available from `http://www.cs.waikato.ac.nz/ml/`

	mood	perceived tempo	complexity	emotion	focus	genre
Baseline	50.00 %	42.53 %	75.66 %	44.46 %	68.92 %	60.48 %
Set from TC02	50.00 %	42.53 %	76.63 %	45.06 %	71.08 %	65.66 %
Some Mpeg7 LLDs	50.00 %	43.13 %	76.14 %	46.75 %	70.00 %	64.94 %
"Large" Set	51.08 %	44.70 %	76.87 %	47.47 %	71.20 %	69.52 %
Best from AP04	50.24 %	48.67 %	78.55 %	57.95 %	75.18 %	70.84 %

Table 4.1: Best classification accuracies for the different categorisations in the small preliminary experiment. TC02 = (Tzanetakis and Cook, 2002); AP04 = (Aucouturier and Pachet, 2004).

These results show that, with the examined techniques, in some cases it is even not possible to get classification accuracies higher than the baseline. For all categorisations except mood, the algorithm from Aucouturier and Pachet (2004) performed better than the other approaches. There is a number of ways in which this experiment could be improved, e.g., by the application of feature selection algorithms or the development of dedicated descriptors for each different task. Still, the results point to some fundamental limitations of the feature-based learning approach; concepts like the emotional quality of a piece of music seem to elude a purely audio-based approach.

4.4 A new direction: Inferring high-level descriptors from extra-musical information

Listening to and making sense of music is much more than decoding and parsing an incoming stream of sound waves into higher-level objects such as onsets, notes, melodies, harmonies, etc. Music is embedded in a rich web of cultural, historical, cultural, and social (and marketing) contexts that influence how music is heard, interpreted, and categorised. That is, many qualities or categorisations attributed to a piece of music by listeners cannot solely be explained by the content of the audio signal itself.

Also, recent research on genre classification clearly shows that purely audio-based approaches to music classification may be hitting a kind of "glass ceiling" (Aucouturier and Pachet, 2004): there seem to be strict limits to the level of classification accuracy that can be obtained with purely audio-based features, no matter how sophisticated the audio descriptors. From a pragmatic point of view, then, it is clear that, if at all, high-quality automatic music annotation and classification can only be achieved by also taking into account and exploiting information sources that are external to the music itself.

The Internet is a rich, albeit unstructured, source of potential information, where millions of music lovers and experts discuss, describe, and exchange music. Possible information sources include personal web pages, music and concert reviews published on the Web, newspaper articles, discussion forums, chat rooms, playlists exchanged through peer-to-peer networks, and many more. A common term for denoting all the musically relevant information that is potentially "out there" is community metadata (Whitman and Lawrence, 2002). Recent approaches to high-level music characterisation try to automatically extract relevant descriptors from the Internet – mostly from general, unstructured web pages –, via the use of information retrieval, text mining, and information extraction techniques (e.g. Baumann and Hummel, 2003; Whitman and Ellis, 2004; Whitman and Lawrence, 2002; Whitman and Smaragdis, 2002). In a sense, this is like learning about music without ever listening to it, by analysing the way people talk about and describe music, rather than what the music actually sounds like.

In the following, two research projects are briefly presented that show in a prototypical way how the Internet can be exploited as a source of information about – in this case – music artists. Section 4.4.1 shows how artists can be probabilistically related to genres via web mining, and section 4.4.2 presents an approach to the hierarchical clustering of music artists, and the automatic labelling of the individual clusters with descriptive terms gleaned from the Web.

4.4.1 Assigning artists to genres via web mining

In this section we will explain how to extract features (words) related to artists from web pages and how to use these features to construct a probabilistic genre classifier. This permits the computer to classify new artists present on the web using the Internet community's collective knowledge. To learn the concept of a genre the method requires a set of typical artists for each genre in advance. Based on these artists and a set of web pages that talk about these artists, a characteristic profile is created for each genre. Using this profile (i.e. a weighted list of typical keywords) any artist can be classified by simple evaluation of word occurrences on related web pages. The following is a simplified account of the basic method; the details can be found in a paper by Knees et al. (2004).

To obtain useful data for genre profile generation, Internet search engines like Google are queried with artist names, along with some constraints (e.g. *+music +review*) that should filter out non-musical pages, and the top ranked pages are retrieved (without these constraints, a search for groups such as *Kiss* would result in many unrelated pages). The retrieved pages tend to be common web pages such as fan pages, reviews from online music magazines, or music retailers. The first N available top-ranked webpages for each query are retrieved, all HTML markup tags are removed, so that only the plain text content is left, and common English stop word lists are used to remove frequent terms (e.g. a, and, or, the).

The features by which artists are characterised are the individual words that occur in any of the pages. In order to identify those words that may indicate what genre an artist belongs to, the next important step is feature weighting. A common method for this comes from the field of Information Retrieval and is known as term frequency × inverse document frequency ($tf \times idf$) (Salton and Buckley, 1988). For each artist a and each term t appearing in the retrieved pages, we count the number of occurrences tf_{ta} (term frequency) of term t in documents related to a, and df_t, the number of pages the term occurred in (document frequency). These are combined by multiplying the term frequency with the inverse document frequency. Basically, the intention of the $tf \times idf$ function is to assign a high score to terms that occur frequently, but also to reduce the score if these terms occur on many different pages and thus do not contain useful information.

In the approach described in the paper by Knees et al. (2004), an additional step is performed to find those terms that are most discriminative for each genre: a χ^2 test is used to select those terms that are least independent of (i.e. are likely to be predictive of) the classes. Selecting the top N terms for each category and scaling all χ^2 values per category so that the score for the top ranked term equals 1.0, gives a list of terms that seem to be typical of a given genre. An example of such a list for the genre heavy metal/hard rock is shown in Table 4.2. Note that neither of the constraint words (music and review) are included (they occur in all the pages, but they do not help in discriminating the genres).

The top 4 words are all (part of) artist names which were queried. However, many artists which are not part of the queries are also in the list, such as Phil Anselmo (Pantera), Hetfield, Hammett, Trujillo (Metallica), and Ozzy Osbourne. Furthermore, related groups such as Slayer, Megadeth, Iron Maiden, and Judas Priest are found as well as album names (Hysteria, Pyromania, ...) and song names (Paranoid, Unforgiven, Snowblind, St. Anger, ...) and other descriptive words such as evil, loud, hard, aggression, and heavy metal.

To classify previously unseen artists, we simply query Google with the artist name, count the occurrences of the characteristic genre terms on the retrieved web pages, and multiply these numbers with their respective scores for each genre. The scores in each genre are summed up, and the probability of membership of an artist to a genre is then computed as the fraction of the achieved score of each genre over the sum of scores over all genres.

In Knees et al. (2004), this procedure was tested using a genre taxonomy of 14 genres, and it was shown that correct genre recognition rates of 80% and better are achievable with this purely web-based approach, which compares very favourably with audio-based classification (see section 4.3.3 above).

On top of this classification system, an interactive demo applet (the GenreCrawler) was implemented that permits the user to experiment with the

1.00 *sabbath	0.26 heavy	0.17 riff	0.12 butler
0.97 *pantera	0.26 ulrich	0.17 leaf	0.12 blackened
0.89 *metallica	0.26 vulgar	0.17 superjoint	0.12 bringin
0.72 *leppard	0.25 megadeth	0.17 maiden	0.12 purple
0.58 metal	0.25 pigs	0.17 armageddon	0.12 foolin
0.56 hetfield	0.24 halford	0.17 gillan	0.12 headless
0.55 hysteria	0.24 dio	0.17 ozzfest	0.12 intensity
0.53 ozzy	0.23 reinventing	0.17 leps	0.12 mob
0.52 iommi	0.23 lange	0.16 slayer	0.12 excitable
0.42 puppets	0.23 newsted	0.15 purify	0.12 ward
0.40 dimebag	0.21 leppards	0.15 judas	0.11 zeppelin
0.40 anselmo	0.21 adrenalize	0.15 hell	0.11 sandman
0.40 pyromania	0.21 mutt	0.15 fairies	0.11 demolition
0.40 paranoid	0.20 kirk	0.15 bands	0.11 sanitarium
0.39 osbourne	0.20 riffs	0.15 iron	0.11 *black
0.37 *def	0.20 s&m	0.14 band	0.11 appice
0.34 euphoria	0.20 trendkill	0.14 reload	0.11 jovi
0.32 geezer	0.20 snowblind	0.14 bassist	0.11 anger
0.29 vinnie	0.19 cowboys	0.14 slang	0.11 rocked
0.28 collen	0.18 darrell	0.13 wizard	0.10 drummer
0.28 hammett	0.18 screams	0.13 vivian	0.10 bass
0.27 bloody	0.18 bites	0.13 elektra	0.09 rocket
0.27 thrash	0.18 unforgiven	0.13 shreds	0.09 evil
0.27 phil	0.18 lars	0.13 aggression	0.09 loud
0.26 lep	0.17 trujillo	0.13 scar	0.09 hard

Table 4.2: The top 100 terms with highest χ^2_{tc} values for genre "heavy metal/hard rock" defined by 4 artists (Black Sabbath, Pantera, Metallica, Def Leppard). Words marked with * are part of the search queries. The values are normalised so that the highest score equals 1.0.

system by typing in arbitrary new artists. In fact, the words to be typed in need not be artist names at all — they could be anything. The learned classifier can relate arbitrary words to genres, if that makes sense at all. For example, a

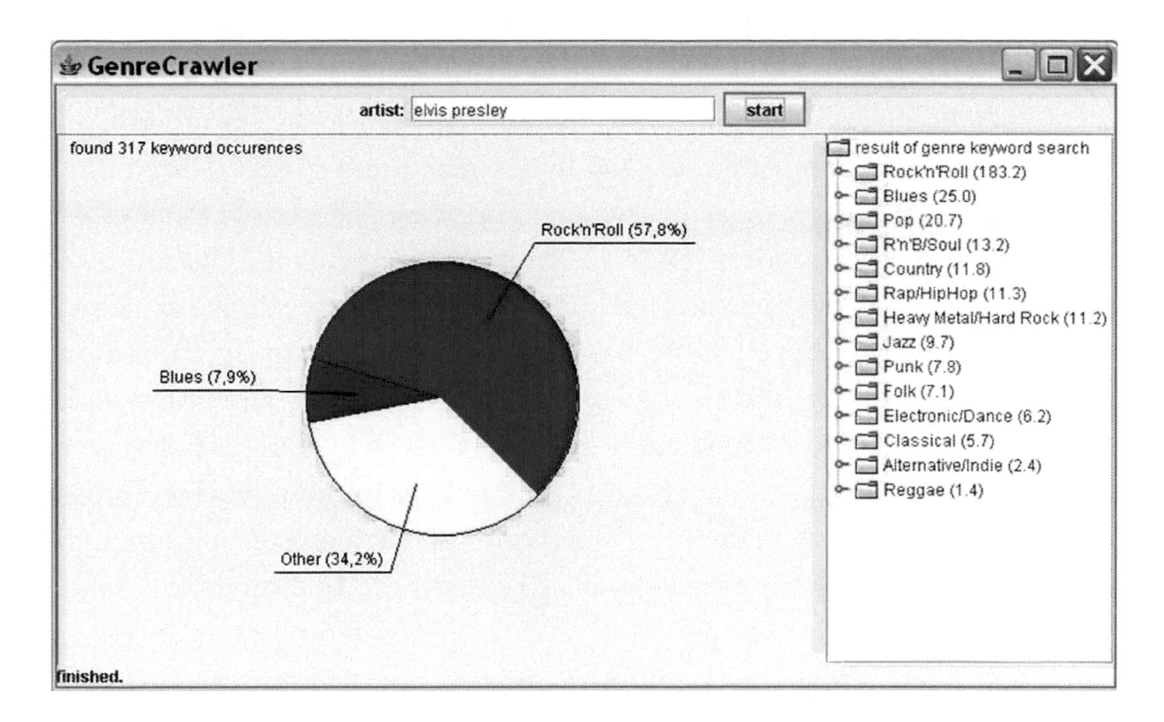

Figure 4.1: The GenreCrawler Knees et al. (2004) trying to classify Elvis Presley.

query for "Pathétique" results in an unambiguous answer: Classical Music. A screenshot of the GenreCrawler at work can be seen in Figure 4.1.

4.4.2 Learning textual characterisations

It is easy to convert the linguistic features (words) identified with the above method into a similarity measure, again using standard methods from information retrieval. Similarity measures have a wide range of applications, and one is presented in this section: learning to group music artists into meaningful categories, and describing these categories with characteristic words. Again, this is exploiting the Internet as an information source and could not be achieved on an audio basis alone.

More precisely, the goal is to find words to describe what a group of artists has in common, or what distinguishes it from other groups. Such information can be used for hierarchical user interfaces to explore music collections at the artist level (Pampalk et al., 2005). A simple text-based interface is shown in Figure 4.2 below.

As a first step, artists must be clustered hierarchically, and then appropriate terms (words) must be selected to describe these clusters. The basis of clustering is a similarity measure, which in our case is based on the linguistic features (characteristic words) extracted from Web pages by the GenreCrawler. There is a multitude of methods for hierarchical clustering. In the system described here (Pampalk et al., 2005), basically, a one-dimensional self organising map (SOM) (Kohonen, 2001) is used, with extensions for hierarchical structuring (Miikkulainen, 1990; Koikkalainen and Oja, 1990). Overlaps between the clusters are permitted, so that an artist may belong to more than one cluster. To obtain a multi-level hierarchical clustering, for each cluster found another one-dimensional SOM is trained (on all artists assigned to the cluster) until the cluster size falls below a certain limit.

The second step is the selection of characteristic terms to describe the individual clusters. The goal is to select those words that best summarise a group of artists. The assumption underlying this application is that the artists are mostly unknown to the user (otherwise we could just label the clusters with the artists' names).

There are a number of approaches to select characteristic words (Pampalk et al., 2005). One of these was developed by Lagus and Kaski (1999) for labelling large document collections organised by SOMs. Lagus and Kaski use only the term frequency tf_{ta} for each term t and artist a. The heuristically motivated ranking formula (higher values are better) is,

$$f_{tc} = (tf_{tc} / \sum_{t'} tf_{t'c}) \cdot \frac{(tf_{tc} / \sum_{t'} tf_{t'c})}{\sum_{c'} (tf_{tc'} / \sum_{t'} tf_{t'c'})}, \tag{4.1}$$

where tf_{tc} is the average term frequency in cluster c. The left side of the product is the importance of t in c defined through the frequency of t relative to the frequency of other terms in c. The right side is the importance of t in c relative to the importance of t in all other clusters.

Figure 4.2 shows a simple HTML interface that permits a user to explore the cluster structure learned by the system. There are two main parts to it: the hierarchy of clusters visualised as a grid of boxed texts and, just to the right of it, a display of a list of artists mapped to the currently selected cluster. The clusters of the first level in the hierarchy are visualised using the five boxes in the first (top) row. After the user selects a cluster, a second row appears which displays the children of the selected cluster. The selected clusters are highlighted in a different color. The hierarchy is displayed in such a way that the user can always see every previously made decision on a higher level. The number of artists mapped to a cluster is visualised by a bar next to the cluster. Inside a text box, at most the top 10 terms are displayed. The value of the ranking function for each term is coded through the color in which the term is displayed. The best term is always black and as the values decrease the color fades out. In the screenshot, at the first level the second node was selected, on the second level the fifth node, and on the third level, the first node. More details about method and experimental results can be found in the paper by Pampalk et al. (2005).

To summarise, the last two sections were meant to illustrate how the Internet can be used as a rich source of information about music. These are just simple first steps, and a lot of research on extracting richer music-related information from the Web can be expected.

A general problem with web-based approaches is that many new and not so well-known artists or music pieces do not appear on web pages. That limits the approach to yesterday's mainstream western culture. Another issue is the dynamics of web contents (e.g Lawrence and Giles, 1999). This has been studied by Knees et al. (2004) and the study was continued in a following paper (Knees, 2004). The experiments reported there indicate that, while the web may indeed be unstable, simple approaches like the ones described here may be highly robust to such fluctuations in web contents. Thus, the web mining approach may turn out to be an important pillar in research on music categorisation, if not music understanding.

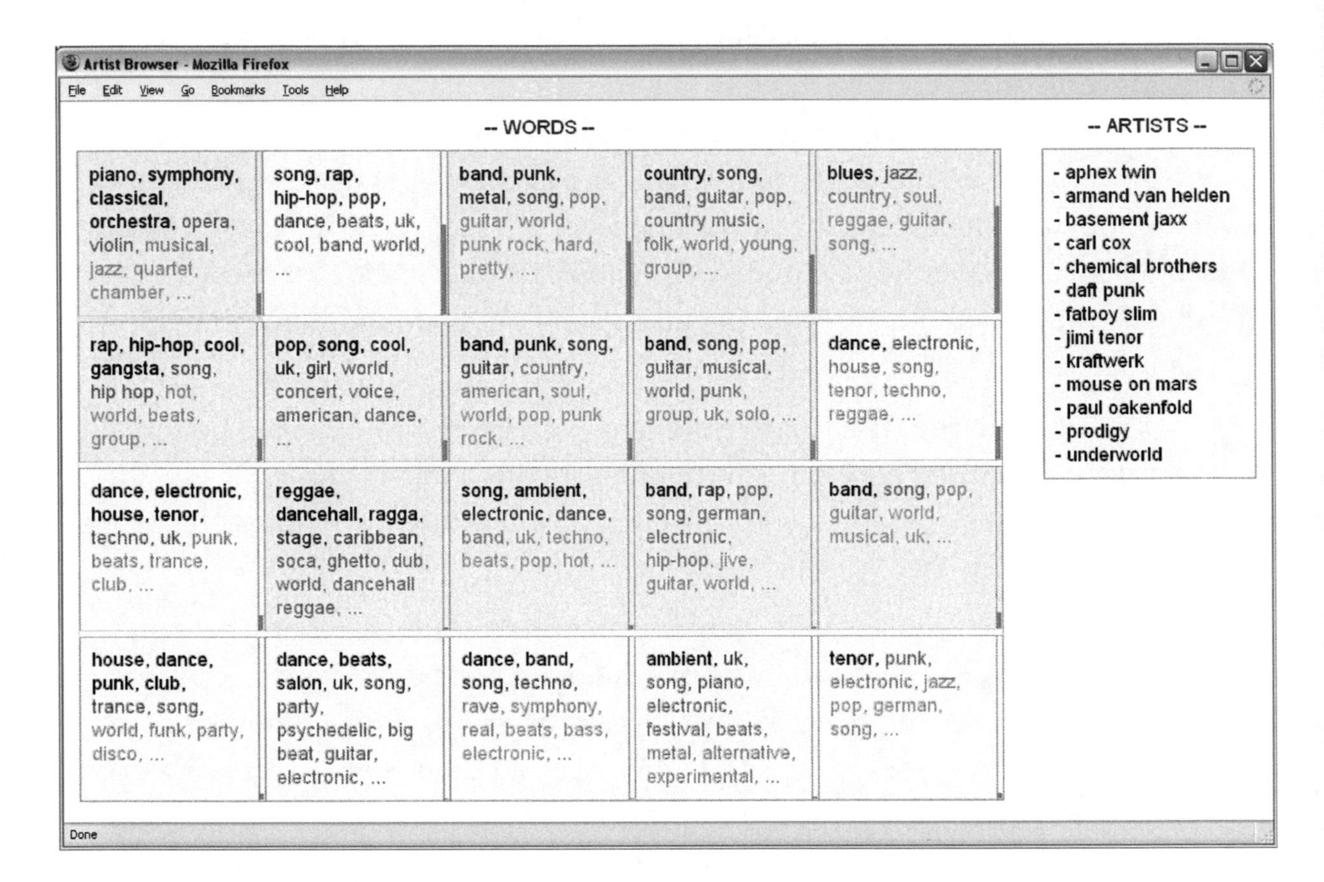

Figure 4.2: Screen shot of the HTML user interface to a system that automatically infers textual characterisations of artist clusters (cf. Pampalk et al., 2005).

4.5 Research and application perspectives

Building computers that can make sense of music has long been a goal topic that inspired scientists, especially in the field of Artificial Intelligence (AI). For the past 20 or so years, research in AI and Music has been aiming at creating systems that could in some way mimic human music perception, or to put it in more technical terms, that could recognise musical structures like melodies, harmonic structure, rhythm, etc. at the same level of competence as human experts. While there has been some success in specialised problems such as beat tracking, most of the truly complex musical capabilities are still outside of the range of computers. For example, no machine is currently capable of correctly transcribing an audio recording of even modest complexity, or of understanding the high-level form of music (e.g. recognising whether a classical piece is in sonata form, identifying a motif and its variations in a Mozart sonata, or unambiguously segmenting a popular piece into verse and chorus and bridge).

The new application field of Music Information Retrieval has led to, or at least contributed to, a shift of expectations: from a practical point of view, the real goal is not so much for a computer to understand music in a human-like way, but simply to have enough intelligence to support intelligent musical services and applications. Perfect musical understanding may not be required here. For instance, genre classification need not reach 100% accuracy to be useful in music recommendation systems. Likewise, a system for quick music browsing (e.g. Goto, 2003) need not perform a perfect segmentation of the music – if it finds roughly those parts in a recording where some of the interesting things are going on, that may be perfectly sufficient. Also, relatively simple capabilities like classifying music recordings into broad categories (genres) or assigning other high-level semantic labels to pieces can be immensely useful.

As has been indicated in this chapter, some of these capabilities are within reach, and indeed, some highly interesting real-world applications of this technology are currently emerging in the music market. From the research point of view, it is quite clear that there is still ample room for improvement, even within the relatively narrow domain of learning to assign high-level descriptors and labels to music recordings, which was the topic of this chapter. For instance, recent work on musical web mining has shown the promise of using extra-musical information for music classification, but little research has so far been performed on integrating different information sources – low-level audio features, higher-level structures automatically extracted from audio, web-based features, and possibly lyrics (which can also be recovered automatically from the Internet as discussed by Knees et al.) – in non-trivial ways.

A concept of central importance to MIR is music similarity measures. These are useful not only for classification, but for a wide variety of practical application scenarios, e.g., the automatic structuring and visualisation of large digital music collections (Pampalk et al., 2002, 2004), automatic playlist generation (e.g. Aucouturier and Pachet, 2002a), automatic music recommendation, and many more. Current music similarity measures are usually based on lower-level descriptors which are somehow averaged over a whole piece, so that a Euclidean distance metric can be applied to them. More complex approaches like clustering and distribution modelling via mixtures give a slightly more detailed account of the contents of a piece, but still ignore the temporal aspect of music. While preliminary experiments with Hidden Markov Models (Aucouturier and Pachet, 2004; Flexer et al., 2005), which do model temporal dependencies, do not seem to lead to improvements when based on low-level timbral features (like MFCCs), there is no reason to assume that the integration of higher-level descriptors (like melody, harmony, etc.) and temporal modelling will not permit substantial improvement. A lot of research on these issues is to be expected in the near future, driven by the sheer practical potential of music similarity measures. To put it simply: computers equipped with good music similarity measures may not be able to make sense of music in any human-like way, but they will be able to do more and more sensible things with music.

Acknowledgments

This work was supported by the European Union in the context of the projects S2S^2 ("Sound to Sense, Sense to Sound", IST-2004-03773) and SIMAC ("Semantic Interaction with Music Audio Contents", FP6 507142). Further support for ÖFAI's research in the area of intelligent music processing was provided by the following institutions: the European Union (project COST 282 KnowlEST "Knowledge Exploration in Science and Technology"); the Austrian *Fonds zur Förderung der Wissenschaftlichen Forschung* (FWF; projects Y99-START "Artificial Intelligence Models of Musical Expression" and L112-N04 "Operational Models of Music Similarity for MIR"); and the Viennese *Wissenschafts-, Forschungs- und Technologiefonds* (WWTF; project CI010 "Interfaces to Music"). The Austrian Research Institute for Artificial Intelligence also acknowledges financial support by the Austrian Federal Ministries of Education, Science and Culture and of Transport, Innovation and Technology.

Bibliography

M. Alonso, B. David, and G. Richard. Tempo and beat estimation of musical signals. In *Proceedings of the 5th International Symposium on Music Information Retrieval (ISMIR 2004)*, pages 158–163, 2004.

J.J. Aucouturier and F. Pachet. Improving timbre similarity: How high is the sky? *Journal of Negative Results in Speech and Audio Sciences*, 1(1), 2004.

J.J. Aucouturier and F. Pachet. Scaling up music playlist generation. In *Proceedings of IEEE International Conference on Multimedia and Expo (ICME 2002)*, pages 105–108, Lausanne, Switzerland, 2002a.

J.J. Aucouturier and F. Pachet. Music similarity measures: What's the use? In *Proceedings of the Third International Symposium on Music Information Retrieval (ISMIR 2002)*, pages 157–163, Paris, France, 2002b.

J.J Aucouturier and F. Pachet. Musical genre: A survey. *Journal of New Music Research*, 32(1):83–93, 2003.

S. Baumann and O. Hummel. Using cultural metadata for artist recommendation. In *Proceedings of the International Conference on Web Delivery of Music (WedelMusic)*, pages 138–141, Leeds, UK, 2003.

J.-J. Burred and A. Lerch. A hierarchical approach to automatic musical genre classification. In *Proceedings of th 6th International Conference on Digital Audio Effects (DAFx-03)*, London, UK, September, 8-11 2003.

C. H. L. Costa, J. D. Valle Jr., and A. L. Koerich. Automatic classification of audio data. In *Proceedings of the 2004 IEEE International Conference on Systems, Man and Cybernetics - SMC*, pages 562–567, Hague, Netherlands, October, 10-13 2004.

N. Cristianini and J. Shawe-Taylor. *An Introduction to Support Vector Machines and other Kernel-based Learning Methods*. Cambridge University Press, 2000.

S. Dixon, E. Pampalk, and G. Widmer. Classification of dance music by periodicity patterns. In *Proceedings of the 4th International Symposium on Music Information Retrieval*, pages 159–165, 2003.

S. Dixon, F. Gouyon, and G. Widmer. Towards characterisation of music via rhythmic patterns. In *Proceedings of the 5th International Symposium on Music Information Retrieval*, pages 509–516, 2004.

J.S. Downie, J. Futrelle, and D. Tcheng. The international music information retrieval systems evaluation laboratory: Governance, access and security. In *Proceedings of the 5th International Symposium on Music Information Retrieval (ISMIR'04)*, Barcelona, Spain, 2004.

R. Duda, P Hart, and D. Stork. *Pattern Classification (2nd Edition)*. John Wiley & Sons, New York, 2001.

J. Eggink and G. Brown. Extracting melody lines from complex audio. In *Proceedings of the 5th International Symposium on Music Information Retrieval (ISMIR 2004)*, pages 84–91, 2004.

A. Flexer, E. Pampalk, and G. Widmer. Hidden markov models for spectral similarity of songs. In *Proceedings of the 8th International Conference on Digital Audio Effects (DAFx'05)*, pages 131–136, Madrid, Spain, 2005.

J. Fürnkranz. Separate-and-conquer rule learning. *Artificial Intelligence Review*, 13(1):3–54, 1999.

E. Gómez and P. Herrera. Extimating the tonality of polyphonic audio files: Cognitive versus machine learning modelling strategies. In *Proceedings of the 5th International Symposium on Music Information Retrieval (ISMIR 2004)*, pages 92–95, 2004.

E. Gómez, A. Klapuri, and B. Meudic. Melody description and extraction in the context of music content processing. *Journal of New Music Research*, 32 (1):23–41, 2003.

M. Goto. Smartmusickiosk: Music listening station with chorus-search function. In *Proceedings of the 16th Annual ACM Symposium on User Interface Software and Technology (UIST 2003)*, pages 31–40, 2003.

M. Goto and S. Hayamizu. A real-time music scene description system: Detecting melody and bass lines in audio signals. In *Working Notes of the IJCAI-99 Workshop on Computational Auditory Scene Analysis*, pages 31–40. International Joint Conference on Artificial Intelligence, 1999.

F. Gouyon and S. Dixon. A review of automatic rhythm description systems. *Computer Music Journal*, 29(1):34–54, 2005.

F. Gouyon, S. Dixon, E. Pampalk, and G. Widmer. Evaluating rhythmic descriptors for musical genre classification. In *Proceedings of the AES 25th International Conference*, London, UK, June 17-19 2004.

T. Hastie, R. Tibshirani, and J. Friedman. *The Elements of Statistical Learning*. Springer Verlag, New York, 2001.

O. Hellmuth, E. Allamanche, J. Herre, T. Kastner, N. Lefebvre, and R. Wistorf. Music genre estimation from low level audio features. In *Proceedings of the AES 25th International Conference*, London, UK, June 17-19 2004.

P. Herrera, G. Peeters, and S. Dubnov. Automatic classification of musical instrument sounds. *Journal of New Music Research*, 32(1):3–22, 2003.

T. Kastner, J. Herre, E. Allamanche, O. Hellmuth, C. Ertel, and M. Schalek. Automatic optimization of a music similarity metric using similarity pairs. In *Proceedings of the AES 25th International Conference*, London, UK, June 17-19 2004.

A. Klapuri. Automatic music transcription as we know it today. *Journal of New Music Research*, 33(3):269–282, 2004.

P. Knees. Automatische Klassifikation von Musikkünstlern basierend auf Web-Daten (automatic classification of music artists based on web-data). Master thesis, Vienna University of Technology, Vienna, 2004.

P. Knees, M. Schedl, and G. Widmer. Multiple lyrics alignment: Automatic retrieval of song lyrics. In *Proceedings of the 5th International Conference on Music Information Retrieval (ISMIR 2004), year = 2005, pages = 564–569, address = London, UK.*

P. Knees, E. Pampalk, and G. Widmer. Artist classification with web-based data. In *Proceedings of the 5th International Symposium on Music Information Retrieval (ISMIR'04)*, Barcelona, Spain, 2004.

T. Kohonen. *Self-Organizing Maps*. Springer Verlag, Berlin, Germany, 2001. Third Edition.

P. Koikkalainen and E. Oja. Self-organizing hierarchical feature maps. In *Proceedings of the International Joint Conference on Neural Networks*, pages 279–284, San Diego, CA, 1990.

K. Lagus and S. Kaski. Keyword selection method for characterizing text document maps. In *Proceedings of ICANN99, Ninth International Conference on Artificial Neural Networks*, volume 1, pages 371–376, London, 1999. IEEE.

S. Lawrence and C. Lee Giles. Accessibility of information on the web. *Nature*, 400(6740):107–109, 1999.

T. Li and M. Ogihara. Detecting emotion in music. In *Proceedings of the International Symposium on Music Information Retrieval (ISMIR'03)*, Baltimore, MD, USA, October 26-30 2003.

D. Liu, L. Lu, and H.-J. Zhang. Automatic mood detection from acoustic music data. In *Proceedings of the International Symposium on Music Information Retrieval (ISMIR'03)*, Baltimore, MD, USA, October 26-30 2003.

B. Logan and A. Salomon. A music similarity function based on signal analysis. In *Proceedings of the IEEE International Conference on Multimedia and Expo*, pages 745–748, 2001.

I. Mierswa and K. Morik. Automatic feature extraction for classifying audio data. *Machine Learning*, 58(2-3):127–149, 2005.

R. Miikkulainen. Script recognition with hierarchical feature maps. *Connection Science*, 2:83–102, 1990.

T.M. Mitchell. *Machine Learning*. McGraw-Hill, New York, 1997.

E. Pampalk, A. Rauber, and D. Merkl. Content-based organization and visualization of music archives. In *Proceedings of the 10th ACM International Conference on Multimedia*, pages 570–579, Juan les Pins, France, 2002.

E. Pampalk, S. Dixon, and G. Widmer. Exploring music collections by browsing different views. *Computer Music Journal*, 28(2):49–62, 2004.

E. Pampalk, A. Flexer, and G. Widmer. Hierarchical organization and description of music collections at the artist level. In *Proceedings of the 9th European Conference on Research and Advanced Technology for Digital Libraries (ECDL 2005)*, pages 37–48, 2005.

J.R. Quinlan. Induction of decision trees. *Machine Learning*, 1(1):81–106, 1986.

C. Roads. *The Computer Music Tutorial.* MIT Press, Cambridge MA, 1996.

G. Salton and C. Buckley. Term-weighting approaches in automatic text retrieval. *Information Processing and Management*, 24(5):513–523, 1988.

G. Tzanetakis and P. Cook. Musical genre classification of audio signals. *IEEE Transactions on Speech and Audio Processing*, 10(5):293–302, 2002.

C. Uhle and C. Dittmar. Drum pattern based genre classification of popular music. In *Proceedings of the AES 25th International Conference*, London, UK, June 17-19 2004.

K. West and S. Cox. Features and classifiers for the automatic classification of musical audio signals. In *Proceedings of the International Symposium on Music Information Retrieval (ISMIR'04)*, Barcelona, Spain, October, 10-14 2004.

B. Whitman and D. Ellis. Automatic record reviews. In *Proceedings of the 5th International Symposium on Music Information Retrieval (ISMIR 2004)*, Barcelona, Spain, 2004.

B. Whitman and S. Lawrence. Inferring descriptions and similarity for music from community metadata. In *Proceedings of the 2002 International Computer Music Conference (ICMC 2002)*, pages 591–598, Göteborg, Sweden, 2002.

B. Whitman and B. Smaragdis. Combining musical and cultural features for intelligent style detection. In *Proceedings of the International Symposium on Music Information Retrieval (ISMIR 2002)*, Paris, France, 2002.

C. Xu, N. C. Maddage, X. Shao, and F. C. Tian. Musical genre classification using support vector machines. In *Proceedings of the International Conference on Acoustics, Speech, and Signal Processing*, pages 429–432, Hong Kong, China, April 6-10 2003.

K. Yoshii, M. Goto, and H. Okuno. Automatic drum sound description for real-world music using template adaptation and matching methods. In *Proceedings of the 5th International Symposium on Music Information Retrieval (ISMIR 2004)*, pages 184–191, 2004.

T. Yoshioka, T. Kitahara, K. Komatani, T. Ogata, and H. Okuno. Automatic chord transcription with concurrent recognition of chord symbols and boundaries. In *Proceedings of the 5th International Symposium on Music Information Retrieval (ISMIR 2004)*, pages 100–105, 2004.

Chapter 5

Sense in Expressive Music Performance: Data Acquisition, Computational Studies, and Models

Werner Goebl[1], Simon Dixon[1], Giovanni De Poli[2], Anders Friberg[3], Roberto Bresin[3], and Gerhard Widmer[1,4]

[1]Austrian Research Institute for Artificial Intelligence (OFAI), Vienna
[2]Department of Information Engineering, University of Padua;
[3]Department of Speech, Music, and Hearing, KTH, Stockholm;
[4]Department of Computational Perception, Johannes Kepler University, Linz

About this Chapter

This chapter gives an introduction to basic directions of current research in expressive music performance. A special focus is given on the various methods to acquire performance data either during a performance (e.g. through computer-monitored instruments) or from audio recordings. We then survey computational approaches to formalise and model the various aspects in expressive music performance. Future challenges and open problems are discussed briefly at the end of the chapter.

5.1 Introduction

Millions of people are regularly attending live music events or listening to recordings of music performances. What drives them to do so is hard to pin down with certainty, and the reasons for it might be manifold. But while enjoying the music, they are all listening to (mostly) human-made music that contains a specific human expression, whatever kind it might be – what they hear makes intuitive sense to them. Without this expressivity the music would not attract people; it is an integral part of the music.

Given the central importance of expressivity (not only in music, but in all communication modes and interaction contexts), it is not surprising that human expression and expressive behaviour have become a domain of intense scientific study. In the domain of music, much research has focused on the act of *expressive music performance*, as it is commonly and most typically found in classical music: the deliberate shaping of the music by the performer, the imposing of expressive qualities onto an otherwise "dead" musical score via controlled variation of parameters such as intensity, tempo, timing, articulation, etc. Early attempts at quantifying this phenomenon date back to the beginning of the 20th century, and even earlier than that.

If we wish to precisely measure and analyse every detail of an expressive music performance (onset timing, timbre and intensity, duration, etc.), we end up with huge amounts of data that quickly become unmanageable. Since the first large-scale, systematic investigations into expression in music performance (usually of classical music) in the 1930s, this has always been a main problem, which was controlled either by reducing the amount of music investigated to some seconds of music, or by limiting the number of performances studied to one or two. Recent approaches try to overcome this problem by using modern computational methods in order to study, model, and understand musical performance in its full complexity.

In the past ten years, some very comprehensive overview papers have been published on the various aspects of music performance research. The probably most cited is Alf Gabrielsson's chapter in Diana Deutsch's book "Psychology of Music" (Gabrielsson, 1999), in which he reviewed over 600 papers in this field published until approximately 1995. In a follow-up paper, he added and discussed another 200 peer-reviewed contributions that appeared until 2002 (Gabrielsson, 2003). A cognitive-psychological review has been contributed by Palmer (1997) summarising empirical research that focuses on cognitive aspects of music performance such as memory retrieval, anticipatory planning, or motor control. The musicologist's perspective is represented by two major edited books devoted exclusively to music performance research (Rink, 1995, 2002). Lately, more introductory chapters highlight the various methodological issues of systematic musicological performance research (Rink, 2003; Clarke, 2004; Cook, 2004; Windsor, 2004). Two recent contributions surveyed the diversity of computational approaches to modelling expressive music performance (De Poli, 2004; Widmer and Goebl, 2004). Parncutt and McPherson (2002) attempted to bridge the gap between research on music performance and music practice by bringing together two authors from each of the two sides for each chapter of their book.

Considering this variety of overview papers, we aim in this chapter to give a systematic overview on the more technological side of accessing, measuring, analysing, studying, and modelling expressive music performances. As a start, we survey the current literature of the past century on various ways of obtaining expression-related data from music performances. Then, we review current computational models of expressive music performance. In a final section we briefly sketch possible future directions and open problems that might be tackled by future research in this field.

5.2 Data acquisition and preparation

This section is devoted to very practical issues of obtaining precise empirical data on expressive performance. We can distinguish basically two different strategies for obtaining information on music performance. The first is to monitor performances during the production process with various measurement devices (MIDI pianos, accelerometers, movement sensors, video systems, etc.). Specific performance parameters can be accessed directly (hammer velocity of each played tone, bow speed, fingering, etc.). The other way is to extract all these relevant data from the recorded audio signal. This method has the disadvantage that some information, easy to extract during performance, is almost impossible to gain from the audio domain (consider, for instance, the sustain pedal on the piano). The advantage, however, is that we now have more than a century of recorded music at our disposal that could serve as a valuable resource for various kinds of scientific investigation. In the following sub-sections, we discuss the various approaches for monitoring and measuring music performance, and survey the major empirical performance studies that used them. As will be seen, by far the largest part of research has been done on piano performances.

5.2.1 Using specially equipped instruments

Before computers and digital measurement devices were invented and readily available for everyone, researchers employed a vast variety of mechanical and electrical measurement apparati to capture all sorts of human or mechanical movements during performance. We will review the most important of them, in chronological order, from rather old to state-of-the-art.

Mechanical and electro-mechanical setups

Among the first to record the movement of piano keys were Binet and Courtier (1895), who used a 6-mm caoutchouc rubber tube placed under the keys that was connected to a cylindric graphical recorder that captured continuous air pressure resulting from striking different keys on the piano. They investigated some basic pianistic tasks such as playing trills, connecting tones, or passing-under of the thumb in scales with exemplary material. In the first of the two contributions of this study, Ebhardt (1898) mounted metal springs on a bar above the strings that closed an electrical shutter when the hammer was about to touch the strings. The electric signal was recorded with a kymograph and timed with a 100-Hz oscillator. He studied the timing precision of simple finger tapping and playing scales. Further tasks with binary and ternary metrum revealed some characteristic timing patterns (e.g. a lengthening of the time interval before an accentuated onset). Onset and offset timing of church hymn performances were investigated by Sears (1902). He equipped a reed organ with mercury contacts that registered key depression of 10 selected keys. This information was recorded on four tracks on the surface of a smoked kymograph drum. He studied several temporal aspects of performances by four organ players, such as duration of the excerpts, bars, and individual note values, accent behavior, or note overlap (articulation).

A multitude of mechanical measurement devices were introduced by Ortmann (1925, 1929) in studies on physiological determinants of piano playing. To investigate the different behaviors of the key, he mounted a tuning fork to the side of one piano key that wrote wave traces into smoked paper which varied with the speed of the key. With this setup, he was one of the first to study the response of the key in different pianistic playing techniques. For assessing finger movements, Ortmann (1929, p. 230) used a custom-built mechanical apparatus with non-flexible aluminum strips that, on one side, were connected to either the finger (proximal phalanx) or the key surface and, on the other side, wrote onto a revolving drum. With this apparatus, continuous displacement of finger and key could be recorded and analysed. Another mechanical system was the "Pantograph" (Ortmann, 1929, p. 164), a parallelogram lever construction to record lateral arm movement. For other types of movement, he used active optical systems. The motion of a tiny light bulb attached to the wrist or the finger left a trace on a photo plate (the room was kept in very subdued light) when the shutter of the photo camera remained open for the entire duration of the movement.

Similar active markers mounted on head, shoulder, elbow, and wrist were used by Bernstein and Popova in their important study in 1930 (reported by Kay et al., 2003) to study the complex interaction and coupling of the limbs in piano playing. They used their "kymocyclographic camera" to record the movements of the active markers. A rotating shutter allowed the light of the markers to impinge on the constantly moving photographic film. With this device they could record up to 600 instances of the movement per second.

Piano rolls as a data source

A special source of expression data are piano rolls for reproducing pianos by different manufacturers (e.g. Welte-Mignon, Hupfeld, Aeolian Duo-Art, Ampico). A number of renowned pianists made recordings on these devices in the early part of the 20th century (Bowers, 1972; Hagmann, 1984). Such pianos were the first means to record and store artistic music performances before the gramophone was invented. Starting in the late 1920s, scientists took advantage of this source of data and investigated various aspects of performance. Heinlein (1929a,b, 1930) used Duo-Art rolls by the Aeolian company to study pedal use of four pianists playing Schumann's *Träumerei*. Rolls of the same company were the basis of Vernon's 1936 study. He investigated vertical synchronisation of the tones in a chord (see Goebl, 2001). Hartmann (1932) used Hupfeld "Animatic Rolls" and provided a very detailed study on tone and bar durations as well as note onset asynchronies in two recordings (by Josef Pembaur and Harold Bauer) of the first movement of Beethoven's "Moonlight Sonata" Op. 27 No. 2. Since the precise recording procedures used by these companies are still unknown (they were deliberately held back for commercial reasons), the authenticity of these rolls is sometimes questionable (Hagmann, 1984; Gottschewski, 1996). For example, the Welte-Mignon system was able to simultaneously control dynamics only for keyboard halves. Hence, emphasising the melody note and playing the rest of the chord tones more softly was only possible when the melody tone was played at a different point in time than the others (Gottschewski, 1996, pp. 26–42). Although we know today that pianists anticipate melody notes (Palmer, 1996b; Repp, 1996c; Goebl, 2001), the Welte-Mignon rolls cannot be taken literally as a source for studying note asynchronies (as done by Vernon, 1936). The interpretation of piano rolls must be done with care, keeping in mind the conditions of their production. There are currently some private attempts to systematically scan piano rolls and transform them into standard symbolic format (e.g. MIDI). However, we are not aware of any scientific project concerned with this.

The Iowa piano camera

During the 1930s, Carl E. Seashore guided a research group that focused on different aspects of music performance, namely the singing voice, violin playing, and piano performance (Seashore, 1932, 1936b,a). They developed various measurement setups for scientific investigation, among them the "Iowa Piano Camera" (Henderson et al., 1936) that optically captured onset and offset times and hammer velocity of each key and additionally the movement of the two pedals. It was therefore a complete and rather precise device that was not topped until the advent of modern computer-controlled pianos (such as the Disklavier or the Bösendorfer SE, see Goebl and Bresin, 2003). Each hammer is equipped with a shutter that controls light exposure of a moving film. The hammer shutter interrupts the light exposure on the film twice: a first time from 24 to 12 mm before the hammer touches the strings, and a second time at hammer–string contact. The average hammer speed of the last 12 mm of the hammer's travel can be inferred from the distance on the film between these two interrupts (today's computer-controlled pianos take the average speed of the final 5 mm). According to Skinner and Seashore (1936), the temporal resolution is around 10 ms. The hammer velocity is quantised into 17 dynamics categories (Henderson, 1936). With this system, the IOWA group performed several studies with professional pianists. Henderson (1936) had two professionals play the middle section of Chopin's Nocturne Op. 15 No. 3. In this very comprehensive study, they examined temporal behavior, phrasing, accentuation, pedalling, and chord asynchronies. Skinner and Seashore (1936) analysed repeated performances of pieces by Beethoven and Chopin and found high timing consistency among the pianists.

Henry Shaffer's photocell Bechstein

After the efforts of Seashore's research group at Iowa, it took over 40 years before a new group of researchers used modern technology to capture piano performance. It was L. Henry Shaffer at Exeter who equipped each of the 88 keys of a Bechstein grand piano with pairs of photocells to capture the essential expressive parameters of piano performance (Shaffer, 1980, 1981, 1984; Shaffer et al., 1985; Shaffer and Todd, 1987; Shaffer, 1992). The optical registration of the action's movements had the advantage of not affecting the playability of the piano. The photocells were mounted in the piano action in pairs, each capturing the moment of the hammer's transit. One was placed to register the instant of hammer-string contact, the other one the resting position of the hammer. The position of the two pedals were monitored by micro switches and stored as 12-bit words on the computer. Each such event was assigned a time stamp rounded to the nearest microsecond. The sensor at the strings yielded the note onset time, the one at the hammer's resting position (when the hammer returns) the note offset time. The time difference between the two sensors was an inverse estimate of the force at which the key was depressed. This technology is in principle identical to the computer-monitored pianos that are commercially available now (e.g. the Yamaha Disklavier series or the Bösendorfer SE).

Studies with synthesiser keyboards or digital pianos

Before computer-monitored acoustic pianos became widely available, simple synthesiser keyboards or digital pianos were used to capture expressive data from music performances. These devices provide timing and loudness data for each performed event through the standardised digital communications protocol MIDI (Musical Instrument Digital Interface) (Huber, 1999). However, such keyboards do not provide a realistic performance setting for advanced pianists, because the response of the keys is very different from an acoustic piano and the synthesised sound (especially with extensive use of the right pedal) does not satisfy the trained ears of highly skilled pianists.

Still, such electronic devices were used for various general expression studies (e.g. Palmer, 1989, 1992; Repp, 1994a,b, 1995c; Desain and Honing, 1994). Bruno Repp later repeated two of his studies that were first performed with data from a digital piano (Repp, 1995c, concerned with legato articulation; Repp, 1996b, concerned with the use of the right pedal) on a computer-controlled grand piano (Repp, 1997c,b, respectively). Interestingly, the results of both pairs of studies were similar to each other, even though the acoustic properties of the digital piano were considerably different from the grand piano.

The Yamaha Disklavier system

Present performance studies dealing with piano performances generally make use of commercially available computer-controlled acoustic pianos. Apart from systems that can be built into a piano (e.g. Autoklav, Pianocorder, see Coenen and Schäfer, 1992), the most common is the Disklavier system by Yamaha. The first computer-controlled grand pianos were available from 1989 onwards. The Mark IV series that is currently available includes also a computer with screen and several high-level functions such as an automatic accompaniment system. From 1998, Yamaha introduced their high-end PRO series of Disklaviers that involves an extended MIDI format to store more than 7-bit velocity information (values from 0 to 127) and information on key release.

There were few attempts to assess the Disklavier's accuracy in recording and reproducing performances. Coenen and Schäfer (1992) compared various reproducing systems (among them a Disklavier DG2RE and a SE225) with respect to their usability for reproducing compositions for mechanical instruments. More systematic tests on recording and reproduction accuracy were performed by Goebl and Bresin (2001, 2003) using accelerometer registration to inspect key and hammer movements during recording and reproduction.

Yamaha delivers both upright and grand piano versions of its Disklavier system. The upright model was used for several performance studies (Palmer and van de Sande, 1993; Palmer and Holleran, 1994; Repp, 1995a,b, 1996c,a,d, 1997d,a). The Yamaha Disklavier grand piano was even more widely used.

Moore (1992) combined data from a Disklavier grand piano with electromyographic recordings of the muscular activity of four performers playing trills. Behne and Wetekam (1994) recorded student performances of the theme from Mozart's K.331 sonata on a Disklavier grand piano and studied systematic timing variations of the Siciliano rhythm. As mentioned above, Repp repeated his work on legato and pedalling on a Disklavier grand piano (Repp, 1997c,b). Juslin and Madison (1999) used a Disklavier grand piano to record and play back different (manipulated) performances of two melodies to assess listeners' ability to recognise simple emotional categories. Bresin and Battel (2000) analysed multiple performances recorded on a Disklavier grand piano of Mozart's K.545 sonata in terms of articulation strategies. Clarke and Windsor (2000) used recordings made on a Disklavier grand piano for perceptual evaluation of real and artificially created performances. A short piece by Beethoven was recorded on a Disklavier grand piano played by one (Windsor et al., 2001) and by 16 professional pianists (Timmers et al., 2002; Timmers, 2002) in different tempi. Timing characteristics of different types of grace notes were investigated. Riley-Butler (2002) used a Disklavier grand piano in educational settings. She presented students with piano roll representations of their performances and observed considerable increase of learning efficiency with this method.

Bösendorfer's SE system

The SE ("Stahnke Electronics") System dates back to the early 1980s when the engineer Wayne Stahnke developed a reproducing system in cooperation with the MIT Artificial Intelligence Laboratory. It was built into a Bösendorfer Imperial grand piano (Roads, 1986; Moog and Rhea, 1990). A first prototype was ready in 1985; the system was officially sold by Kimball (at that time owner of Bösendorfer) starting from summer 1986. This system was very expensive and only few academic institutions could afford it. Until the end of its production, only about three dozen of these systems have been built and sold. In principle, the SE works like the Disklavier system (optical sensors register hammershank speed and key release, and linear motors reproduce final hammer velocity). However, its recording and reproducing capabilities are superior even compared with other much younger systems (Goebl and Bresin, 2003). Despite its rare occurrence in academic institutions, it was used for performance research in some cases.

Palmer and Brown (1991) performed basic tests on the relationship between hammer velocity and peak amplitude of the resulting sound. Repp (1993) tried to estimate peak sound level of piano tones from the two lowest partials as measured in the spectrogram and compared a digital piano, a Disklavier MX100A upright piano, with the Bösendorfer SE. Studies in music performance were performed at Ohio State University (Palmer and van de Sande, 1995; Palmer, 1996b,a), at the Musikhochschule Karlsruhe (e.g. Mazzola and Beran, 1998; Mazzola, 2002, p. 833), and on the grand piano located at the Bösendorfer company in Vienna (Goebl, 2001; Widmer, 2001, 2002b, 2003; Goebl and Bresin, 2003; Widmer, 2005).

Very recently (2006), the Bösendorfer company in Vienna has finished development of a new computer-controlled reproducing piano called "CEUS" that includes, among other features, sensors that register the continuous motion of each key. These data might be extremely valuable for studies regarding pianists' touch and tone control.

5.2.2 Measuring audio by hand

An alternative to measuring music expression during performance through sensors placed in or around the performer or the instrument is to analyse the recorded sound of music performances. This has the essential advantage that any type of recording may serve as a basis for investigation, e.g. commercially available CDs, historic recordings, or recordings from ethnomusicological research. One could just simply go into a record store and buy all the performances by the great pianists of the past century.[1]

However, extracting precise performance information from audio is difficult and sometimes impossible. The straight-forward method is to inspect

[1] In analysing recordings the researcher has to be aware that almost all records are glued together from several takes so the analysed performance might never have taken place in this particular rendition (see also Clarke, 2004, p. 88).

the waveform of the audio signal with computer software and mark manually with a cursor the onset times of selected musical events. Though this method is time-consuming, it delivers timing information with a reasonable precision. Dynamics is a more difficult issue. Overall dynamics (loudness) can be measured (e.g. by reading peak energy values from the root-mean-square of the signal averaged over a certain time window), but we are not aware of a successful procedure to extract individual dynamics of simultaneous tones (for an attempt, see Repp, 1993). Many other signal processing problems have not been solved either (e.g. extracting pedal information, tone length and articulation, etc., see also McAdams et al., 2004).

First studies that extracted timing information directly from sound used oscillogram filming (e.g. Bengtsson and Gabrielsson, 1977; for more references see Gabrielsson, 1999, p. 533). Povel (1977) analysed gramophone records of three performances of Johann Sebastian Bach's first prelude of The Well-Tempered Clavier, Vol. I. He determined the note onsets "by eye" from two differently obtained oscillograms of the recordings (which were transferred onto analog tape). He reported a temporal precision of 1–2 ms (!). Recordings of the same piece were investigated by Cook (1987), who obtained timing (and intensity) data with a computational method. Onset detection was automated by a threshold procedure applied to the digitised sound signal (8 bit, 4 kHz) and post-corrected by hand. He reported a timing resolution of 10 ms. He also stored intensity values, but did not specify in more detail what exactly was measured there.

Gabrielsson et al. (1983) analysed timing patterns of performances from 28 different monophonic melodies played by 5 performers. The timing data were measured from the audio recordings with a precision of ±5 ms (p. 196). In a later study, Gabrielsson (1987) extracted both timing and (overall) intensity data from the theme of Mozart's sonata K.331. In this study, a digital sampling system was used that allowed a temporal precision of 1–10 ms. The dynamics was estimated by reading peak amplitudes of each score event (in voltages). Nakamura (1987) used a Brüel & Kjær level recorder to register dynamics of solo performances played on a violin, oboe, and recorder. He analysed the produced dynamics in relation to the perceived intensity of the music.

The first larger corpus of recordings was measured by Repp (1990) who fed 19 recordings of the third movement of Beethoven's piano sonata Op. 31 No. 3 into a VAX 11/780 computer and read off the note onsets from waveform displays. In cases of doubt, he played the sound up to the onset and moved the cursor stepwise back in time, until the following note was no longer audible (Repp, 1990, p. 625). He measured the performances at the quarter-note level[2] and reported an absolute mean error of 6.5 ms for repeated measurements (equivalent to 1% of the inter-onset intervals, p. 626). In a further study, Repp (1992) collected 28 recordings of Schumann's "Träumerei" by 24 renowned pianists. He used a standard waveform editing program to hand-measure the 10-kHz sampled audio files. The rest of the procedure was identical (aural control of ambiguous onsets). He reported an average absolute measurement error of 4.3 ms (or less than 1%). In his later troika on the "microcosm of musical expression" (Repp, 1998, 1999a,b), he applied the same measurement procedure on 115 performances of the first five bars of Chopin's Etude Op. 10 No. 3 collected from libraries and record stores. He also extracted overall intensity information (Repp, 1999a) by taking the peak sound levels (pSPL in dB) extracted from the root-mean-square (RMS) integrated sound signal (over a rectangular window of 30 ms).

Nettheim (2001) measured parts of recordings of four historical performances of Chopin's e-minor Nocturne Op. 72 No. 1 (Pachmann, Godowsky, Rubinstein, Horowitz). He used a time-stretching software to reduce the playback speed by a factor of 7 (without changing the pitch of the music). He then simply took the onset times from a time display during playback. Tone onsets of all individual tones were measured with this method.[3] In repeated measurements, he reported an accuracy of around 14 ms. In addition to note onset timing, he assigned arbitrary intensity values to each tone ranging from 1 to 100 by ear.

In recent contributions on timing and synchronisation in jazz performances, the timing of the various instruments of jazz ensembles was inves-

[2]In the second part of this paper, he measured and analysed eight-note and sixteenth-note values as well.

[3]Obviously, the chosen excerpts were slow pieces with a comparatively low note density.

tigated. Friberg and Sundström (2002) measured cymbal onsets from spectrogram displays with a reported precision of ±3 ms. Ashley (2002) studied the synchronisation of the melody instruments with the double bass line. He repeatedly measured onsets of both lines from waveform plots of the digitised signal with usual differences between the measurements of 3–5 ms. About the same level of consistency (typically 2 ms) was achieved by Collier and Collier (2002) through a similar measurement procedure (manual annotation of physical onsets in trumpet solos). Lisboa et al. (2005) used a wave editor to extract onset timing in solo cello performances; Moelants (2004) made use of a speech transcription software ("Praat") to assess trill and ornament timing in solo string performances.

In a recent commercial enterprise, John Q. Walker and colleagues have been trying to extract the complete performance information out of historical (audio) recordings in order to play them back on a modern Disklavier.[4] Their commercial aim is to re-sell old recordings with modern sound quality or live performance feel. They computationally extract as much performance information as possible and add the missing information (e.g. tone length, pedalling) to an artificially-created MIDI file. They use it to control a modern Disklavier grand piano and compare this performance to the original recording. Then they modify the added information in the MIDI files and play it back again and repeat this process iteratively until the Disklavier's reproduction sounds "identical" to the original recording (see also Midgette, 2005).

Another way of assessing temporal content of recordings is by tapping along with the music recording e.g. on a MIDI drum pad or a keyboard, and recording this information (Cook, 1995; Bowen, 1996; Bachmann, 1999). This is a comparably fast method to gain rough timing data at a tappable beat level. However, perceptual studies on tapping along with expressive music showed that tappers – even after repeatedly tapping along with the same short piece of music – still underestimate abrupt tempo changes or systematic variations (Dixon et al., 2005).

[4]`http://www.zenph.com`

5.2.3 Computational extraction of expression from audio

The most general approach to extracting performance-related data directly from audio recordings would be fully automatic transcription, but such systems are currently not robust enough to provide the level of precision required for analysis of expression (Klapuri, 2004). However, more specialised systems were developed with the specific goal of expression extraction, in an attempt to support the painstaking effort of manual annotation (e.g. Dixon, 2000). Since the score is often available for the performances being analysed, Scheirer (1997) recognised that much better performance could be obtained by incorporating score information into the audio analysis algorithms, but his system was never developed to be sufficiently general or robust to be used in practice. One thing that was lacking from music analysis software was an interface for interactive editing of partially correct automatic annotations, without which the use of the software was not significantly more efficient than manual annotation.

The first system with such an interface was BeatRoot (Dixon, 2001a,b), an automatic beat tracking system with a graphical user interface which visualised (and auralised) the audio and derived beat times, allowing the user to edit the output and retrack the audio data based on the corrections. BeatRoot produces a list of beat times, from which tempo curves and other representations can be computed. Although it has its drawbacks, this system has been used extensively in studies of musical expression (Goebl and Dixon, 2001; Dixon et al., 2002; Widmer, 2002a; Widmer et al., 2003; Goebl et al., 2004). Recently, Gouyon et al. (2004) implemented a subset of BeatRoot as a plug-in for the audio editor WaveSurfer (Sjölander and Beskow, 2000).

A similar methodology was applied in the development of JTranscriber (Dixon, 2004), which was written as a front end for an existing transcription system (Dixon, 2000). The graphical interface shows a spectrogram scaled to a semitone frequency scale, with the transcribed notes superimposed over the spectrogram in piano roll notation. The automatically generated output can be edited with simple mouse-based operations, with audio playback of the original and the transcription, together or separately.

These tools provide a better approach than manual annotation, but since they have no access to score information, they still require a significant amount of interactive correction, so that they are not suitable for very large scale studies. An alternative approach is to use existing knowledge, such as from previous annotations of other performances of the same piece of music, and to transfer the metadata after aligning the audio files. The audio alignment system MATCH (Dixon and Widmer, 2005) finds optimal alignments between pairs of recordings, and is then able to transfer annotations from one recording to the corresponding time points in the second. This proves to be a much more efficient method of annotating multiple performances of the same piece, since manual annotation needs to be performed only once. Further, audio alignment algorithms are generally much more accurate than techniques for direct extraction of expressive information from audio data, so the amount of subsequent correction for each matched file is much less.

Taking this idea one step further, the initial annotation phase can be avoided entirely if the musical score is available in a symbolic format, by synthesising a mechanical performance from the score and matching the audio recordings to the synthetic performance. For analysis of expression in audio, e.g. tempo measurements, the performance data must be matched to the score, so that the relationship between actual and nominal durations can be computed. Several score-performance alignment systems have been developed for various types of music (Cano et al., 1999; Soulez et al., 2003; Turetsky and Ellis, 2003; Shalev-Shwartz et al., 2004).

Other relevant work is the on-line version of the MATCH algorithm, which can be used for tracking live performances with high accuracy (Dixon, 2005a,b). This system is being developed for real time visualisation of performance expression. The technical issues are similar to those faced by score-following systems, such as those used for automatic accompaniment (Dannenberg, 1984; Orio and Déchelle, 2001; Raphael, 2004), although the goals are somewhat different. Matching involving purely symbolic data has also been explored. Cambouropoulos developed a system for extracting score files from expressive performances in MIDI format (Cambouropoulos, 2000). After manual correction, the matched MIDI and score files were used in detailed studies of musical expression. Various other approaches to symbolic score-performance matching are reviewed by Heijink et al. (2000b,a).

5.2.4 Extracting expression from performers' movements

While the previous sections dealt with the extraction of expression contained in music performances, this section is devoted to expression as represented in all kinds of movements that occur when performers interact with their instruments during performance (for an overview, see Davidson and Correia, 2002; Clarke, 2004). Performers' movements are a powerful communication channel of expression to the audience, sometimes even overriding the acoustic information (Behne, 1990; Davidson, 1994).

There are several ways to monitor performers' movements. One possibility is to connect mechanical devices to the playing apparatus of the performer (Ortmann, 1929), but that has the disadvantage of inhibiting the free execution of the movements. More common are optical tracking systems that either simply video-tape a performer's movements or record special passive or active markers placed on particular joints of the performer's body. We already mentioned an early study by Berstein and Poppova (1930), who introduced an active photographical tracking system (Kay et al., 2003). Such systems use light-emitting markers placed on the various limbs and body parts of the performer. They are recorded by video cameras and tracked by software that extracts the position of the markers (e.g. the Selspot System, as used by Dahl, 2004, 2005). The disadvantage of these systems is that the participants need to be cabled, which is a time-consuming process. Also, the cables might inhibit the participants to move as they would normally move. Passive systems use reflective markers that are illuminated by external lamps. In order to create a three-dimensional picture of movement, the data from several cameras are coupled by software (Palmer and Dalla Bella, 2004).

Even less intrusive are video systems that simply record performance movements without any particular marking of the performer's limbs. Elaborated software systems (e.g. EyesWeb[5], see Camurri et al., 2004, 2005) are able to track defined body joints directly from the plain video signal (see Camurri

[5]http://www.megaproject.org

and Volpe, 2004, for an overview on gesture-related research). Perception studies on communication of expression through performers' gestures use simpler point-light video recordings (reflective markers on body joints recorded in a darkened room) to present them to participants for ratings (Davidson, 1993).

5.2.5 Extraction of emotional content from MIDI and audio

For listeners and musicians, an important aspect of music is its ability to express emotions (Juslin and Laukka, 2004). An important research question has been to investigate the coupling between emotional expression and the underlying musical parameters. Two important distinctions have to be made. The first distinction is between perceived emotional expression ("what is communicated") and induced emotion ("what you feel"). Here, we will concentrate on the perceived emotion which has been the focus of most of the research in the past. The second distinction is between compositional parameters (pitch, melody, harmony, rhythm) and performance parameters (tempo, phrasing, articulation, accents). The influence of compositional parameters has been investigated for a long time starting with the important work of Hevner (1937). A comprehensive summary is given in Gabrielsson and Lindström (2001). The influence of performance parameters has recently been investigated in a number of studies (for overviews see Juslin and Sloboda, 2001; Juslin, 2003). These studies indicate that for basic emotions such as happy, sad or angry, there is a simple and consistent relationship between the emotional description and the parameter values. For example, a sad expression is generally characterised by slow tempo, low sound level, legato articulation, and a happy expression is often characterised by fast tempo, moderate sound level and staccato articulation.

Predicting the emotional expression is usually done in a two-step process (Lindström et al., 2005). The first step extracts the basic parameters from the incoming signal. The selection of parameters is a trade-off between what is needed in terms of emotion-mapping and what is possible. MIDI performances are the simplest case in which the basic information in terms of notes, dynamics and articulation is already available. From this data it is possible to deduce for example the tempo using beat-tracking methods as described above. Audio from monophonic music performances can also be analyzed at the note level, which gives similar parameters as in the MIDI case (with some errors). In addition, using audio, a few extra parameters are available such as the spectral content and the attack velocity. The CUEX algorithm by Friberg et al. (2005) was specifically designed for prediction of emotional expression; it determines eight different parameters for each recognised note. Polyphonic audio is the most difficult case which has only recently been considered. One possibility is to first perform note extraction using polyphonic transcription (e.g. Klapuri, 2004) and then extract the parameters. Due to the lack of precision of polyphonic transcription there will be many errors. However, this may not be too problematic for the prediction of the emotion if the mapping is redundant and insensitive to small errors in the parameters. A more straightforward approach is to extract overall parameters directly from audio, such as using auditory-based measures for pitch, rhythm and timbre (Leman et al., 2004; Liu et al., 2003).

The second step is the mapping from the extracted parameters to the emotion character. A typical data-driven method is to use listener ratings (the "right" answer) for a set of performances to train a model. Common statistical/mathematical models are used such as regression (Leman et al., 2004; Juslin, 2000), Bayesian networks (Canazza et al., 2003), or Hidden Markov Models (Dillon, 2003).

5.3 Computational models of music performance

As the preceding sections have demonstrated, a large amount of empirical data about expressive performance has been gathered and analysed (mostly using statistical methods). The ultimate goal of this research is to arrive at an understanding of the relationships between the various factors involved in performance that can be formulated in a general model. Models describe relations among different kinds of observable (and often measurable) information about a phenomenon, discarding details that are felt to be irrelevant. They serve to generalise empirical findings and have both a descriptive and predictive value. Often the information is quantitative and we can distinguish input data, supposedly known, and output data, which are inferred by the model. In this case, inputs can be considered as the causes, and outputs the effects of the phenomenon. Computational models – models that are implemented on a computer – can compute the values of output data corresponding to the provided values of inputs. This process is called simulation and is widely used to predict the behaviour of the phenomenon in different circumstances. This can be used to validate the model, by comparing the predicted results with actual observations.

5.3.1 Modelling strategies

We can distinguish several strategies for developing the structure of the model and finding its parameters. The most prevalent ones are analysis-by-measurement and analysis-by-synthesis. Recently also methods from artificial intelligence have been employed: machine learning and case based reasoning. One can distinguish local models, which operate at the note level and try to explain the observed facts in a local context, and global models that take into account the higher level of the musical structure or more abstract expression patterns. The two approaches often require different modelling strategies and structures. In certain cases, it is possible to devise a combination of both approaches. The composed models are built by several components, each one aiming to explain different sources of expression. However, a good combination of the different parts is still quite a challenging research problem.

Analysis by measurement

The first strategy, analysis-by-measurement, is based on the analysis of deviations from the musical notation measured in recorded human performances. The goal is to recognise regularities in the deviation patterns and to describe them by means of a mathematical model, relating score to expressive values (see Gabrielsson 1999 and Gabrielsson 2003, for an overview of the main results). The method starts by selecting the performances to be analyzed. Often rather small sets of carefully selected performances are used. The physical properties of every note are measured using the methods seen in section 5.2 and the data so obtained are checked for reliability and consistency. The most relevant variables are selected and analysed. The analysis assumes an interpretation model that can be confirmed or modified by the results of the measurements. Often the assumption is made that patterns deriving from different sources or hierarchical levels can be separated and then added. This assumption helps the modelling phase, but may be overly simplistic. The whole repertoire of statistical data analysis techniques is then available to fit descriptive or predictive models onto the empirical data – from regression analysis to linear vector space theory to neural networks or fuzzy logic.

Many models address very specific aspects of expressive performance, for example, the final ritard and its relation to human motion (Kronman and Sundberg, 1987; Todd, 1995; Friberg and Sundberg, 1999; Sundberg, 2000; Friberg et al., 2000b); the timing of grace notes (Timmers et al., 2002); vibrato (Desain and Honing, 1996; Schoonderwaldt and Friberg, 2001); melody lead (Goebl, 2001, 2003); legato (Bresin and Battel, 2000); or staccato and its relation to local musical context (Bresin and Widmer, 2000; Bresin, 2001).

A global approach was pursued by Todd in his phrasing model (Todd, 1992, 1995). This model assumes that the structure of a musical piece can be decomposed into a hierarchy of meaningful segments (phrases), where each phase is in turn composed of a sequence of sub-phrases. The fundamental assumption of the model is that performers emphasise the hierarchical structure by an accelerando-ritardando pattern and a crescendo-decrescendo pattern for each phrase, and that these patterns are superimposed (summed) onto each other to give the actually observed complex performance. It has recently been shown empirically on a substantial corpus of Mozart performances (Tobudic and Widmer, 2006) that this model may be appropriate to explain (in part, at least) the shaping of dynamics by a performer, but less so as a model of expressive timing and tempo.

Analysis by synthesis

While analysis by measurement develops models that best fit quantitative data, the analysis-by-synthesis paradigm takes into account the human perception and subjective factors. First, the analysis of real performances and the intuition of expert musicians suggest hypotheses that are formalised as rules. The rules are tested by producing synthetic performances of many pieces and then evaluated by listeners. As a result the hypotheses are refined, accepted or rejected. This method avoids the difficult problem of objective comparison of performances, including subjective and perceptual elements in the development loop. On the other hand, it depends very much on the personal competence and taste of a few experts.

The most important model developed in this way is the KTH rule system (Friberg, 1991, 1995; Friberg et al., 1998, 2000a; Sundberg et al., 1983, 1989, 1991). In the KTH system, a set of rules describe quantitatively the deviations to be applied to a musical score, in order to produce a more attractive and human-like performance than the mechanical one that results from a literal playing of the score. Every rule tries to predict (and to explain with musical or psychoacoustic principles) some deviations that a human performer is likely to apply. Many rules are based on a low-level structural analysis of the musical score. The KTH rules can be grouped according to the purposes that they apparently have in music communication. For instance, differentiation rules appear to facilitate categorisation of pitch and duration, whereas grouping rules appear to facilitate grouping of notes, both at micro and macro levels.

Machine learning

In the "traditional" way of developing models, the researcher normally makes some hypothesis on the performance aspects s/he wishes to model and then tries to establish the empirical validity of the model by testing it on real data or on synthetic performances. An alternative approach, pursued by Widmer and coworkers (Widmer, 1995a,b, 1996, 2000, 2002b; Widmer and Tobudic, 2003; Widmer, 2003; Widmer et al., 2003; Widmer, 2005; Tobudic and Widmer, 2006), tries to extract new and potentially interesting regularities and performance principles from many performance examples, by using machine learning and data mining algorithms (see also Chapter 4 in this book). The aim of these methods is to search for and discover complex dependencies on very large data sets, without a specific preliminary hypothesis. A possible advantage is that machine learning algorithms may discover new (and possibly interesting) knowledge, avoiding any musical expectation or assumption. Moreover, some algorithms induce models in the form of rules that are directly intelligible and can be analysed and discussed with musicologists. This was demonstrated in a large-scale experiment (Widmer, 2002b), where a machine learning system analysed a large corpus of performance data (recordings of 13 complete Mozart piano sonatas by a concert pianist), and autonomously discovered a concise set of predictive rules for note-level timing, dynamics, and articulation. Some of these rules turned out to describe regularities similar to those incorporated in the KTH performance rule set (see above), but a few discovered rules actually contradicted some common hypotheses and thus pointed to potential shortcomings of existing theories.

The note-level model represented by these learned rules was later combined with a machine learning system that learned to expressively shape timing and dynamics at various higher levels of the phrase hierarchy (in a similar way as described in Todd's 1989; 1992 structure-level models), to yield a multi-level model of expressive phrasing and articulation (Widmer and Tobudic, 2003). A computer performance of a (part of a) Mozart piano sonata generated by this model was submitted to the International Performance Rendering Contest (RENCON) in Tokyo, 2002, where it won the Second Prize behind a rule-based rendering system that had been carefully tuned by hand. The rating was done by a jury of human listeners. This can be taken as a piece of evidence of the musical adequacy of the model. However, as an explanatory model, this system has a serious shortcoming: in contrast to the note-level rules, the phrase-level performance model is not interpretable, as it is based on a kind of case-based learning (see also below). More research into learning structured, interpretable models from empirical data will be required.

Case-based reasoning

An alternative approach, closer to the observation-imitation-experimentation process observed in humans, is that of directly using the knowledge implicit in human performances. Case-based reasoning (CBR) is based on the idea of solving new problems by using (often with some kind of adaptation) similar previously solved problems. An example in this direction is the SaxEx system for expressive performance of jazz ballads (Arcos et al., 1998; López de Mántaras and Arcos, 2002),which predicts expressive transformations to recordings of saxophone phrases by looking at how other, similar phrases were played by a human musician. The success of this approach greatly depends on the availability of a large amount of well-distributed previously solved problems, which are not easy to collect.

Mathematical theory approach

A rather different model, based mainly on mathematical considerations, is the Mazzola model (Mazzola, 1990; Mazzola and Zahorka, 1994; Mazzola et al., 1995; Mazzola, 2002; Mazzola and Göller, 2002). This model basically consists of a musical structure analysis part and a performance part. The analysis part involves computer-aided analysis tools, for various aspects of the music structure, that assign particular weights to each note in a symbolic score. The performance part, that transforms structural features into an artificial performance, is theoretically anchored in the so-called Stemma Theory and Operator Theory (a sort of additive rule-based structure-to-performance mapping). It iteratively modifies the performance vector fields, each of which controls a single expressive parameter of a synthesised performance.

The Mazzola model has found a number of followers who studied and used the model to generate artificial performances of various pieces. Unfortunately, there has been little interaction or critical exchange between this "school" and other parts of the performance research community, so that the relation between this model and other performance theories, and also the empirical validity of the model, are still rather unclear.

5.3.2 Perspectives

Computer-based modelling of expressive performance has shown its promise over the past years and has established itself as an accepted methodology. However, there are still numerous open questions related both to the technology, and to the questions that could be studied with it. Two prototypical ones are briefly discussed here.

Comparing performances and models

A problem that naturally arises in quantitative performance research is how performances can be compared. In subjective comparison often a supposed "ideal" performance is used as a reference by the evaluator. In other cases, an actual reference performance can be assumed. Of course subjects with different background may have dissimilar preferences that are not easily made explicit.

When we consider computational models, objective numerical comparisons would be desirable. In this case, performances are represented by sets of values. Various similarity or distance measures (e.g. absolute difference, Euclidean distance, etc.) can be defined over these, and it is not at all clear which of these is most appropriate musically. Likewise, it is not clear how to weight individual components or aspects (e.g. timing vs. dynamics), or how these "objective" differences relate to subjectively perceived differences. Agreed-upon methods for performance comparison would be highly important for further fruitful research in this field.

Common principles vs. differences

The models discussed in the previous sections aim at explaining and simulating general principles that seem to govern expressive performance, that is, those aspects of the relation between score and performance that seem predictable and more or less common to different performances and artists. Recently research has also started to pay attention to aspects that differentiate performances and performers' styles (Repp, 1992; Widmer, 2003). The same piece of music can be performed trying to convey different expressive intentions (Gabrielsson and Lindström, 2001), sometimes changing the character of the performance drastically. The CARO model (Canazza et al., 2004) is able to modify a neutral performance (i.e. played without any specific expressive intention) in order to convey different expressive intentions. Bresin and Friberg (2000) developed some macro rules for selecting appropriate values for the parameters of the KTH rule system in order to convey different emotions. The question of the boundary between predictability and individuality in performance remains a challenging one.

5.4 Conclusions

Quantitative research on expressive human performance has been developing quickly during the past decade, and our knowledge of this complex phenomenon has improved considerably. There is ample room for further investigations, and the field of computational performance research continues to be active. As the present survey shows, the computer has become a central player in this kind of research, both in the context of measuring and extracting expression-related information from performances, and in analysing and modelling the empirical data so obtained. Intelligent computational methods are thus helping us advance our understanding of a complex and deeply human ability and phenomenon. In addition, operational computer models of music performance will also find many applications in music education and entertainment – think, for instance, of expressive music generation or interactive expressive music control in multimedia applications or games, of quasi-autonomous systems for interactive music performance, of new types of musical instruments or interfaces that provide novel means of conveying expressive intentions or emotions, or of intelligent tutoring or teaching support systems in music education.

Still, there are fundamental limits that will probably be very hard to overcome for music performance research, whether computer-based or not. The very idea of a creative activity being predictable and, more specifically, the notion of a direct quasi-causal relation between the content of the music and the performance has obvious limitations. The person and personality of the artist as a mediator between music and listener is totally neglected in basically all models discussed above. There are some severe general limits to what any predictive model can describe. For instance, very often performers intentionally play the repetition of the same phrase or section totally differently the second time around. Being able to model and predict this would presuppose models of aspects that are outside the music itself, such as performance context, artistic intentions, personal experiences, listeners' expectations, etc.

Although it may sound quaint, there are concrete attempts at elaborating computational models of expressive performance to a level of complexity where they are able to compete with human performers. Since 2002, a scientific initiative brings together scientists from all over the world for a competition of artificially created performances (RENCON, contest for performance rendering systems[6]). Their aim is to construct computational systems that are able to pass a kind of expressive performance Turing Test (that is, an artificial performance sounds indistinguishable from a human performance, Hiraga et al., 2004). The very ambitious goal proclaimed by the RENCON initiative is for a computer to win the Chopin competition by 2050 (Hiraga et al., 2004). It is hard to imagine that this will ever be possible, not only because the organisers of such a competition will probably not permit a computer to participate, but also because a computational model would have to take into account the complex social and cognitive contexts in which, like any human intellectual and artistic activity, a music performance is situated. But even if complete predictive models of such phenomena are strictly impossible, they advance our understanding and appreciation of the complexity of artistic behaviour, and it remains an intellectual and scientific challenge to probe the limits of formal modelling and rational characterisation.

[6] `http://shouchan.ei.tuat.ac.jp/~rencon/`

Acknowledgments

This research was supported by the European Union (project FP6 IST-2004-03773 S2S2 "Sound to Sense, Sense to Sound"); the Austrian Fonds zur Förderung der Wissenschaftlichen Forschung (FWF; START project Y99-INF "Computer-Based Music Research: Artificial Intelligence Models of Musical Expression"); and the Viennese Science and Technology Fund (WWTF; project CI010 "Interfaces to Music"). The Austrian Research Institute for Artificial Intelligence (OFAI) acknowledges basic financial support by the Austrian Federal Ministries for Education, Science, and Culture, and for Transport, Innovation and Technology.

Bibliography

J. L. Arcos, R. López de Màntaras, and X. Serra. SaxEx: A case-based reasoning system for generating expressive performances. *Journal of New Music Research*, 27(3):194–210, 1998.

R. Ashley. Do[n't] change a hair for me: The art of Jazz Rubato. *Music Perception*, 19(3):311–332, 2002.

K. Bachmann. *Das Verhältnis der Tempi in mehrsätzigen Musikwerken: ein Beitrag zur musikalischen Aufführungsanalyse am Beispiel der Symphonien Ludwig van Beethovens*. Unpublished doctoral thesis, Institut für Musikwissenschaft, Universität Salzburg, Salzburg, 1999.

K.-E. Behne. "Blicken Sie auf die Pianisten?!" Zur bildbeeinflußten Beurteilung von Klaviermusik im Fernsehen. *Medienpsychologie*, 2(2):115–131, 1990.

K.-E. Behne and B. Wetekam. Musikpsychologische Interpretationsforschung: Individualität und Intention. In Klaus-Ernst Behne, Günter Kleinen, and Helga de la Motte-Haber, editors, *Musikpsychologie. Empirische Forschungen, ästhetische Experimente*, volume 10, pages 24–32. Noetzel, Wilhelmshaven, 1994.

I. Bengtsson and A. Gabrielsson. Rhythm research in Uppsala. In *Music, Room, Acoustics*, volume 17, pages 19–56. Publications issued by the Royal Swedish Academy of Music, Stockholm, 1977.

A. Binet and J. Courtier. Recherches graphiques sur la musique. *L'Année Psychologique*, 2:201–222, 1895. URL `http://www.musica-mechana.de`. Available also in a German translation by Schmitz, H.-W. (1994), Das Mechanische Musikinstrument 61, 16–24.

J. A. Bowen. Tempo, duration, and flexibility: Techniques in the analysis of performance. *Journal of Musicological Research*, 16(2):111–156, 1996.

Q. D. Bowers. *Encyclopedia of Automatic Musical Instruments*. Vestal Press Ltd., New York, 13th edition, 1972.

R. Bresin. Articulation rules for automatic music performance. In A. Schloss and R. Dannenberg, editors, *Proceedings of the 2001 International Computer Music Conference, Havana, Cuba*, pages 294–297. International Computer Music Association, San Francisco, 2001.

R. Bresin and G. U. Battel. Articulation strategies in expressive piano performance. *Journal of New Music Research*, 29(3):211–224, 2000.

R. Bresin and A. Friberg. Emotional coloring of computer-controlled music performances. *Computer Music Journal*, 24(4):44–63, 2000.

R. Bresin and G. Widmer. Production of staccato articulation in Mozart sonatas played on a grand piano. Preliminary results. *Speech, Music, and Hearing. Quarterly Progress and Status Report*, 2000(4):1–6, 2000.

E. Cambouropoulos. Score Extraction from MIDI Files. In *In Proceedings of the 13th Colloquium on Musical Informatics (CIM'2000)*. L'Aquila, Italy, 2000.

A. Camurri and G. Volpe, editors. *Gesture-Based Communication in Human-Computer Interaction*. Springer, Berlin, 2004. LNAI 2915.

A. Camurri, C. L. Krumhansl, B. Mazzarino, and G. Volpe. An exploratory study of aniticipation human movement in dance. In *Proceedings of the 2nd International Symposium on Measurement, Analysis, and Modeling of Human Functions*. Genova, Italy, 2004.

A. Camurri, G. Volpe, G. De Poli, and M. Leman. Communicating expressiveness and affect in multimodal interactive systems. *IEEE Multimedia*, 12(1): 43–53, 2005.

S. Canazza, G. De Poli, G. Mion, A. Rodà, A. Vidolin, and P. Zanon. Expressive classifiers at CSC: An overview of the main research streams. In *Proceedings of the XIV Colloquium on Musical Informatics (XIV CIM 2003) May 8–10*. Firenze, 2003.

S. Canazza, G. De Poli, C. Drioli, A. Rodà, and A. Vidolin. Modeling and control of expressiveness in music performance. *Proceedings of the IEEE*, 92 (4):686–701, 2004.

P. Cano, A. Loscos, and J. Bonada. Score-performance matching using HMMs. In *Proceedings of the International Computer Music Conference*, pages 441–444. International Computer Music Association, 1999.

E. F. Clarke. Empirical methods in the study of performance. In Eric F. Clarke and Nicholas Cook, editors, *Empirical Musicology. Aims, Methods, and Prospects*, pages 77–102. University Press, Oxford, 2004.

E. F. Clarke and W. L. Windsor. Real and simulated expression: A listening study. *Music Perception*, 17(3):277–313, 2000.

A. Coenen and S. Schäfer. Computer-controlled player pianos. *Computer Music Journal*, 16(4):104–111, 1992.

G. L. Collier and J. L. Collier. A study of timing in two Louis Armstrong solos. *Music Perception*, 19(3):463–483, 2002.

N. Cook. Computational and comparative Musicology. In Eric F. Clarke and Nicholas Cook, editors, *Empirical Musicology. Aims, Methods, and Prospects*, pages 103–126. University Press, Oxford, 2004.

N. Cook. Structure and performance timing in Bach's C major prelude (WTC I): An empirical study. *Music Analysis*, 6(3):100–114, 1987.

N. Cook. The conductor and the theorist: Furtwängler, Schenker and the first movement of Beethoven's Ninth Symphony. In John Rink, editor, *The Practice of Performance*, pages 105–125. Cambridge University Press, Cambridge, UK, 1995.

S. Dahl. Playing the accent – comparing striking velocity and timing in an ostinato rhythm performed by four drummers. *Acta Acustica*, 90(4):762–776, 2004.

S. Dahl. Movements and analysis of drumming. In Eckart Altenmüller, J. Kesselring, and M. Wiesendanger, editors, *Music, Motor Control and the Brain*, page in press. University Press, Oxford, 2005.

R.B. Dannenberg. An on-line algorithm for real-time accompaniment. In *Proceedings of the International Computer Music Conference*, pages 193–198, 1984.

J. W. Davidson. Visual perception of performance manner in the movements of solo musicians. *Psychology of Music*, 21(2):103–113, 1993.

J. W. Davidson. What type of information is conveyed in the body movements of solo musician performers? *Journal of Human Movement Studies*, 26(6): 279–301, 1994.

J. W. Davidson and J. S. Correia. Body movement. In Richard Parncutt and Gary McPherson, editors, *The Science and Psychology of Music Performance. Creating Strategies for Teaching and Learning*, pages 237–250. University Press, Oxford, 2002.

G. De Poli. Methodologies for expressiveness modelling of and for music performance. *Journal of New Music Research*, 33(3):189–202, 2004.

P. Desain and H. Honing. Does expressive timing in music performance scale proportionally with tempo? *Psychological Research*, 56:285–292, 1994.

P. Desain and H. Honing. Modeling continuous aspects of music performance: Vibrato and Portamento. In Bruce Pennycook and Eugenia Costa-Giomi, editors, *Proceedings of the 4th International Conference on Music Perception and Cognition (ICMPC'96)*. Faculty of Music, McGill University, Montreal, Canada, 1996.

R. Dillon. A statistical approach to expressive intention recognition in violin performances. In Roberto Bresin, editor, *Proceedings of the Stockholm Music Acoustics Conference (SMAC'03), August 6–9, 2003*, pages 529–532. Department of Speech, Music, and Hearing, Royal Institute of Technology, Stockholm, Sweden, 2003.

S. Dixon. Extraction of musical performance parameters from audio data. In *Proceedings of the First IEEE Pacific-Rim Conference on Multimedia*, pages 42–45, Sydney, 2000. University of Sydney.

S. Dixon. Analysis of musical content in digital audio. In J. DiMarco, editor, *Computer Graphics and Multimedia: Applications, Problems, and Solutions*, pages 214–235. Idea Group, Hershey PA, 2004.

S. Dixon. An on-line time warping algorithm for tracking musical performances. In *Proceedings of the International Joint Conference on Artificial Intelligence*, 2005a. to appear.

S. Dixon. Live tracking of musical performances using on-line time warping. 2005b. Submitted.

S. Dixon. Automatic extraction of tempo and beat from expressive performances. *Journal of New Music Research*, 30(1):39–58, 2001a.

S. Dixon. Learning to detect onsets of acoustic piano tones. In Claudia Lomeli Buyoli and Ramon Loureiro, editors, *MOSART Workshop on current research directions in computer music, November 15–17, 2001*, pages 147–151. Audiovisual Institute, Pompeu Fabra University, Barcelona, Spain, 2001b.

S. Dixon and G. Widmer. MATCH: A music alignment tool chest. In *6th International Conference on Music Information Retrieval*, page Submitted, 2005.

S. Dixon, W. Goebl, and G. Widmer. The Performance Worm: Real time visualisation based on Langner's representation. In Mats Nordahl, editor, *Proceedings of the 2002 International Computer Music Conference, Göteborg, Sweden*, pages 361–364. International Computer Music Association, San Francisco, 2002.

S. Dixon, W. Goebl, and E. Cambouropoulos. Smoothed tempo perception of expressively performed music. *Music Perception*, 23:in press, 2005.

K. Ebhardt. Zwei Beiträge zur Psychologie des Rhythmus und des Tempo. *Zeitschrift für Psychologie und Physiologie der Sinnesorgane*, 18:99–154, 1898.

A. Friberg. Generative rules for music performance. *Computer Music Journal*, 15(2):56–71, 1991.

A. Friberg. *A Quantitative Rule System for Musical Performance.* Doctoral dissertation, Department of Speech, Music and Hearing, Royal Institute of Technology, Stockholm, 1995.

A. Friberg and J. Sundberg. Does music performance allude to locomotion? A model of final ritardandi derived from measurements of stopping runners. *Journal of the Acoustical Society of America*, 105(3):1469–1484, 1999.

A. Friberg and A. Sundström. Swing ratios and ensemble timing in jazz performance: Evidence for a common rhythmic pattern. *Music Perception*, 19 (3):333–349, 2002.

A. Friberg, R. Bresin, L. Frydén, and J. Sundberg. Musical punctuation on the mircolevel: Automatic identification and performance of small melodic units. *Journal of New Music Research*, 27(3):271–292, 1998.

A. Friberg, V. Colombo, L. Frydén, and J. Sundberg. Generating musical performances with Director Musices. *Computer Music Journal*, 24(3):23–29, 2000a.

A. Friberg, J. Sundberg, and Frydén. Music from motion: Sound level envelopes of tones expressing human locomotion. *Journal of New Music Research*, 29(3):199–210, 2000b.

A. Friberg, E. Schoonderwaldt, and P. N. Juslin. CUEX: An algorithm for extracting expressive tone variables from audio recordings. *Acoustica united with Acta Acoustica*, in press, 2005.

A. Gabrielsson. Music performance research at the millenium. *Psychology of Music*, 31(3):221–272, 2003.

A. Gabrielsson. Once again: The Theme from Mozart's Piano Sonata in A Major (K.331). In Alf Gabrielsson, editor, *Action and Perception in Rhythm and Music*, volume 55, pages 81–103. Publications issued by the Royal Swedish Academy of Music, Stockholm, Sweden, 1987.

A. Gabrielsson. Music Performance. In Diana Deutsch, editor, *Psychology of Music*, pages 501–602. Academic Press, San Diego, 2nd edition, 1999.

A. Gabrielsson and E. Lindström. The influence of musical structure on emotional expression. In Patrik N. Juslin and John A. Sloboda, editors, *Music and Emotion: Theory and Research*, pages 223–248. Oxford University Press, New York, 2001.

A. Gabrielsson, I. Bengtsson, and B. Gabrielsson. Performance of musical rhythm in 3/4 and 6/8 meter. *Scandinavian Journal of Psychology*, 24:193–213, 1983.

W. Goebl. Melody lead in piano performance: Expressive device or artifact? *Journal of the Acoustical Society of America*, 110(1):563–572, 2001.

W. Goebl. *The Role of Timing and Intensity in the Production and Perception of Melody in Expressive Piano Performance.* Doctoral thesis, Institut für Musikwissenschaft, Karl-Franzens-Universität Graz, Graz, Austria, 2003. available online at `http://www.ofai.at/music`.

W. Goebl and R. Bresin. Are computer-controlled pianos a reliable tool in music performance research? Recording and reproduction precision of a Yamaha Disklavier grand piano. In Claudia Lomeli Buyoli and Ramon Loureiro, editors, *MOSART Workshop on Current Research Directions in Computer Music, November 15–17, 2001*, pages 45–50. Audiovisual Institute, Pompeu Fabra University, Barcelona, Spain, 2001.

W. Goebl and R. Bresin. Measurement and reproduction accuracy of computer-controlled grand pianos. *Journal of the Acoustical Society of America*, 114(4): 2273–2283, 2003.

W. Goebl and S. Dixon. Analyses of tempo classes in performances of Mozart piano sonatas. In Henna Lappalainen, editor, *Proceedings of the Seventh In-*

ternational Symposium on Systematic and Comparative Musicology, Third International Conference on Cognitive Musicology, August 16–19, 2001, pages 65–76. University of Jyväskylä, Jyväskylä, Finland, 2001.

W. Goebl, E. Pampalk, and G. Widmer. Exploring expressive performance trajectories: Six famous pianists play six Chopin pieces. In Scott D. Lipscomp, Richard Ashley, Robert O. Gjerdingen, and Peter Webster, editors, *Proceedings of the 8th International Conference on Music Perception and Cognition, Evanston, IL, 2004 (ICMPC8)*, pages 505–509. Causal Productions, Adelaide, Australia, 2004. CD-ROM.

H. Gottschewski. *Die Interpretation als Kunstwerk. Musikalische Zeitgestaltung und ihre Analyse am Beispiel von Welte-Mignon-Klavieraufnahmen aus dem Jahre 1905*. Freiburger Beiträge zur Musikwissenschaft, Bd. 5. Laaber-Verlag, Laaber, 1996.

F. Gouyon, N. Wack, and S. Dixon. An open source tool for semi-automatic rhythmic annotation. In *Proceedings of the 7th International Conference on Digital Audio Effects*, pages 193–196, 2004.

P. Hagmann. *Das Welte-Mignon-Klavier, die Welte-Philharmonie-Orgel und die Anfänge der Reproduktion von Musik*. Europäische Hochschulschriften: Reihe 35, Musikwissenschaft, Bd. 10. Peter Lang, Bern, Frankfurt am Main, New York, 1984. URL `http://www.freidok.uni-freiburg.de/volltexte/608/`. available at `http://www.freidok.uni-freiburg.de/volltexte/608/`.

A. Hartmann. Untersuchungen über das metrische Verhalten in musikalischen Interpretationsvarianten. *Archiv für die gesamte Psychologie*, 84:103–192, 1932.

H. Heijink, P. Desain, H. Honing, and L. Windsor. Make me a match: An evaluation of different approaches to score-performance matching. *Computer Music Journal*, 24(1):43–56, 2000a.

H. Heijink, L. Windsor, and P. Desain. Data processing in music performance research: Using structural information to improve score-performance matching. *Behavior Research Methods, Instruments and Computers*, 32(4):546–554, 2000b.

C. P. Heinlein. A discussion of the nature of pianoforte damper-pedalling together with an experimental study of some individual differences in pedal performance. *Journal of General Psychology*, 2:489–508, 1929a.

C. P. Heinlein. The functional role of finger touch and damper-pedalling in the appreciation of pianoforte music. *Journal of General Psychology*, 2:462–469, 1929b.

C. P. Heinlein. Pianoforte damper-pedalling under ten different experimental conditions. *Journal of General Psychology*, 3:511–528, 1930.

M. T. Henderson. Rhythmic organization in artistic piano performance. In Carl Emil Seashore, editor, *Objective Analysis of Musical Performance*, volume IV of *University of Iowa Studies in the Psychology of Music*, pages 281–305. University Press, Iowa City, 1936.

M. T. Henderson, J. Tiffin, and C. E. Seashore. The Iowa piano camera and its use. In Carl Emil Seashore, editor, *Objective Analysis of Musical Performance*, volume IV, pages 252–262. University Press, Iowa City, 1936.

K. Hevner. The affective value of pitch and tempo in music. *American Journal of Psychology*, 49:621–630, 1937.

R. Hiraga, R. Bresin, K. Hirata, and H. Katayose. Rencon 2004: Turing Test for musical expression. In *Proceedings of the 2004 Conference on New Interfaces for Musical Expression (NIME04)*, pages 120–123. Hamamatsu, Japan, 2004.

D. M. Huber. *The MIDI Manual*. Butterworth-Heinemann, Boston, MA, 1999.

P. N. Juslin. Cue utilization in communication of emotion in music performance: Relating performance to perception. *Journal of Experimental Psychology: Human Perception and Performance*, 26(1797–1813), 2000.

P. N. Juslin. Studies of music performance: A theoretical analysis of empirical findings. In Roberto Bresin, editor, *Proceedings of the Stockholm Music Acoustics Conference (SMAC'03), August 6–9, 2003*, volume II, pages 513–516. Department of Speech, Music, and Hearing, Royal Institute of Technology, Stockholm, Sweden, 2003.

P. N. Juslin and P. Laukka. Expression, perception, and induction of musical emotions: A review and a questionnaire study of everyday listening. *Journal of New Music Research*, 33(3):217–238, 2004.

P. N. Juslin and G. Madison. The role of timing patterns in recognition of emotional expression from musical performance. *Music Perception*, 17(2): 197–221, 1999.

P. N. Juslin and J. A. Sloboda. *Music and Emotion: Theory and Research*. Oxford University Press, New York, 2001.

B. A. Kay, M. T. Turvey, and O. G. Meijer. An early oscillator model: Studies on the biodynamics of the piano strike (Bernstein & Popova, 1930). *Motor Control*, 7(1):1–45, 2003.

A. Klapuri. Automatic music transcription as we know it today. *Journal of New Music Research*, 33(3):269–282, 2004.

U. Kronman and J. Sundberg. Is the musical retard an allusion to physical motion? In Alf Gabrielsson, editor, *Action and Perception in Rhythm and Music*, volume 55, pages 57–68. Publications issued by the Royal Swedish Academy of Music, Stockholm, Sweden, 1987.

M. Leman, V. Vermeulen, L. De Voogdt, J. Taelman, D. Moelants, and M. Lesaffre. Correlation of gestural musical audio cues and perceived expressive qualities. In Antonio Camurri and Gualterio Volpe, editors, *Gesture-based Communication in Human-Computer Interaction*, pages xx–yy. Springer, Berlin, 2004. LNAI 2915.

E. Lindström, A. Camurri, A. Friberg, G. Volpe, and M.-L. Rinman. Affect, attitude and evaluation of multi-sensory performances. *Journal of New Music Research*, in press, 2005.

T. Lisboa, A. Williamon, M. Zicari, and H. Eiholzer. Mastery through imitation: A preliminary study. *Musicae Scientiae*, 9(1), 2005.

D. Liu, L. Lie, and H.-J. Zhang. Automatic mood detection from acoustic music data. In *Proceedings of the International Symposium on Music Information Retrieval*. 2003.

R. López de Mántaras and Josep L. A. Arcos. AI and music: From composition to expressive performances. *AI Magazine*, 23(3):43–57, 2002.

G. Mazzola, editor. *The Topos of Music – Geometric Logic of Concepts, Theory, and Performance*. Birkhäuser Verlag, Basel, 2002.

G. Mazzola. *Geometrie der Töne. Elemente der Mathematischen Musiktheorie*. Birkhäuser Verlag, Basel, 1990.

G. Mazzola and J. Beran. Rational composition of performance. In Reinhard Kopiez and Wolfgang Auhagen, editors, *Controlling Creative Processes in Music*, volume 12 of *Schriften zur Musikpsychologie and Musikästhetik, Bd. 12*, pages 37–68. Lang, Frankfurt/M., 1998.

G. Mazzola and S. Göller. Performance and interpretation. *Journal of New Music Research*, 31(3):221–232, 2002.

G. Mazzola and O. Zahorka. Tempo curves revisited: Hierarchies of performance fields. *Computer Music Journal*, 18(1):40–52, 1994.

G. Mazzola, O. Zahorka, and J. Stange-Elbe. Analysis and performance of a dream. In Anders Friberg and Johan Sundberg, editors, *Proceedings of the KTH Symposion on Grammars for Music Performance*, pages 59–68. Department of Speech Communication and Music Acoustics, Stockholm, Sweden, 1995.

S. McAdams, P. Depalle, and E. F. Clarke. Analyzing musical sound. In Eric F. Clarke and Nicholas Cook, editors, *Empirical Musicology. Aims, Methods, and Prospects*, pages 157–196. University Press, Oxford, 2004.

A. Midgette. Play it again, Vladimir (via computer). *The New York Times*, June 5 2005.

D. Moelants. Temporal aspects of instrumentalists' performance of tremolo, trills, and vibrato. In *Proceedings of the International Symposium on Musical Acoustics (ISMA'04)*, pages 281–284. The Acoustical Society of Japan, Nara, Japan, 2004.

R. A. Moog and T. L. Rhea. Evolution of the keyboard interface: The Bösendorfer 290 SE recording piano and the Moog multiply-touch-sensitive keyboards. *Computer Music Journal*, 14(2):52–60, 1990.

G. P. Moore. Piano trills. *Music Perception*, 9(3):351–359, 1992.

T. Nakamura. The communication of dynamics between musicians and listeners through musical performance. *Perception and Psychophysics*, 41(6): 525–533, 1987.

N. Nettheim. A musical microscope applied to the piano playing of Vladimir de Pachmann, 2001. http://users.bigpond.net.au/nettheim/pachmic/microsc.htm.

N. Orio and F. Déchelle. Score following using spectral analysis and Hidden Markov Models. In *Proceedings of the International Computer Music Conference*, pages 151–154, 2001.

O. Ortmann. *The Physical Basis of Piano Touch and Tone*. Kegan Paul, Trench, Trubner; J. Curwen; E. P. Dutton, London, New York, 1925.

O. Ortmann. *The Physiological Mechanics of Piano Technique*. Kegan Paul, Trench, Trubner, E. P. Dutton, London, New York, 1929. Paperback reprint: New York: E. P. Dutton 1962.

C. Palmer. Mapping musical thought to musical performance. *Journal of Experimental Psychology: Human Perception and Performance*, 15(12):331–346, 1989.

C. Palmer. The role of interpretive preferences in music performance. In Mari Riess Jones and Susan Holleran, editors, *Cognitive Bases of Musical Communication*, pages 249–262. American Psychological Association, Washington DC, 1992.

C. Palmer. Anatomy of a performance: Sources of musical expression. *Music Perception*, 13(3):433–453, 1996a.

C. Palmer. On the assignment of structure in music performance. *Music Perception*, 14(1):23–56, 1996b.

C. Palmer. Music performance. *Annual Review of Psychology*, 48:115–138, 1997.

C. Palmer and J. C. Brown. Investigations in the amplitude of sounded piano tones. *Journal of the Acoustical Society of America*, 90(1):60–66, 1991.

C. Palmer and S. Dalla Bella. Movement amplitude and tempo change in piano performance. *Journal of the Acoustical Society of America*, 115(5):2590, 2004.

C. Palmer and S. Holleran. Harmonic, melodic, and frequency height influences in the perception of multivoiced music. *Perception and Psychophysics*, 56(3):301–312, 1994.

C. Palmer and C. van de Sande. Units of knowledge in music performance. *Journal of Experimental Psychology: Learning, Memory, and Cognition*, 19(2): 457–470, 1993.

C. Palmer and C. van de Sande. Range of planning in music performance. *Journal of Experimental Psychology: Human Perception and Performance*, 21(5): 947–962, 1995.

R. Parncutt and G. McPherson, editors. *The Science and Psychology of Music Performance. Creating Strategies for Teaching and Learning*. University Press, Oxford, New York, 2002.

D.-J. Povel. Temporal structure of performed music: Some preliminary observations. *Acta Psychologica*, 41(4):309–320, 1977.

C. Raphael. A hybrid graphical model for aligning polyphonic audio with musical scores. In *Proceedings of the 5th International Conference on Musical Information Retrieval*, pages 387–394, 2004.

B. H. Repp. Patterns of expressive timing in performances of a Beethoven minuet by nineteen famous pianists. *Journal of the Acoustical Society of America*, 88(2):622–641, 1990.

B. H. Repp. Diversity and commonality in music performance: An analysis of timing microstructure in Schumann's "Träumerei". *Journal of the Acoustical Society of America*, 92(5):2546–2568, 1992.

B. H. Repp. Some empirical observations on sound level properties of recorded piano tones. *Journal of the Acoustical Society of America*, 93(2):1136–1144, 1993.

B. H. Repp. On determining the basic tempo of an expressive music performance. *Psychology of Music*, 22:157–167, 1994a.

B. H. Repp. Relational invariance of expressive microstructure across global tempo changes in music performance: an exploratory study. *Psychological Research*, 56(4):269–284, 1994b.

B. H. Repp. Detectability of duration and intensity increments in melody tones: a partial connection between music perception and performance. *Perception and Psychophysics*, 57(8):1217–1232, 1995a.

B. H. Repp. Expressive timing in Schumann's "Träumerei:" An analysis of performances by graduate student pianists. *Journal of the Acoustical Society of America*, 98(5):2413–2427, 1995b.

B. H. Repp. Acoustics, perception, and production of legato articulation on a digital piano. *Journal of the Acoustical Society of America*, 97(6):3862–3874, 1995c.

B. H. Repp. The art of inaccuracy: Why pianists' errors are difficult to hear. *Music Perception*, 14(2):161–184, 1996a.

B. H. Repp. Pedal timing and tempo in expressive piano performance: A preliminary investigation. *Psychology of Music*, 24(2):199–221, 1996b.

B. H. Repp. Patterns of note onset asynchronies in expressive piano performance. *Journal of the Acoustical Society of America*, 100(6):3917–3932, 1996c.

B. H. Repp. The dynamics of expressive piano performance: Schumann's "Träumerei" revisited. *Journal of the Acoustical Society of America*, 100(1): 641–650, 1996d.

B. H. Repp. The Aesthetic Quality of a Quantitatively Average Music Performance: Two Preliminary Experiments. *Music Perception*, 14(4):419–444, 1997a.

B. H. Repp. The effect of tempo on pedal timing in piano performance. *Psychological Research*, 60(3):164–172, 1997b.

B. H. Repp. Acoustics, perception, and production of legato articulation on a computer-controlled grand piano. *Journal of the Acoustical Society of America*, 102(3):1878–1890, 1997c.

B. H. Repp. Expressive timing in a Debussy Prelude: A comparison of student and expert pianists. *Musicae Scientiae*, 1(2):257–268, 1997d.

B. H. Repp. A microcosm of musical expression. I. Quantitative analysis of pianists' timing in the initial measures of Chopin's Etude in E major. *Journal of the Acoustical Society of America*, 104(2):1085–1100, 1998.

B. H. Repp. A microcosm of musical expression: II. Quantitative analysis of pianists' dynamics in the initial measures of Chopin's Etude in E major. *Journal of the Acoustical Society of America*, 105(3):1972–1988, 1999a.

B. H. Repp. A microcosm of musical expression: III. Contributions of timing and dynamics to the aesthetic impression of pianists' performances of the initial measures of Chopin's Etude in E major. *Journal of the Acoustical Society of America*, 106(1):469–478, 1999b.

K. Riley-Butler. Teaching expressivity: An aural-visual feedback-replication model. In *ESCOM 10th Anniversary Conference on Musical Creativity, April 5–8, 2002*. Université de Liège, Liège, Belgium, 2002. CD-ROM.

J. Rink, editor. *Musical Performance. A Guide to Understanding*. Cambridge University Press, Cambridge, UK, 2002.

J. Rink, editor. *The Practice of Performance: Studies in Musical Interpretation*. University Press, Cambridge UK, 1995.

J. Rink. In respect of performance: The view from Musicology. *Psychology of Music*, 31(3):303–323, 2003.

C. Roads. Bösendorfer 290 SE computer-based piano. *Computer Music Journal*, 10(3):102–103, 1986.

E.D. Scheirer. Using musical knowledge to extract expressive performance information from audio recordings. In H. Okuno and D. Rosenthal, editors, *Readings in Computational Auditory Scene Analysis*. Lawrence Erlbaum, 1997.

E. Schoonderwaldt and A. Friberg. Towards a rule-based model for violin vibrato. In Claudia Lomeli Buyoli and Ramon Loureiro, editors, *MOSART Workshop on Current Research Directions in Computer Music, November 15–17, 2001*, pages 61–64. Audiovisual Institute, Pompeu Fabra University, Barcelona, Spain, 2001.

C. H. Sears. A contribution to the psychology of rhythm. *American Journal of Psychology*, 13(1):28–61, 1902.

C. E. Seashore, editor. *The Vibrato*, volume I of *University of Iowa Studies in the Psychology of Music*. University Press, Iowa City, 1932.

C. E. Seashore, editor. *Psychology of the Vibrato in Voice and Instrument*, volume III of *University of Iowa Studies. Studies in the Psychology of Music Volume III*. University Press, Iowa City, 1936a.

C. E. Seashore, editor. *Objective Analysis of Musical Performance*, volume IV of *University of Iowa Studies in the Psychology of Music*. University Press, Iowa City, 1936b.

L. H. Shaffer. Analysing piano performance. In George E. Stelmach and Jean Requin, editors, *Tutorials in Motor Behavior*. North-Holland, Amsterdam, 1980.

L. H. Shaffer. Performances of Chopin, Bach and Bartòk: Studies in motor programming. *Cognitive Psychology*, 13(3):326–376, 1981.

L. H. Shaffer. Timing in solo and duet piano performances. *Quarterly Journal of Experimental Psychology: Human Experimental Psychology*, 36A(4):577–595, 1984.

L. H. Shaffer. How to interpret music. In Mari Riess Jones and Susan Holleran, editors, *Cognitive Bases of Musical Communication*, pages 263–278. American Psychological Association, Washington DC, 1992.

L. H. Shaffer and N. P. Todd. The interpretative component in musical performance. In Alf Gabrielsson, editor, *Action and Perception in Rhythm and Music*, volume 55, pages 139–152. Publications issued by the Royal Swedish Academy of Music, Stockholm, Sweden, 1987.

L. H. Shaffer, E. F. Clarke, and N. P. Todd. Metre and Rhythm in Piano playing. *Cognition*, 20(1):61–77, 1985.

S. Shalev-Shwartz, J. Keshet, and Y. Singer. Learning to align polyphonic music. In *5th International Conference on Music Information Retrieval*, pages 381–386, 2004.

K. Sjölander and J. Beskow. WaveSurfer – an open source speech tool. In *Proceedings of the International Conference on Spoken Language Processing*, 2000.

L. Skinner and C. E. Seashore. A musical pattern score of the first movement of the Beethoven Sonata, Opus 27, No. 2. In Carl Emil Seashore, editor, *Objective Analysis of Musical Performance*, volume IV of *Studies in the Psychology of Music*, pages 263–279. University Press, Iowa City, 1936.

F. Soulez, X. Rodet, and D. Schwarz. Improving polyphonic and poly-instrumental music to score alignment. In *4th International Conference on Music Information Retrieval*, pages 143–148, 2003.

J. Sundberg. Four years of research on music and motion. *Journal of New Music Research*, 29(3):183–185, 2000.

J. Sundberg, A. Askenfelt, and L. Frydén. Musical performance. A synthesis-by-rule approach. *Computer Music Journal*, 7:37–43, 1983.

J. Sundberg, A. Friberg, and L. Frydén. Rules for automated performance of ensemble music. *Contemporary Music Review*, 3:89–109, 1989.

J. Sundberg, A. Friberg, and L. Frydén. Threshold and preference quantities of rules for music performance. *Music Perception*, 9(1):71–92, 1991.

R. Timmers. *Freedom and Constraints in Timing and Ornamentation*. Shaker Publishing, Maastricht, 2002.

R. Timmers, R. Ashley, P. Desain, H. Honing, and L. W. Windsor. Timing of ornaments in the theme of Beethoven's Paisello Variations: Empirical data and a model. *Music Perception*, 20(1):3–33, 2002.

A. Tobudic and G. Widmer. Relational IBL in classical music. *Machine Learning*, 64:5–24, 2006.

N. P. Todd. A computational model of Rubato. *Contemporary Music Review*, 3: 69–88, 1989.

N. P. Todd. The dynamics of dynamics: A model of musical expression. *Journal of the Acoustical Society of America*, 91(6):3540–3550, 1992.

N. P. Todd. The kinematics of musical expression. *Journal of the Acoustical Society of America*, 97(3):1940–1949, 1995.

R. Turetsky and D. Ellis. Ground-truth transcriptions of real music from force-aligned MIDI syntheses. In *4th International Conference on Music Information Retrieval*, pages 135–141, 2003.

L. N. Vernon. Synchronization of chords in artistic piano music. In Carl Emil Seashore, editor, *Objective Analysis of Musical Performance*, volume IV of *Studies in the Psychology of Music*, pages 306–345. University Press, Iowa City, 1936.

G. Widmer. Large-scale induction of expressive performance rules: first quantitative results. In Ioannis Zannos, editor, *Proceedings of the 2000 International Computer Music Conference, Berlin, Germany*, pages 344–347. International Computer Music Association, San Francisco, 2000.

G. Widmer. Using AI and machine learning to study expressive music performance: Project survey and first report. *AI Communications*, 14(3):149–162, 2001.

G. Widmer. In search of the Horowitz factor: Interim report on a musical discovery project. In *Proceedings of the 5th International Conference on Discovery Science (DS'02), Lübeck, Germany*. Springer, Berlin, 2002a.

G. Widmer. Machine discoveries: A few simple, robust local expression principles. *Journal of New Music Research*, 31(1):37–50, 2002b.

G. Widmer. Discovering simple rules in complex data: A meta-learning algorithm and some surprising musical discoveries. *Artificial Intelligence*, 146(2): 129–148, 2003.

G. Widmer. Studying a creative act with computers: Music performance studies with automated discovery methods. *Musicae Scientiae*, 9(1):11–30, 2005.

G. Widmer. A machine learning analysis of expressive timing in pianists' performances of Schumann' "Träumerei". In Anders Friberg and Johan Sundberg, editors, *Proceedings of the KTH Symposion on Grammars for Music Performance*, pages 69–81. Department of Speech Communication and Music Acoustics, Stockholm, Sweden, 1995a.

G. Widmer. Modeling rational basis for musical expression. *Computer Music Journal*, 19(2):76–96, 1995b.

G. Widmer. Learning expressive performance: The structure-level approach. *Journal of New Music Research*, 25(2):179–205, 1996.

G. Widmer and W. Goebl. Computational models of expressive music performance: The state of the art. *Journal of New Music Research*, 33(3):203–216, 2004.

G. Widmer and A. Tobudic. Playing Mozart by analogy: Learning multi-level timing and dynamics strategies. *Journal of New Music Research*, 32(3):259–268, 2003.

G. Widmer, S. Dixon, W. Goebl, E. Pampalk, and A. Tobudic. In search of the Horowitz factor. *AI Magazine*, 24(3):111–130, 2003.

L. Windsor. Data collection, experimental design, and statistics in musical research. In Eric F. Clarke and Nicholas Cook, editors, *Empirical Musicology. Aims, Methods, and Prospects*, pages 197–222. University Press, Oxford, 2004.

L. Windsor, P. Desain, R. Aarts, H. Heijink, and R. Timmers. The timing of grace notes in skilled musical performance at different tempi. *Psychology of Music*, 29(2):149–169, 2001.

Chapter 6

Controlling Sound with Senses: Multimodal and Cross-Modal Approaches to Control of Interactive Systems

Antonio Camurri[1], Carlo Drioli[2], Barbara Mazzarino[1], Gualtiero Volpe[1]

[1]InfoMus Lab, DIST, University of Genoa
[2]Department of Computer Science, University of Verona

About this chapter

This chapter briefly surveys some relevant aspects of current research into control of interactive (music) systems, putting into evidence research issues, achieved results, and problems that are still open for the future. A particular focus is on multimodal and cross-modal techniques for expressive control of sound and music processing and synthesis. The chapter will discuss a conceptual framework, the methodological aspects, the research perspectives.

It will also present concrete examples and tools such as the EyesWeb XMI platform and the EyesWeb Expressive Gesture Processing Library.

6.1 Introduction

The problem of effectively controlling sound generation and processing has always been relevant for music research in general and for Sound and Music Computing in particular. Research into control concerns perceptual, cognitive, affective aspects. It ranges from the study of the mechanisms involved in playing traditional acoustic instruments to the novel opportunities offered by modern Digital Music Instruments. More recently, the problem of defining effective strategies for real-time control of multimodal interactive systems, with particular reference to music but not limited to it, is attracting growing interest from the scientific community because of its relevance also for future research and applications in broader fields of human-computer interaction.

In this framework, research into control extends its scope to include for example analysis of human movement and gesture (not only gestures of musicians playing an instrument but also gestures of subjects interacting with computer systems), analysis of the perceptual and cognitive mechanisms of gesture interpretation, analysis of the communication of non-verbal expressive and emotional content through gesture, multimodality and cross-modality, identification of strategies for mapping the information obtained from gesture analysis onto real-time control of sound and music output including high-level information (e.g. real-time control of expressive sound and music output).

A key issue in this research is its cross-disciplinary nature. Research can highly benefit from cross-fertilisation between scientific and technical knowledge on the one side, and art and humanities on the other side. Such need of cross-fertilisation opens new perspectives to research in both fields: if from the one side scientific and technological research can benefit from models and theories borrowed from psychology, social science, art and humanities, on the other side these disciplines can take advantage of the tools that technology can provide for their own research, i.e. for investigating the hidden subtleties of human beings at a depth that was hard to reach before. The convergence of different research communities such as musicology, computer science, computer engineering, mathematics, psychology, neuroscience, arts and humanities as well as of theoretical and empirical approaches bears witness to the need and the importance of such cross-fertilisation.

This chapter briefly surveys some relevant aspects of current research on control, putting into evidence research issues, achieved results, and problems that are still open for the future.

A particular focus is on multimodal and cross-modal techniques for expressive control. Multimodal analysis enables the integrated analysis of information coming from different multimedia streams (audio, video) and affecting different sensorial modalities (auditory, visual). Cross-modal analysis enables exploiting potential similarities in the approach for analyzing different multimedia streams: so, for example techniques developed for analysis in a given modality (e.g. audio) can also be used for analysis in another modality (e.g. video); further, commonalities at mid- and high-level in representations of different sensory channels are an important perspective for developing models for control and mapping based on a-modal, converging representations.

A first aspect concerns the definition of a conceptual framework envisaging control at different levels, from low-level analysis of audio signals to feature extraction, to identification and analysis of significant musical structures (note groups, phrases), up to high-level association of semantic descriptions including affective, emotional content (see also Chapters 3 and 4). Such conceptual framework is not limited to the music domain. It can be fruitfully applied to other modalities (e.g. movement and gesture) too, enabling multimodal and cross-modal processing. This includes a level of abstraction such that features at that level do not belong to a given modality, rather they emerge as shared, a-modal representations from the different modalities and can contribute to model and explain the mapping strategies between modalities. A comprehensive definition of such a high-level framework embedding cross-modality and multimodality is still an open research issue deserving particular attention in the future. Section 6.2 presents a conceptual framework worked out in the EU-IST Project MEGA (Multisensory Expressive Gesture Applications) that can be considered as a starting point for research on this direction.

A second aspect is related to the definition of suitable scientific methodologies for investigating – within the conceptual framework – the subtleties involved in sound and music control under a multimodal and cross-modal perspective. For example, an important topic for control research is gesture analysis of both performers and interacting subjects. Such analysis can be performed at different layers, from the tracking of the positions of given body parts, to the interpretation and classification of gestures in term of expressive, emotional content. Section 6.3 provides an overview of some consolidated scientific methodologies for gesture analysis with particular focus on performing arts (dance and music performers) in a multimodal framework.

Moving from these foundational issues, in Section 6.4 we present some concrete examples of multimodal and cross-modal processing of sound and movement information. Availability of such information enables the development of suitable strategies for controlling and/or generating sound and music output in real-time. Further examples are reported in chapter 7 and Appendix A. Chapter 7 focuses on control of music performance with a particular emphasis on the role of the affective, emotional information. Appendix A deals with control issues related with sound production involving control of both traditional acoustic and digital musical instruments and control of sounding objects.

In Section 6.5 the new version of the EyesWeb XMI open platform for multimodal interaction is briefly introduced as an example of software tool for the design and development of multimodal control strategies. Finally, in Section 6.6 some future research perspectives are discussed.

As a final remark, it is worth noticing that the control of sound generation and processing directly involves artistic choices by the designer of the performance. How many degrees of freedom a designer wishes to leave to an automatic system? In other words, control issues are often intrinsically connected to the role of technology in the artwork and, in a certain way, to the concept of artwork itself.

6.2 A conceptual framework

A relevant foundational aspect for research in sound and music control concerns the definition of a conceptual framework envisaging control at different levels under a multimodal and cross-modal perspective: from low-level analysis of audio signals, toward high-level semantic descriptions including affective, emotional content.

This Section presents a conceptual framework worked out in the EU-IST Project MEGA (2000-2003) that can be considered as a starting point for research on this direction.

Research in the MEGA project moved from the assumption that the physical stimuli that make up an artistic environment contain information about expressiveness that can, to some extent, be extracted and communicated. With multiple modalities (music, video, computer animation) this allows the transmission of expressiveness parameters from one domain to another domain, for example from music to computer animation, or from dance to music. That is, expressive parameters are an example of parameters emerging from modalities and independent from them. In other words, expressive parameters define a cross-modal control space that is at a higher level with respect to the single modalities.

A main question in MEGA research thus relates to the nature of the physical cues that carry expressiveness, and a second question is how to set up cross-modal interchanges (as well as person/machine interchanges) of expressiveness. These questions necessitated the development of a layered conceptual framework for affect processing that splits up the problem into different sub-problems. The conceptual framework aims at clarifying the possible links between physical properties of a particular modality, and the affective/emotive/expressive (AEE) meaning that is typically associated with these properties. Figure 6.1 sketches the conceptual framework in terms of (i) a syntactical layer that stands for the analysis and synthesis of physical properties (bottom), (ii) a semantic layer that contains descriptions of affects, emotions, and expressiveness (top), and (iii) a layer of AEE mappings and spaces that link the syntactical layer with the semantic layer (middle).

The syntactical layer contains different modalities, in particular audio, movement, and animation and arrows point to flows of information. Communication of expressiveness in the cross-modal sense could work in the following way. First, (in the upward direction) physical properties of the musical audio are extracted and the mapping onto an AEE-space allows the description of the affective content in the semantic layer. Starting from this description (in the downward direction), a particular AEE-mapping may be selected that is then used to synthesise physical properties of that affect in another modality, such as animation. This path is followed, for example, when sadness is expressed in a piece of music, and correspondingly an avatar is displaying this sadness in his posture.

6.2.1 Syntactic layer

The syntactic layer is about the extraction of the physical features that are relevant for affect, emotion and expressiveness processing. In the domain of musical audio processing, Lesaffre and colleagues worked out a useful taxonomy of concepts that gives a structured understanding of this layer in terms of a number of justified distinctions (Lesaffre et al., 2003). A distinction is made between low-level, mid-level, and high-level descriptors of musical signals. In this viewpoint, the low-level features are related to very local temporal and spatial characteristics of sound. They deal with the physical categories of frequency, duration, spectrum, intensity, and with the perceptual categories of pitch, time, timbre, and perceived loudness. Low-level features are extracted and processed (in the statistical sense) in order to carry out a subsequent analysis related to expression. For example, in the audio domain, these low-level features are related to tempo (i.e. number of beats per minute), tempo variability, sound level, sound level variability, spectral shape (which is related to the timbre characteristics of the sound), articulation (features such as legato, staccato), articulation variability, attack velocity (which is related to the onset characteristics which can be fast or slow), pitch, pitch density, degree of accent on structural important notes, periodicity, dynamics (intensity), roughness (or sensory dissonance), tonal tension (or the correlation between local pitch patterns and global or contextual pitch patterns), and so on (see also Chapter

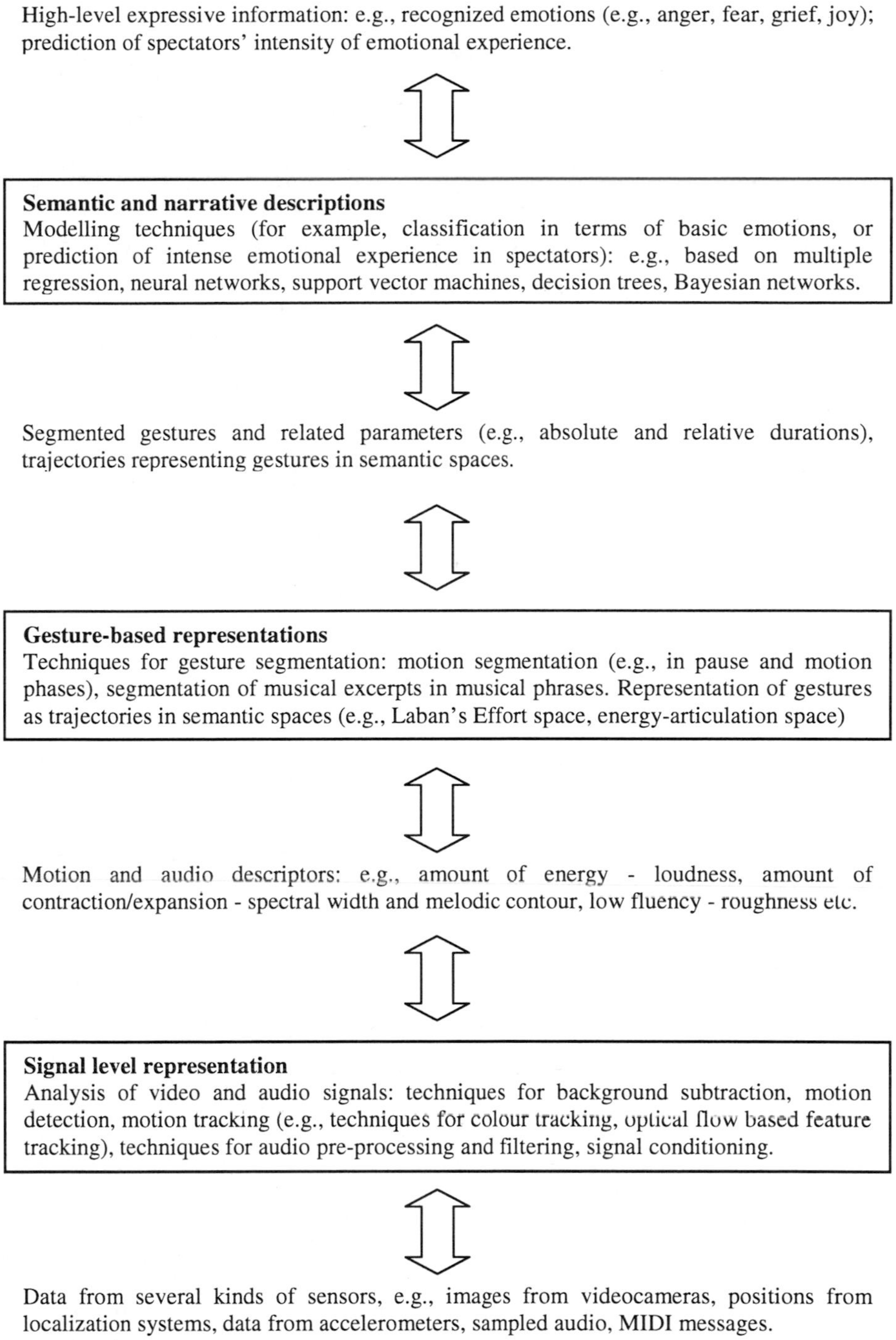

High-level expressive information: e.g., recognized emotions (e.g., anger, fear, grief, joy); prediction of spectators' intensity of emotional experience.

Semantic and narrative descriptions
Modelling techniques (for example, classification in terms of basic emotions, or prediction of intense emotional experience in spectators): e.g., based on multiple regression, neural networks, support vector machines, decision trees, Bayesian networks.

Segmented gestures and related parameters (e.g., absolute and relative durations), trajectories representing gestures in semantic spaces.

Gesture-based representations
Techniques for gesture segmentation: motion segmentation (e.g., in pause and motion phases), segmentation of musical excerpts in musical phrases. Representation of gestures as trajectories in semantic spaces (e.g., Laban's Effort space, energy-articulation space)

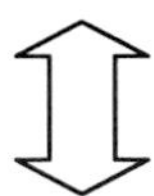

Motion and audio descriptors: e.g., amount of energy - loudness, amount of contraction/expansion - spectral width and melodic contour, low fluency - roughness etc.

Signal level representation
Analysis of video and audio signals: techniques for background subtraction, motion detection, motion tracking (e.g., techniques for colour tracking, optical flow based feature tracking), techniques for audio pre-processing and filtering, signal conditioning.

Data from several kinds of sensors, e.g., images from videocameras, positions from localization systems, data from accelerometers, sampled audio, MIDI messages.

Figure 6.1: The layered conceptual framework makes a distinction between syntax and semantics, and in between, a connection layer that consists of affect / emotion / expressiveness (AEE) spaces and mappings.

3).

When more context information is involved (typically in musical sequences that are longer than 3 seconds), then other categories emerge, in particular, categories related to melody, harmony, rhythm, source, and dynamics. Each of these categories has several distinct specifications, related to an increasing complexity, increasing use of contextual information, and increasing use of top-down knowledge. The highest category is called the expressive category. This layer can in fact be developed into a separate layer because it involves affective, emotive and expressive meanings that cannot be directly extracted from audio structures. Figure 6.1 introduced this layer as a separate layer that is connected with the syntactical cues using a middle layer of mappings and spaces. Examples of mappings and spaces will be given below. Whether all these features are relevant in a context of affect processing and communication of expressiveness is another matter. The experiments discussed in the next sections tried to shed some light on this issue.

In the domain of movement (dance) analysis, a similar approach can be envisaged that leans on a distinction between features calculated on different time scales. In this context also, it makes sense to distinguish between (i) low-level features, calculated on a time interval of a few milliseconds (e.g. one or a few frames coming from a video camera), (ii) mid-level features, calculated on a movement stroke (in the following also referred as "motion phase"), i.e. on time durations of a few seconds, and (iii) high-level features that are related to the conveyed expressive content (but also to cognitive aspects) and referring to sequences of movement strokes or motion (and pause) phases. An example of a low-level feature is the amount of contraction/expansion that can be calculated on just one frame Camurri et al. (2003), i.e. on 40 ms with the common sample rate of 25 fps. Other examples of low-level features are the detected amount of movement, kinematical measures (e.g. velocity and acceleration of body parts), measures related to the occupation of the space surrounding the body. Examples of mid-level descriptors are the overall direction of the movement in the stroke (e.g. upward or downward) or its directness (i.e. how much the movement followed direct paths), motion impulsiveness, and fluency. At this level it is possible to obtain a first segmentation of movement in strokes that can be employed for developing an event-based representation of movement. In fact, strokes or motion phases can be characterised by a beginning, an end, and a collection of descriptors including both mid-level features calculated on the stroke and statistical summaries (e.g. average, standard deviation), performed on the stroke, of low-level features (e.g. average body contraction/expansion during the stroke).

The distinction between low-level, mid-level, and high-level descriptors will be further discussed in Section 6.3 as a possible perspective for gesture analysis.

6.2.2 Semantic layer

The semantic layer is about the experienced meaning of affective, emotive, expressive processing. Apart from aesthetic theories of affect processing in music and in dance, experimental studies were set up that aim at depicting the underlying structure of affect attribution in performing arts (see next sections). Affect semantics in music has been studied by allowing a large number of listeners to use adjectives (either on a completely free basis, or taken from an elaborate list) to specify the affective content of musical excerpts. Afterwards, the data are analyzed and clustered into categories. The early results (Hevner, 1936) showed that listeners tend to use eight different categories of affect attribution. For a recent overview, see the work by Sloboda and Juslin (2001). There seems to be a considerable agreement about two fundamental dimensions of musical affect processing, namely Valence and Activity. Valence is about positively or negatively valued affects, while Activity is about the force of these affects. A third dimension is often noticed, but its meaning is less clearly specified. These results provided the basis for the experiments performed along the project.

6.2.3 Connecting syntax and semantics: Maps and spaces

Different types of maps and spaces can be considered for connecting syntax and semantics. One type is called the semantic map because it relates the meaning of affective/emotive/expressive concepts with physical cues of a certain modality. In the domain of music, for example, several cues have been identified and related to affect processing. For example, tempo is considered to be the most important factor affecting emotional expression in music. Fast tempo is associated with various expressions of activity/excitement, happiness, potency, anger and fear while slow tempo with various expressions of sadness, calmness, solemnity, dignity. Loud music may be determinant for the perception of expressions of intensity, power, anger and joy whereas soft music may be associated with tenderness, sadness, solemnity, and fear. High pitch may be associated with expressions such as happy, graceful, exciting, angry, fearful and active, and low pitch may suggest sadness, dignity, excitement as well as boredom and pleasantness, and so on (see the overviews by Juslin and Laukka, 2003; Gabrielsson and Lindström, 2001). Leman and colleagues show that certain automatically extracted low-level features can be determinants of affect attribution and that maps can be designed that connect audio features with affect/emotion/expression descriptors (Leman et al., 2005). Bresin and Friberg (2000) synthesised music performances starting from a semantic map representing basic emotions. Using qualitative cue descriptions from previous experiments, as listed above, each emotional expression was modelled in terms of a set of rule parameters in a performance rule system. This yielded a fine control of performance parameters relating to performance principles used by musicians such as phrasing and microtiming. A listening experiment was carried out confirming the ability of the synthesised performances to convey the different emotional expressions. Kinaesthetic spaces or energy-velocity spaces are another important type of space. They have been successfully used for the analysis and synthesis of the musical performance (Canazza et al., 2003). This space is derived from factor analysis of perceptual evaluation of different expressive music performances. Listeners tend to use these coordinates as mid level evaluation criteria. The most evident correlation of energy-velocity dimensions with syntactical features is legato-staccato versus tempo. The robustness of this space is confirmed in the synthesis of different and varying expressive intentions in a musical performance. The MIDI parameters typically control tempo and key velocity. The audio-parameters control legato, loudness, brightness, attack time, vibrato, and envelope shape.

In human movement and dance the relationship between syntactical features and affect semantics has been investigated in several studies. For example, in the tradition of the work by Johansson (1973), it has been shown that it is possible for human observers to perceive emotions in dance from point light displays (Walk and Homan, 1984; Dittrich et al., 1996). Pollick et al. (2001) analyzed recognition of emotion in everyday movements (e.g. drinking, knocking) and found significant correlations between motion kinematics (in particular speed) and the activation axis in the two-dimensional space having as axes activation and valence as described by Russell (1980) with respect to his circumplex structure of affect. Wallbott (2001) in his paper dealing with measurement of human expression, after reviewing a collection of works concerning movement features related with expressiveness and techniques to extract them (either manually or automatically), classified these features by considering six different aspects: spatial aspects, temporal aspects, spatio-temporal aspects, aspects related to "force" of a movement, "gestalt" aspects, categorical approaches. Boone and Cunningham (1998), starting from previous studies by Meijer (1989), identified six expressive cues involved in the recognition of the four basic emotions anger, fear, grief, and happiness, and further tested the ability of children in recognizing emotions in expressive body movement through these cues. Such six cues are "frequency of upward arm movement, the duration of time arms were kept close to the body, the amount of muscle tension, the duration of time an individual leaned forward, the number of directional changes in face and torso, and the number of tempo changes an individual made in a given action sequence" (Boone and Cunningham, 1998).

6.3 Methodologies of analysis

The definition of suitable scientific methodologies for investigating – within the conceptual framework and under a multimodal perspective – the subtleties involved in sound and music control is a key issue. An important topic for control research is gesture analysis of both performers and interacting subjects. Gestures are an easy and natural way for controlling sound generation and processing. For these reasons, this section discusses methodologies and approaches focusing on full-body movement and gesture. Nevertheless, the concepts here discussed can be easily generalised to include other modalities. Discovering the key factors that characterise gesture, and in particular expressive gesture, in a general framework is a challenging task. When considering such an unstructured scenario one often has to face the problem of the poor or noisy characterisation of most movements in terms of expressive content. Thus, a common approach consists in starting research from a constrained framework where expressiveness in movement can be exploited to its maximum extent. One such scenario is dance (Camurri et al., 2004c). Another scenario is music performance (Dahl and Friberg, 2004).

6.3.1 Bottom-up approach

Let us consider the dance scenario (consider, however, that what we are going to say also applies to music performance). A possible methodology for designing repeatable experiments is to have a dancer performing a series of dance movements (choreographies) that are distinguished by their expressive content. We use the term "micro-dance" for a short fragment of choreography having a typical duration in the range of 15-90 s. A microdance is conceived as a potential carrier of expressive information, and it is not strongly related to a given emotion (i.e. the choreography has no explicit gestures denoting emotional states). Therefore, different performances of the same micro-dance can convey different expressive or emotional content to spectators: e.g. light/heavy, fluent/rigid, happy/sad, emotional engagement, or evoked emotional strength. Human testers/spectators judge each micro-dance performance. Spectators' ratings are used for evaluation and compared with the output of developed computational models (e.g. for the analysis of expressiveness). Moreover, micro-dances can also be used for testing feature extraction algorithms by comparing the outputs of the algorithms with spectators' ratings of the same micro-dance performance (see for example the work by Camurri et al. (2004b) on spectators' expectation with respect to the motion of the body center of gravity). In case of music performances, we have musical phrases (corresponding to micro-dances above) and the same approach can be applied.

6.3.2 Subtractive approach

Micro-dances can be useful to isolate factors related to expressiveness and to help in providing experimental evidence with respect to the cues that choreographers and psychologists identified. This is obtained by the analysis of differences and invariants in the same micro-dance performed with different expressive intentions. With the same goal, another approach is based on the live observation of genuinely artistic performances, and their corresponding audiovisual recordings. A reference archive of artistic performances has to be carefully defined for this method, chosen after a strict intensive interaction with composers and performers. Image (audio) processing techniques are utilised to gradually subtract information from the recordings. For example, parts of the dancer's body could be progressively hidden until only a set of moving points remain, deforming filters could be applied (e.g. blur), the frame rate could be slowed down, etc. Each time information is reduced, spectators are asked to rate the intensity of their emotional engagement in a scale ranging from negative to positive values (a negative value meaning that the video fragment would rise some negative feeling in the spectator). The transitions between positive and negatives ratings and a zero-rating (i.e. no expressiveness was found by the spectator in the analyzed video sequence) would help to identify what are the movement features carrying expressive information. An intensive interaction is needed between the image processing phase (i.e. the decisions on which information has to be subtracted) and the rating phase. This subtractive approach is different from the already mentioned studies by Johansson (1973) and from more recent results (Cowie et al., 2001) where it is demonstrated that a limited number of visible points on human joints allow an observer to recognise information on movement, including certain emotional content.

6.4 Examples of multimodal and cross-modal analysis

Here we provide some concrete examples of multimodal and cross-modal analysis with reference to the above mentioned conceptual framework. Multimodal and cross-modal analysis can be applied both in a bottom-up approach and in a subtractive approach. In the latter, they are used for extracting and comparing features among subsequent subtraction steps.

6.4.1 Analysis of human full-body movement

A major step in multimodal analysis of human full-body movement is the extraction of a collection of motion descriptors. With respect to the approaches discussed above, such descriptors can be used in the bottom-up approach for characterizing motion (e.g. micro-dances). The top-down approach can be used for validating the descriptors with respect to their role and contribute in conveying expressive content.

With respect to the conceptual framework, at Layer 1 consolidated computer vision techniques (e.g. background subtraction, motion detection, motion tracking) are applied to the incoming video frames. Two kinds of outputs are usually generated: trajectories of points on the dancers' bodies (motion trajectories) and processed images. As an example Figure 6.2 shows the extraction of a Silhouette Motion Image (SMI). A SMI is an image carrying information about variations of the shape and position of the dancer's silhouette in the last few frames. SMIs are inspired by MEI and MHI (Bobick and Davis, 2001). We also use an extension of SMIs taking into account the internal motion in silhouettes.

From such outputs a collection of motion descriptors are extracted including:

- Cues related to the amount of movement (energy) and in particular what we call Quantity of Motion (QoM). QoM is computed as the area (i.e.

Figure 6.2: The SMI is represented as the red area in the picture.

number of pixels) of a SMI. It can be considered as an overall measure of the amount of detected motion, involving velocity and force.

- Cues related to body contraction/expansion and in particular the Contraction Index (CI), conceived as a measure, ranging from 0 to 1, of how the dancer's body uses the space surrounding it. The algorithm to compute the CI (Camurri et al., 2003) combines two different techniques: the individuation of an ellipse approximating the body silhouette and computations based on the bounding region.

- Cues derived from psychological studies (Boone and Cunningham, 1998) such as amount of upward movement, dynamics of the Contraction Index (i.e. how much CI was over a given threshold along a time unit);

- Cues related to the use of space, such as length and overall direction of motion trajectories.

- Kinematical cues, such as velocity and acceleration of motion trajectories.

A relevant task for Layer 2 is motion segmentation. A possible technique for motion segmentation is based on the measured QoM. The evolution in time of the QoM resembles the evolution of velocity of biological motion, which can be roughly described as a sequence of bell-shaped curves (motion bells, see Figure 6.3). In order to segment motion by identifying the component gestures, a list of these motion bells and their features (e.g. peak value and duration) is extracted. An empirical threshold is defined to perform segmentation: the dancer is considered to be moving if the QoM is greater than 2.5% of the total area of the silhouette. It is interesting to notice that the motion bell approach can also be applied to sound signal analysis.

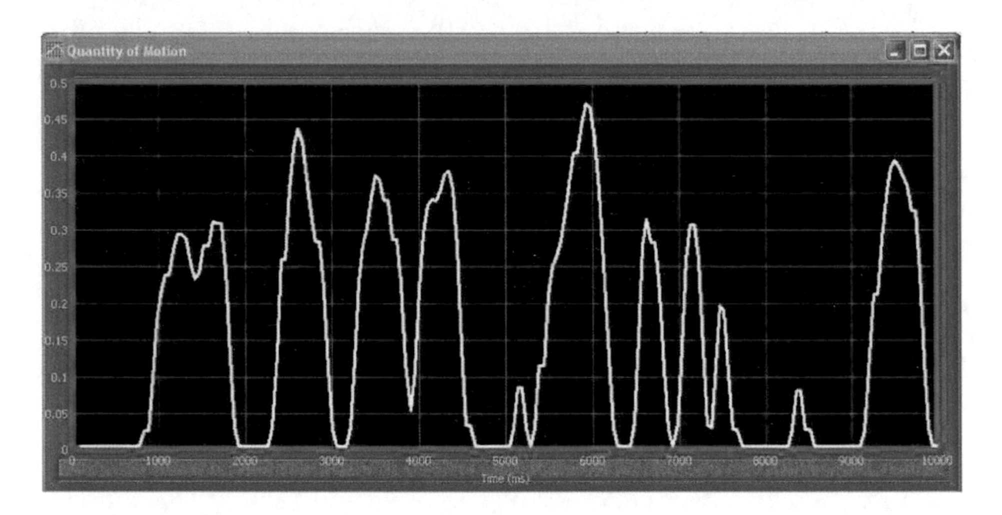

Figure 6.3: Motion bells and motion segmentation (Time on the x axis, QoM on the y axis).

Segmentation allows extracting further higher-level cues at Level 2. A concrete example is the Directness Index (DI), calculated as the ratio between the length of the straight trajectory connecting the first and the last point of a motion trajectory and the sum of the lengths of each segment constituting the trajectory. Furthermore, motion fluency and impulsiveness can be evaluated. Fluency can be estimated from an analysis of the temporal sequence of motion bells. A dance fragment performed with frequent stops and restarts will result less fluent than the same movement performed in a continuous, "harmonic" way. The hesitating, bounded performance will be characterised by a higher percentage of acceleration and deceleration in the time unit (due to the frequent stops and restarts). A first measure of impulsiveness can be obtained from the shape of a motion bell. In fact, since QoM is directly related to the amount of detected movement, a short motion bell having a high pick value will be the result of an impulsive movement (i.e. a movement in which speed rapidly moves from a value near or equal to zero, to a peak and back to zero). On the other hand, a sustained, continuous movement will show a motion bell characterised by a relatively long time period in which the QoM values have little fluctuations around the average value (i.e. the speed is more or less constant during the movement).

One of the tasks of Layer 4 is to classify dances with respect to their emotional/expressive content. For example, in a study carried in the framework of the EU-IST Project MEGA results were obtained on the classification of expressive gestures with respect to their four basic emotions (anger, fear, grief, joy). In an experiment on analysis of dance performances carried out in collaboration with the Department of Psychology of Uppsala University (Sweden), a collection of 20 micro-dances (5 dancers per 4 basic emotions) was rated by subjects and classified by an automatic system based on decision trees. Five decision tree models were trained for classification on five training sets (85% of the available data) and tested on five test sets (15% of the available data). The samples for the training and test sets were randomly extracted from the data set and were uniformly distributed along the four classes and the five dancers. The data set included 18 variables extracted from the dance performances. The outcomes of the experiment show a rate of correct classification for the automatic system (35.6%) in between chance level (25%) and spectators' rate of correct classification (56%) (Camurri et al., 2004c).

6.4.2 Cross-modal analysis of acoustic patterns

An example of cross-modal processing consists in the analysis by means of computer vision techniques of acoustic patterns extracted from an audio signal.

Analysis is performed in EyesWeb (see section 6.5) by means of a library providing the whole auditory processing chain, i.e. cochlear filter banks, hair cell models, and auditory representations including excitation pattern, cochleogram, and correlogram (Camurri et al., 2005). The design of the cochlear filter banks relies on the Matlab Auditory Toolbox (Slaney, 1994). To date, a filter bank configuration can be exported in XML format and loaded into the EyesWeb plugin (see Figure 6.4). For example the cochleogram of a voice sound is depicted in Figure 6.5.

The cochleogram images can be analyzed by image processing techniques to extract information that is not so easily accessible through audio analysis (e.g. activation of particular regions in the image, pattern matching with template images).

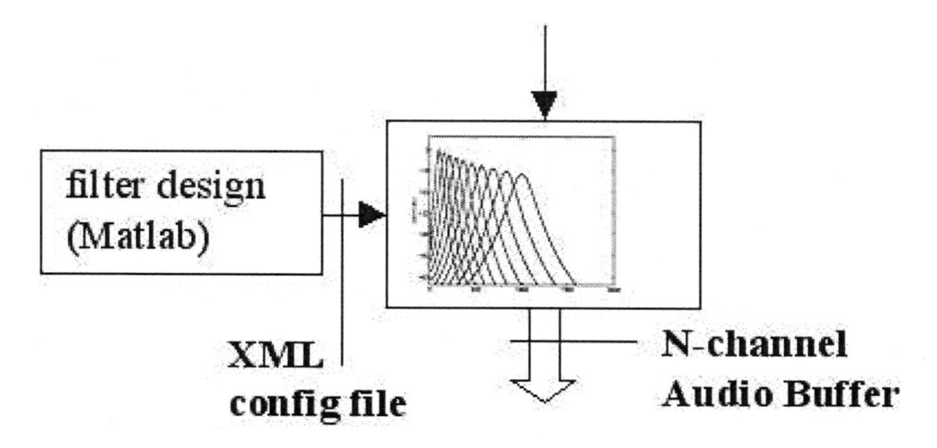

Figure 6.4: Design of the auditory filter bank through the Matlab Auditory Toolbox

In this first example of cross-modal techniques the cochleogram images are analyzed by applying to them the techniques for motion analysis included in the EyesWeb Gesture Processing Library (Camurri et al.). For example, in order to quantify the variation of the cochleogram, i.e. the variance over time of the spectral components in the audio signal, Silhouette Motion Images (SMIs) and Quantity of Motion (QoM) (Camurri et al., 2003) are used. Figure 6.6 shows the SMI of a cochleogram (red shadow). It represents the combined variation of the audio signal over time and frequency in the last 200 ms. The area (i.e. number of pixels) of the SMI (that in motion analysis is usually referred to as Quantity of Motion, i.e. the amount of detected overall motion) summarises such variation of the audio signal. It can be considered as the detected amount of variation of the audio signal both along time and along

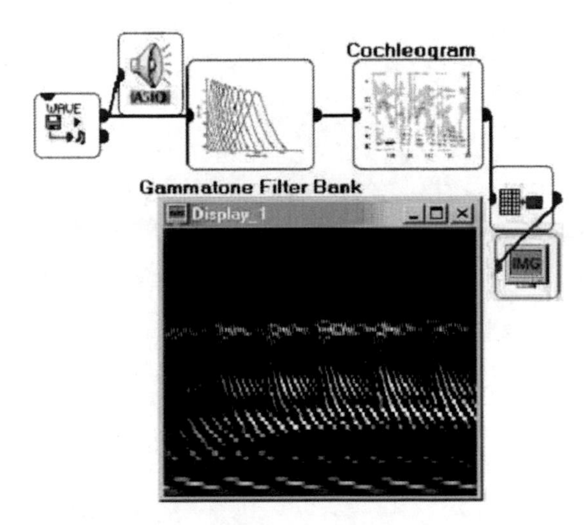

Figure 6.5: Cochleogram of a voice sound obtained through the auditory model blocks

frequency in the time interval over which the corresponding SMI is computed (200 ms in this example).

From a first analysis of the data obtained with this approach it seems that the QoM obtained from the SMIs of the cochleograms can be employed for onset detection especially at the phrase level, i.e. it can be used for detection of phrase boundaries. In speech analysis the same technique can be used for segmenting words. Current research includes performance analysis and comparison with state-of-the-art standard techniques.

6.4.3 Cross-modal processing: auditory-based algorithms for motion analysis

Cross-modal processing applications can also be designed in which the analysis of movement and gestures is inspired by audio analysis algorithms. An example is the patch shown in Figure 6.7, in which a pitch detector is used to measure the frequency of periodic patterns in human gestures: the vertical displacement of a moving hand, measured from the video input signal and rescaled, is converted into the audio domain through an interpolation block, and then analyzed through a pitch detector based on the autocorrelation func-

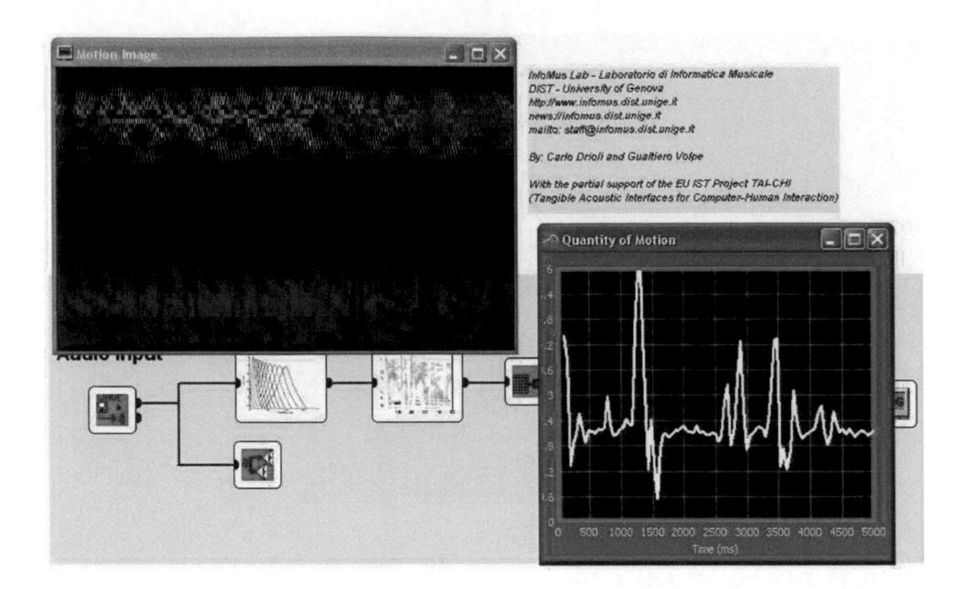

Figure 6.6: SMI of a cochleogram (red shadow) and graph of the corresponding QoM

tion.

Motion-derived signals and audio signals differ in terms of sampling rate and band characteristics. The conversion from a motion-derived signal to one in the audio domain can be performed in principle by upsampling and interpolating the input signal, and a dedicated conversion block is available to perform this operation. If m_{i-1} and m_i are the previous and present input values respectively, and t_i is the initial time of the audio frame in seconds, the audio-rate samples are computed by linear interpolation as

$$s(t_i + \frac{n}{F_s}) = m_{i-1} + n\frac{(m_i - m_{i-1})}{N_s}, n = 1 \ldots N_s$$

where N_s is a selected audio frame length at a given audio sampling rate F_s. However, often sound analysis algorithms are designed to operate in frequency ranges that are much higher if compared to those related to the velocity of body movements. For this reason, the conversion block also provides amplitude modulation (AM) and frequency modulation (FM) functions to shift the original signal band along the frequency axis.

If $c(t) = A_c \cos(2\pi f_c t)$ is a sinusoidal carrier wave with carrier amplitude

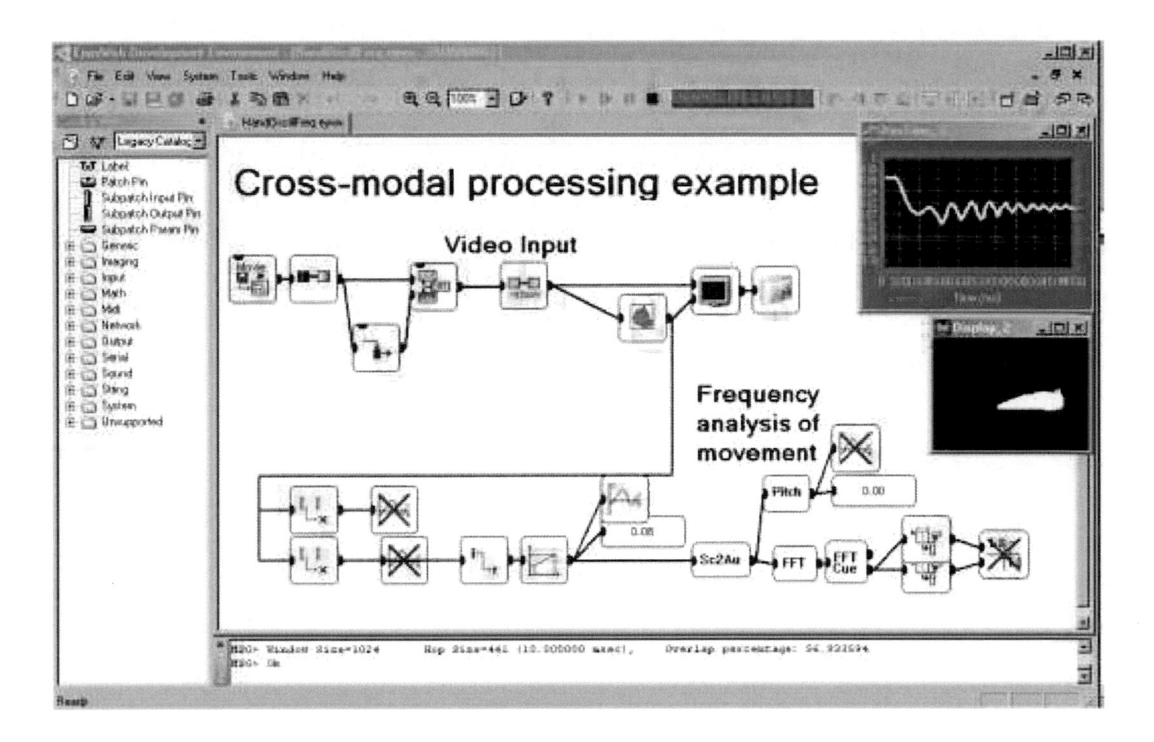

Figure 6.7: An example of EyesWeb application for cross-modal analysis of movement: the hand vertical displacement, measured from the video signal, is converted into the audio domain and analyzed through a pitch detector.

A_c and carrier frequency f_c, an AM audio-rate signal can be computed as

$$s_m(t) = A_c s(t)\cos(2\pi f_c t),$$

and an FM signal as

$$s_m(t) = A_c \cos(2\pi f_c t + 2\pi \int_0^t s(t)dt).$$

The approach to motion analysis by algorithms inspired by acoustic and/or musical cues extraction can be explored further. A possible application is, for example, the control of a digital score reproduction (e.g. a MIDI file) through the detection of tempo, onset, IOI, and other similar musical parameters from the arm and hand movements.

6.4.4 Multimodal analysis of touch gestures

As an example of multimodal analysis of gestural information let us consider an experimental application for the analysis of touch gesture based on Tangible Acoustic Interfaces (TAIs).

Designing and developing TAIs consists of exploring how physical objects, augmented surfaces, and spaces can be transformed into tangible-acoustic embodiments of natural seamless unrestricted interfaces. TAIs can employ physical objects and space as media to bridge the gap between the virtual and physical worlds and to make information accessible through large size touchable objects as well as through ambient media. Research on TAI is carried out for example in the framework of the EU-IST project TAI-CHI (Tangible Acoustic Interfaces for Computer-Human Interaction).

The aim of the sample application here described is twofold: (i) locate where on a TAI the touch gesture takes place, and (ii) analyze how touching is performed (i.e. individuating the expressive qualities of the touching action, such as for example whether the touching action is light and delicate, or heavy and impulsive).

The approach to analysis is multimodal since both the information extracted from the acoustic signal generated by the touching action on the TAI and the information extracted from a video-camera toward the touching position are used.

Localisation is based on two algorithms for in-solid localisation of touching positions developed by the partners in the TAI-CHI project. The first algorithm, developed by the Image and Sound Processing Group at Politecnico di Milano employs 4 sensors and is based on the computation of the Time Delay of Arrival (TDOA) of the acoustical waves to the sensors (Polotti et al., 2005). The second algorithm developed by the Laboratoire Ondes et Acoustique at the Institut pour le Developement de la Science, l'Education et la Technologie, Paris, France, employs just one sensor and is based on pattern matching of the sound patterns generated by the touching action against a collection of stored patterns. In order to increase the reliability of the detected touching position we developed an EyesWeb application integrating the two methods and compensating the possible weakness of one method with the outcomes of the other one.

The position and time of contact information obtained from audio analysis can be employed to trigger and control in a more precise way the video-based gesture analysis process: we are testing hi-speed and hi-res videocameras in EyesWeb XMI in which it is also possible to select the portion of the active ccd area using (x,y) information from a TAI interface.

Video-based analysis (possibly combined with information extracted from the sound generated by the touching action, e.g. the sound level) is then used for extraction of expressive qualities. Gesture analysis is based on hand detection and tracking and builds upon the extraction of information concerning both static and dynamic aspects. As for the static aspects, we developed a collection of EyesWeb modules for real-time classification of hand postures. Classification employs machine learning techniques (namely, Support Vector Machines). As for the dynamic aspects, we used the expressive features currently available in the EyesWeb Expressive Gesture Processing Library (e.g. Quantity of Motion, Contraction/Expansion, Directness Index etc.). Figure 6.8 shows for example the output of an EyesWeb module for the extraction of the hand skeleton.

In other words, while the contact position is detected through an acoustic based localisation system, visual information is employed to get information on how the hand approaches and touches the interface (e.g. with a fluent movement, or in a hesitating way, or in a direct and quick way etc.).

6.5 Tools

6.5.1 The EyesWeb open platform

The EyesWeb open platform has been designed at DIST-InfoMus Lab with a special focus on the multimodal analysis and processing of non-verbal expressive gesture in human movement and music signals (Camurri et al., 2000). Since the starting of the EyesWeb project in 1997, the focus has been on the

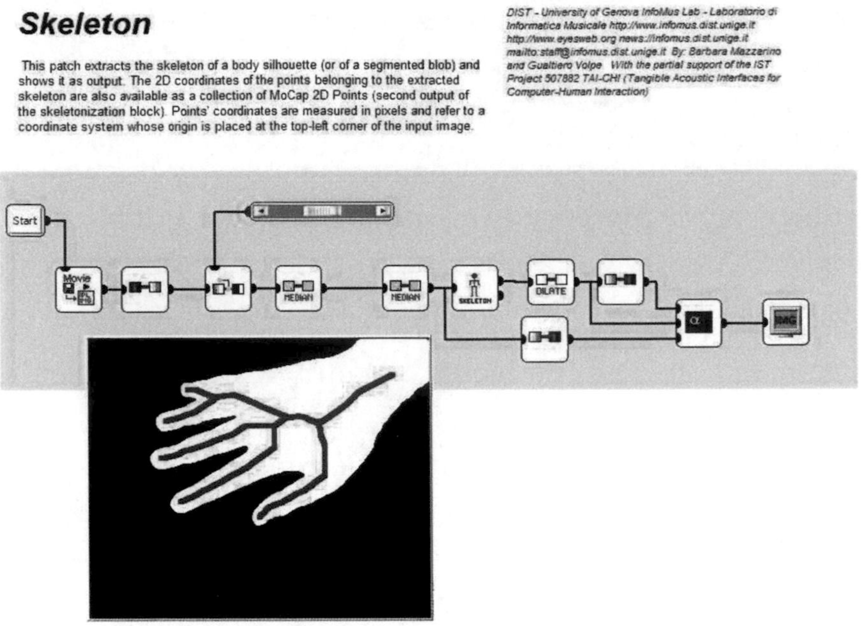

Figure 6.8: An EyesWeb 4 patch extracting the skeleton of a hand touching a TAI

development of a system supporting on the one hand multimodal processing both in its conceptual and technical aspects, and allowing on the other hand fast development of robust application prototypes for use in artistic performances and interactive multimedia installations. In 2001 the platform has been made freely available on the Internet[1] and the number of users has rapidly grown. In recent years, EyesWeb has been satisfactorily used by the DIST-InfoMus Lab both for research purposes and for several kinds of applications, e.g. in museum exhibits and in the field of performing arts. It has also been adopted as standard in several EU funded research projects (in the IST Program: projects MEGA, CARE-HERE, MEDIATE, TAI-CHI) and thousands of users currently employ it in universities, public and private research centers, and companies. Recently, the EyesWeb platform has been reconceived in order to fulfill new requirements coming from the continuously enlarging EyesWeb community. Such process led to the development of another platform (EyesWeb version 4.0) which is completely new with respect to its predecessors in the way it

[1]http://www.eyesweb.org

deals with the conceptual issues involved in multimodal processing, in how it supports and implements multimodality, in the additional features it provides to users (Camurri et al., 2004a). The first beta version of EyesWeb 4.0 was publicly released in September 2004.

The EyesWeb open platform consists of a number of integrated hardware and software modules that can be easily interconnected and extended in a visual environment. The EyesWeb software includes a development environment and a set of libraries of reusable software components that can be assembled by the user in a visual language to build patches as in common computer music languages inspired to analog synthesisers. EyesWeb supports the user in experimenting computational models of non-verbal expressive communication and in mapping, at different levels, gestures from different modalities (e.g. human full-body movement, music) onto real-time generation of multimedia output (sound, music, visual media, mobile scenery). It allows fast development and experiment cycles of interactive performance setups. EyesWeb is a Win32 multi-thread application. At run-time, an original real-time patch scheduler supports several modalities of activation for modules in order to support and optimise management and integration of multimodal streams. A patch is automatically splitted by the scheduler according to its topology and possible synchronization needs. Asynchronous modules having an internal dynamics are also supported. They receive inputs as any other kind of modules but their outputs are asynchronous with respect to their inputs. For example, an "emotional resonator" able to react to the perceived expressive content of a dance performance, embedding an internal dynamics, may have a delay in activating its outputs due to its actual internal state, memory of past events. This is one of the mechanisms explicitly supported by the system to implement interaction metaphors beyond the "musical instrument" and to support interactive narrative structures. It should be noted that usually the user does not have to care about activation mechanisms and scheduling of the modules, since EyesWeb directly manages these aspects. The user is therefore free to take care of higher-level tasks, such as the interactive narrative structure and dynamic evolution of patches in timelines or execution graphs. EyesWeb supports the integrated processing of different streams of (expressive) data, such as music audio, video, and, in general, gestural information.

A set of open libraries of basic modules is available including the following:

- Input and output modules: support for frame grabbers (from webcams to professional frame grabbers), wireless on-body sensors (e.g. accelerometers), live audio input, video and audio players (several different video and audio format supported), OSC (OpenSoundControl), Steinberg ASIO, MIDI, input devices (mouse, keyboard, joystick, data gloves, etc.), audio, video, and numeric output both live and recorded on files (avi, wav, text, etc.).

- Math and filters: modules for basic mathematical operations (both on scalars and matrices), pre-processing, signal conditioning, signal processing in the time and frequency domains.

- Imaging: processing and conversions of images, computer vision techniques, blob extraction and analysis, graphic primitives, support to FreeFrame plug-ins.

- Sound and MIDI libraries: audio processing, extraction of audio features in the time and frequency domains, extraction of features from MIDI, support to VST plug-ins.

- Communication: TCP/IP, serial, OSC, MIDI, Microsoft DCOM.

Users can also build new EyesWeb modules and use them in patches. In order to help programmers in developing blocks, the EyesWeb Wizard software tool has been developed and is available. Users can develop (possibly independently from EyesWeb) the algorithms and the basic software skeletons of their own modules. Then, the Wizard supports them in the process of transforming algorithms in integrated EyesWeb modules. Multiple versions of modules (versioning mechanism) are supported by the system, thus allowing the use in patches of different versions of the same data-type or module. The compatibility with future versions of the systems, in order to preserve the existing work (i.e. modules and patches) in the future is supported.

EyesWeb has been the basic platform of the MEGA EU IST project. In the EU V Framework Program it has also been adopted in the IST CARE HERE (Creating Aesthetically Resonant Environments for the Handicapped, Elderly and Rehabilitation) and IST MEDIATE (A Multisensory Environment Design for an Interface between Autistic and Typical Expressiveness) projects on therapy and rehabilitation and by the MOSART (Music Orchestration Systems in Algorithmic Research and Technology) network for training of young researchers. In the EU VI Framework Program EyesWeb has been adopted and extended to the new version 4.0 in the TAI-CHI project (Tangible Acoustic Interfaces for Computer-Human Interaction). Some partners in the EU Networks of Excellence ENACTIVE (Enactive Interfaces) and HUMAINE (Human-Machine Interaction Network on Emotion) adopted EyesWeb for research. EyesWeb is fully available at its website. Public newsgroups also exist and are daily managed to support the EyesWeb community.

6.5.2 The EyesWeb expressive gesture processing library

Many of the algorithms for extracting the motion descriptors illustrated above have been implemented as software modules for the EyesWeb open platform. Such modules are included in the EyesWeb Expressive Gesture Processing Library.

The EyesWeb Expressive Gesture Processing Library includes a collection of software modules and patches (interconnections of modules) contained in three main sub-libraries:

- The EyesWeb Motion Analysis Library: a collection of modules for real-time motion tracking and extraction of movement cues from human full-body movement. It is based on one or more video cameras and other sensor systems.
- The EyesWeb Space Analysis Library: a collection of modules for analysis of occupation of 2D (real as well as virtual) spaces. If from the one hand this sub-library can be used to extract low-level motion cues (e.g. how much time a given position in the space has been occupied), on the other

hand it can also be used to carry out analyses of gesture in semantic, abstract spaces.

- The EyesWeb Trajectory Analysis Library: a collection of modules for extraction of features from trajectories in 2D (real as well as virtual) spaces. These spaces may again be either physical spaces or semantic and expressive spaces.

The EyesWeb Motion Analysis Library (some parts of this library can be downloaded for research and educational purposes from the EyesWeb website) applies computer vision, statistical, and signal processing techniques to extract expressive motion features (expressive cues) from human full-body movement. At the level of processing of incoming visual inputs the library provides modules including background subtraction techniques for segmenting the body silhouette, techniques for individuating and tracking motion in the images from one or more video cameras, algorithms based on searching for body centroids and on optical flow techniques (e.g. the Lucas and Kanade tracking algorithm), algorithms for segmenting the body silhouette in subregions using spatio-temporal projection patterns, modules for extracting a silhouette's contour and computing its convex hull. At the level of extraction of motion descriptors a collection of parameters is available. They include the above mentioned Quantity of Motion, Contraction Index, Stability Index, Asymmetry Index, Silhouette shape, and direction of body parts. The EyesWeb Motion Analysis Library also includes blocks and patches extracting measures related to the temporal dynamics of movement. A main issue is the segmentation of movement in pause and motion phases. Several movement descriptors can be measured after segmenting motion: for example, blocks are available for calculating durations of pause and motion phases and inter-onset intervals as the time interval between the beginning of two subsequent motion phases.

The EyesWeb Space Analysis Library is based on a model considering a collection of discrete potential functions defined on a 2D space. The space is divided into active cells forming a grid. A point moving in the space is considered and tracked. Three main kinds of potential functions are considered: (i) potential functions not depending on the current position of the tracked point,

(ii) potential functions depending on the current position of the tracked point, (iii) potential functions depending on the definition of regions inside the space. Objects and subjects in the space can be modelled by time-varying potentials. Regions in the space can also be defined. A certain number of "meaningful" regions (i.e. regions on which a particular focus is placed) can be defined and cues can be measured on them (e.g. how much time a tracked subject occupied a given region). The metaphor can be applied both to real spaces (e.g. scenery and actors on a stage, the dancer's General Space as described by Rudolf Laban) and to virtual, semantic, expressive spaces (e.g. a space of parameters where gestures are represented as trajectories). For example, if, from the one hand, the tracked point is a dancer on a stage, a measure of the time duration along which the dancer was in the scope of a given light can be obtained; on the other hand, if the tracked point represents a position in a semantic, expressive space where regions corresponds to basic emotions, the time duration along which a given emotion has been recognised can also be obtained. The EyesWeb Space Analysis Library implements the model and includes blocks allowing the definition of interacting discrete potentials on 2D spaces, the definition of regions, and the extraction of cues (such as, for example, the occupation rates of regions in the space).

The EyesWeb Trajectory Analysis Library contains a collection of blocks and patches for extraction of features from trajectories in 2D (real or virtual) spaces. It complements the EyesWeb Space Analysis Library and it can be used together with the EyesWeb Motion Analysis Library. Blocks can deal with many trajectories at the same time, for example trajectories of body joints (head, hands, and feet tracked by means of color tracking techniques – occlusions are not dealt with at this stage) or trajectories of points tracked using the Lucas and Kanade (1981) feature tracker available in the Motion Analysis Library. Features that can be extracted include geometric and kinematics measures. They include directness index, trajectory length, trajectory local and average direction, velocity, acceleration, and curvature. Descriptive statistic measures can also be computed both along time (for example, average and peak values of features calculated either on running windows or on all of the samples between two subsequent commands such as the average velocity of the hand of a dancer during a given motion phase) and among trajectories (for example, average velocity of groups of trajectories available at the same time such as the average instantaneous velocity of all of the tracked points located on the arm of a dancer). Trajectories can be real trajectories coming from tracking algorithms in the real world (e.g. the trajectory of the head of a dancer tracked using a tracker included in the EyesWeb Motion Analysis Library) or trajectories in virtual, semantic spaces (e.g. a trajectory representing a gesture in a semantic, expressive space).

6.6 Perspectives

Multimodal and cross-modal approaches for integrated analysis of multimedia streams offers an interesting challenge and opens novel perspectives for control of interactive music systems. Moreover, they can be exploited in the broader fields of multimedia content analysis, multimodal interactive systems, innovative natural and expressive interfaces.

This chapter presented a conceptual framework, research methodologies and concrete examples of cross-modal and multimodal techniques for control of interactive music systems. Preliminary results indicate the potential of such approach: cross-modal techniques enable to adapt to the analysis in a given modality approaches originally conceived for another modality, allowing in this way the development of novel and original techniques. Multimodality allows integration of features and use of complementary information, e.g. use of information in a given modality for supplementing lack of information in another modality or for reinforcing the results obtained by analysis in another modality.

While these preliminary results are encouraging, further research is needed for fully exploiting cross-modality and multimodality. For example, an open problem which is currently under investigation at DIST – InfoMus Lab concerns the development of high-level models allowing the definition of cross-modal features. That is, while the examples in this chapter concern cross-modal algorithms, a research challenge consists of identifying a collection of features that, being at a higher-level of abstraction with respect to modal features, are in fact independent of modalities and can be considered cross-modal since they can be extracted from and applied to data coming from different modalities. Such cross-modal features are abstracted from the currently available modal features and define higher-level feature spaces allowing for multimodal mapping of data from one modality to another.

Another, more general, open research issue is how to exploit the information obtained from multimodal and cross-modal techniques for effective control of future interactive music systems. That is, how to define suitable strategies for mapping the information obtained from the analysis of users' behavior (e.g. performer's expressive gestures) onto real-time generation of expressive outputs (e.g. expressive sound and music output). This issue includes the development of mapping strategies integrating both fast adaptive and reactive behavior and more high-level decision-making processes. Current state-of-the-art control strategies often consist of direct associations, without any dynamics, of features of analyzed (expressive) gestures with parameters of synthesised (expressive) gestures (e.g. the actual position of a dancer on the stage may be mapped onto the reproduction of a given sound). Such direct associations are usually employed for implementing statically reactive behavior. The objective is to develop high-level indirect strategies, including reasoning and decision-making processes, and related to rational and cognitive processes. Indirect strategies implement adaptive and dynamic behavior and are usually characterised by a state evolving over time and decisional processes. Production systems and decision-making algorithms may be employed to implement this kind of strategies. Multimodal interactive systems based on a dialogical paradigm may employ indirect strategies only or a suitable mix of direct and indirect strategies.

As a final remark, it should be noticed that control issues in the Sound and Music Computing field are often related to aesthetic, artistic choices. To which extent can a multimodal interactive (music) system make autonomous decisions? That is, does the system have to follow the instructions given by the director, the choreographer, the composer, (in general the creator of a performance or of an installation) or is it allowed to have some degree of freedom in its behavior? The expressive autonomy of a multimodal interactive system is defined as the amount of degrees of freedom that a director, a choreographer, a composer (or in general the designer of an application involving communication of expressive content) leaves to the system in order to make decisions about the most suitable expressive content to convey in a given moment and about the way to convey it. In general, a multimodal interactive system can have different degrees of expressive autonomy and the required degree of expressive autonomy is crucial for the development of its multimodal and cross-modal control strategies.

Bibliography

A.F Bobick and J. Davis. The recognition of human movement using temporal templates. *IEEE Transactions on Pattern Analysis and Machine Intelligence*, 23 (3):257–267, 2001.

R.T. Boone and J.G. Cunningham. Children's decoding of emotion in expressive body movement: The development of cue attunement. *Developmental Psychology*, 34(5):1007–1016, 1998.

R. Bresin and A. Friberg. Emotional coloring of computer-controlled music performances. *Computer Music Journal*, 24(4):44–63, 2000.

A. Camurri, B. Mazzarino, and G. Volpe. Analysis of expressive gesture: The Eyesweb expressive gesture processing library. In A. Camurri and G. Volpe, editors, *Gesture-based Communication in Human-Computer Interaction, LNAI 2915*. Springer Verlag.

A. Camurri, S. Hashimoto, M. Ricchetti, R. Trocca, K. Suzuki, and G. Volpe. EyesWeb: Toward gesture and affect recognition in interactive dance and music systems. *Computer Music Journal*, 24(1):941–952, Spring 2000.

A. Camurri, I. Lagerlöf, and G. Volpe. Recognizing emotion from dance movement: Comparison of spectator recognition and automated techniques. *International Journal of Human-Computer Studies*, 59(1):213–225, July 2003.

A. Camurri, P. Coletta, A. Massari, B. Mazzarino, M. Peri, M. Ricchetti, A. Ricci, and G. Volpe. Toward real-time multimodal processing: Eyesweb 4.0. In

Proc. AISB 2004 Convention: Motion, Emotion and Cognition, pages 22–26, Leeds, UK, March 2004a.

A. Camurri, C. L. Krumhansl, B. Mazzarino, and G. Volpe. An exploratory study of anticipating human movement in dance. In *Proc. 2nd International Symposium on Measurement, Analysis and Modeling of Human Functions*, Genova, Italy, June 2004b.

A. Camurri, B. Mazzarino, M. Ricchetti, R. Timmers, and G. Volpe. Multimodal analysis of expressive gesture in music and dance performances. In A. Camurri and G. Volpe, editors, *Gesture-based Communication in Human-Computer Interaction, LNAI 2915*, pages 20–39. Springer Verlag, February 2004c.

A. Camurri, P. Coletta, C. Drioli, A. Massari, and G. Volpe. Audio processing in a multimodal framework. In *Proc. 118th AES Convention*, Barcelona, Spain, May 2005.

S. Canazza, G. De Poli, A. Rodà, and A. Vidolin. An abstract control space for communication of sensory expressive intentions in music performance. *Journal of New Music Research*, 32(3):281–294, 2003.

R. Cowie, E. Douglas-Cowie, N. Tsapatsoulis, G. Votsis, S. Kollias, W. Fellenz, and J. Taylor. Emotion recognition in human-computer interaction. *IEEE Signal Processing Magazine*, 18(1):32–80, January 2001.

S. Dahl and A. Friberg. Expressiveness of musician's body movements in performances on marimba. In A. Camurri and G. Volpe, editors, *Gesture-based Communication in Human-Computer Interaction, LNAI 2915*. Springer Verlag, February 2004.

W.H. Dittrich, T. Troscianko, S.E.G. Lea, and D. Morgan. Perception of emotion from dynamic point-light displays represented in dance. *Perception*, 25:727–738, 1996.

A. Gabrielsson and E. Lindström. The influence of musical structure on emotion. In P.N. Juslin and J.A. Sloboda, editors, *Music and emotion: Theory and research*. Oxford University Press, 2001.

K. Hevner. Experimental studies of the elements of expression in music. *American Journal of Psychology*, 48:246–268, 1936.

G. Johansson. Visual perception of biological motion and a model for its analysis. *Perception and Psychophysics*, 14:201–211, 1973.

P.N. Juslin and J. Laukka. Communication of emotions in vocal expression and music performance: Different channels, same code? *Psychological Bulletin*, 129(5):770–814, 2003.

M. Leman, V. Vermeulen, L. De Voogdt, and D. Moelants. Prediction of musical affect attribution using a combination of structural cues extracted from musical audio. *Journal of New Music Research*, 34(1):39–67, January 2005.

M. Lesaffre, M. Leman, K. Tanghe, B. De Baets, H. De Meyer, and J.P. Martens. User-dependent taxonomy of musical features as a conceptual framework for musical audio-mining technology. In Roberto Bresin, editor, *Proc. Stockholm Music Acoustics Conference 2003*, pages 635–638, August 2003.

B. Lucas and T. Kanade. An iterative image registration technique with an application to stereo vision. In *Proc. of the International Joint Conference on Artificial Intelligence*, 1981.

M. De Meijer. The contribution of general features of body movement to the attribution of emotions. *Journal of Nonverbal Behavior*, 13:247–268, 1989.

F.E. Pollick, A. Bruderlin, and A.J. Sanford. Perceiving affect from arm movement. *Cognition*, 82:B51–B61, 2001.

P. Polotti, M. Sampietro, A. Sarti, S. Tubaro, and A. Crevoisier. Acoustic localization of tactile interactions for the development of novel tangible interfaces. In *Proc. of the 8th Int. Conference on Digital Audio Effects (DAFX-05)*, September 2005.

J.A. Russell. A circumplex model of affect. *Journal of Personality and Social Psychology*, 39:1161–1178, 1980.

M. Slaney. Auditory toolbox documentation. Technical report, Apple Computers Inc., 1994.

J.A. Sloboda and P.N. Juslin, editors. *Music and Emotion: Theory and Research*, 2001. Oxford University Press.

R.D. Walk and C.P. Homan. Emotion and dance in dynamic light displays. *Bulletin of the Psychonomic Society*, 22:437–440, 1984.

H.G. Wallbott. The measurement of human expressions. In Walbunga von Rallfer-Engel, editor, *Aspects of communications*, pages 203–228. 2001.

Chapter 7

Real-Time Control of Music Performance

Anders Friberg and Roberto Bresin

Department of Speech, Music and Hearing, KTH, Stockholm

About this chapter

In this chapter we will look at the real-time control of music performance on a higher level dealing with semantic/gestural descriptions rather than the control of each note as in a musical instrument. It is similar to the role of the conductor in a traditional orchestra. The conductor controls the overall interpretation of the piece but leaves the execution of the notes to the musicians. A computer-based music performance system typically consists of a human controller using gestures that are tracked and analysed by a computer generating the performance. An alternative could be to use audio input. In this case the system would follow a musician or even computer-generated music.

7.1 Introduction

What do we mean by higher level control? The methods for controlling a music performance can be divided in three different categories: (1) Tempo/dynamics. A simple case is to control the instantaneous values of tempo and dynamics of a performance. (2) Performance models. Using performance models for musical structure, such as the KTH rule system (see also Section 7.2.1), it is possible to control performance details such as how to perform phrasing, articulation, accents and other aspect of a musical performance. (3) Semantic descriptions. These descriptions can be an emotional expression such as aggressive, dreamy, melancholic or typical performance instructions (often referring to motion) such as andante or allegretto. The input gestures/audio can by analysed in different ways roughly similar to the three control categories above. However, the level of detail obtained by using the performance models cannot in the general case be deduced from a gesture/audio input. Therefore, the analysis has to be based on average performance parameters. A short overview of audio analysis including emotion descriptions is found in Section 7.3.1. The analysis of gesture cues is described in Chapter 6.

Several conductor systems using control of tempo and dynamics (thus mostly category 1) have been constructed in the past. The Radio Baton system, designed by Mathews (1989), was one of the first systems and it is still used both for conducting a score as well as a general controller. The Radio Baton controller consists of two sticks (2 radio senders) and a rectangular plate (the receiving antenna). The 3D position of each stick above the plate is measured. Typically one stick is used for beating the time and the other stick is used for controlling dynamics. Using the *Conductor* software, a symbolic score (a converted MIDI file) is played through a MIDI synthesiser. The system is very precise in the sense that the position of each beat is exactly given by the downbeat gesture of the stick. This allows for very accurate control of tempo but also requires practice - even for an experienced conductor! A more recent system controlling both audio and video is the *Personal Orchestra* developed by Borchers et al. (2004) and its further development in *You're the Conductor* (see Lee et al., 2004). These systems are conducted using a wireless baton with infrared light for estimating the baton position in two dimensions. The Personal Orchestra is an installation in House of Music in Vienna, Austria, where the user can conduct real recordings of the Vienna Philharmonic Orchestra. The tempo of both the audio and the video as well as the dynamics of the audio can be controlled yielding a very realistic experience. Due to restrictions in the time manipulation model, tempo is only controlled in discrete steps. The installation *You're the conductor* is also a museum exhibit but aimed for children rather than adults. Therefore it was carefully designed to be intuitive and easily used. This time it is recordings of the Boston Pops orchestra that are conducted. A new time stretching algorithm was developed allowing any temporal changes of the original recording. From the experience with children users they found that the most efficient interface was a simple mapping of gesture speed to tempo and gesture size to volume. Several other conducting systems have been constructed. For example, the Conductor's jacket by Marrin Nakra (2000) senses several body parameters such as muscle tension and respiration that is translated to musical expression. The Virtual Orchestra is a graphical 3D simulation of an orchestra controlled by a baton interface developed by Ilmonen (2000).

A general scheme of a computer-based system for the real-time control of musical performance can be idealised as made by a "controller" and a "mapper". The controller is based on the analysis of audio or gesture input (i.e. the musician gestures). The analysis provides parameters (i.e. speed and size of the movements) which can be mapped into acoustic parameters (i.e. tempo and sound level) responsible for expressive deviations in the musical performance.

In the following we will look more closely at the mapping between expressive control gestures and acoustic cues by using music performance models and semantic descriptions, with special focus on systems which we have been developing at KTH during the years.

7.2 Control in musical performance

7.2.1 Control parameters

Expressive music performance implies the control of a set of acoustical parameters as extensively described in Chapter 5. Once these parameters are identified, it is important to make models which allow their manipulation in a musically and aesthetically meaningful way. One approach to this problem is that provided by the KTH performance rule system. This system is the result of an on-going long-term research project about music performance initiated by Johan Sundberg (e.g. Sundberg et al., 1983; Sundberg, 1993; Friberg, 1991; Friberg and Battel, 2002). The idea of the rule system is to model the variations introduced by the musician when playing a score. The rule system contains currently about 30 rules modelling many performance aspects such as different types of phrasing, accents, timing patterns and intonation (see Table 7.1). Each rule introduces variations in one or several of the performance variables: IOI (Inter-Onset Interval), articulation, tempo, sound level, vibrato rate, vibrato extent as well as modifications of sound level and vibrato envelopes. Most rules operate on the "raw" score using only note values as input. However, some of the rules for phrasing as well as for harmonic, melodic charge need a phrase analysis and a harmonic analysis provided in the score. This means that the rule system does not in general contain analysis models. This is a separate and complicated research issue. One exception is the punctuation rule which includes a melodic grouping analysis (Friberg et al., 1998).

Table 7.1: Most of the rules in Director Musices (Friberg et al., 2000), showing the affected performance variables (sl = sound level, dr = interonset duration, dro = offset to onset duration, va = vibrato amplitude, dc = cent deviation from equal temperament in cents).

MARKING PITCH CONTEXT

Rule Name	**Performance Variables**	**Short Description**
High-loud	sl	The higher the pitch, the louder
Melodic-charge	sl dr va	Emphasis on notes remote from current chord
Harmonic-charge	sl dr	Emphasis on chords remote from current key
Chromatic-charge	dr sl	Emphasis on notes closer in pitch; primarily used for atonal music
Faster-uphill	dr	Decrease duration for notes in uphill motion
Leap-tone-duration	dr	Shorten first note of an up-leap and lengthen first note of a down-leap
Leap-articulation-dro	dro	Micropauses in leaps
Repetition-articulation-dro	dro	Micropauses in tone repetitions

MARKING DURATION AND METER CONTEXT

Rule Name	**Performance Variables**	**Short Description**

Continued on next page

Table 7.1: (continued)

Duration-contrast	dr sl	The longer the note, the longer and louder; and the shorter the note, the shorter and softer
Duration-contrast-art	dro	The shorter the note, the longer the micropause
Score-legato-art	dro	Notes marked legato in scores are played with duration overlapping with interonset duration of next note; resulting onset to offset duration is dr+dro
Score-staccato-art	dro	Notes marked staccato in scores are played with micropause; resulting onset to offset duration is dr-dro
Double-duration	dr	Decrease duration contrast for two notes with duration relation 2:1
Social-duration-care	dr	Increase duration for extremely short notes
Inegales	dr	Long-short patterns of consecutive eighth notes; also called swing eighth notes
Ensemble-swing	dr	Model different timing and swing ratios in an ensemble proportional to tempo
Offbeat-sl	sl	Increase sound level at offbeats
	Intonation	

Continued on next page

Table 7.1: (continued)

Rule Name	**Performance Variables**	**Short Description**
High-sharp	dc	The higher the pitch, the sharper
Mixed-intonation	dc	Ensemble intonation combining both melodic and harmonic intonation
Harmonic-intonation	dc	Beat-free intonation of chords relative to root
Melodic-intonation	dc	Close to Pythagorean tuning, e.g. with sharp leading tones

PHRASING

Rule Name	**Performance Variables**	**Short Description**
Punctuation	dr dro	Automatically locates small tone groups and marks them with lengthening of last note and a following micropause
Phrase-articulation	dro dr	Micropauses after phrase and subphrase boundaries, and lengthening of last note in phrases
Phrase-arch	dr sl	Each phrase performed with arch-like tempo curve: starting slow, faster in middle, and ritardando towards end; sound level is coupled so that slow tempo corresponds to low sound level

Continued on next page

Table 7.1: (continued)

Final-ritard	dr	Ritardando at end of piece, modelled from stopping runners

SYNCHRONISATION

Rule Name	Performance Variables	Short Description
Melodic-sync	dr	Generates new track consisting of all tone onsets in all tracks; at simultaneous onsets, note with maximum melodic charge is selected; all rules applied on this sync track, and resulting durations are transferred back to original tracks
Bar-sync	dr	Synchronise tracks on each bar line

The rules are designed using two methods, (1) the analysis-by-synthesis method, and (2) the analysis-by-measurements method. In the first method, the musical expert, Lars Frydén in the case of the KTH performance rules, tells the scientist how a particular performance principle functions (see 5.3.1). The scientist implements it, e.g. by implementing a function in lisp code. The expert musician tests the new rules by listening to its effect produced on a musical score. Eventually the expert asks the scientist to change or calibrate the functioning of the rule. This process is iterated until the expert is satisfied with the results. An example of a rule obtained by applying the analysis-by-synthesis method is the Duration Contrast rule in which shorter notes are shortened and longer notes are lengthened (Friberg, 1991). The analysis-by-measurements method consists of extracting new rules by analyzing databases of performances (see 5.3.1). For example two databases have been used for the design of the articulation rules. One database consisted in the same piece of music[1] performed by five pianists with nine different expressive intentions. The second database was made by thirteen Mozart piano sonatas performed by a professional pianist. The performances of both databases were all made on computer-monitored grand pianos, a Yamaha Disklavier for the first database, and a Bösendorfer SE for the second one (Bresin and Battel, 2000; Bresin and Widmer, 2000).

For each rule there is one main parameter k which controls the overall rule amount. When $k = 0$ there is no effect of the rule and when $k = 1$ the effect of the rule is considered normal. However, this "normal" value is selected arbitrarily by the researchers and should be used only for the guidance of parameter selection. By making a selection of rules and k values, different performance styles and performer variations can be simulated. Therefore, the rule system should be considered as a musician's toolbox rather than providing a fixed interpretation (see Figure 7.1).

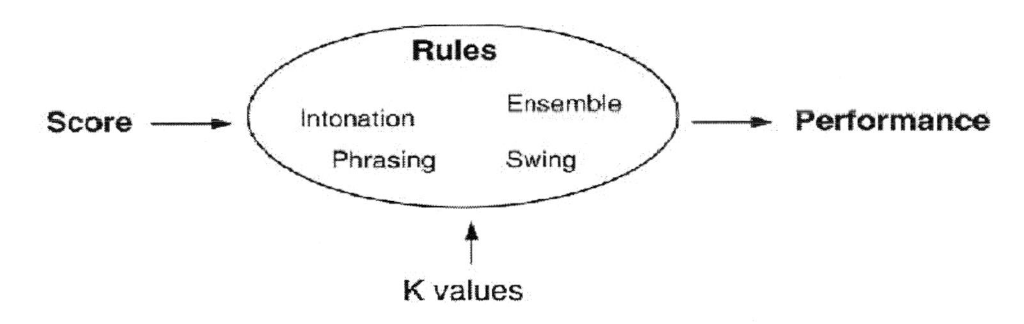

Figure 7.1: Functioning scheme of the KTH performance rule system.

A main feature of the rule system is that most rules are related to the performance of different structural elements in the music (Friberg and Battel, 2002). Thus, for example, the phrasing rules enhance the division in phrases already apparent in the score. This indicates an interesting limitation for the freedom of expressive control: it is not possible to violate the inherent

[1] Andante movement of Mozart's sonata in G major, K 545.

musical structure. One example would be to make ritardandi and accelerandi in the middle of a phrase. From our experience with the rule system such a violation will inevitably not be perceived as musical. However, this toolbox for marking structural elements in the music can also be used for modelling musical expression on the higher semantic level.

Director Musices[2] (DM) is the main implementation of the rule system and is a stand-alone lisp program available for Windows, MacOS, and GNU/Linux documented in (Friberg et al., 2000) and (Bresin et al., 2002).

7.2.2 Mapping: from acoustic cues to high-level descriptors

Emotional expressive music performances can easily be modelled using different selections of KTH rules and their parameters as demonstrated by Bresin and Friberg (2000). Studies in psychology of music have shown that it is possible to communicate different emotional intentions by manipulating the acoustical parameters which characterise a specific musical instrument (Juslin, 2001). For instance in piano performance it is possible to control duration and sound level of each note. In string and blowing instruments it is also possible to control attack time, the vibrato and spectral energy. Table 7.2 shows a possible organisation of rules and their k parameters for obtaining performances with different expressions anger, happiness and sadness.

Table 7.2: Cue profiles for emotions Anger, Happiness and Sadness, as outlined by Juslin (2001), and compared with the rule set-up utilised for the synthesis of expressive performances with Director Musices (DM)

ANGER

Expressive Cue	Juslin	Macro-Rule in DM

Continued on next page

[2] http://www.speech.kth.se/music/performance/download/dm-download.html

Table 7.2: (continued)

Tempo	Fast	Tone IOI is shortened by 20%
Sound level	High	Sound level is increased by 8 dB
	Abrupt tone attacks	Phrase arch rule applied on phrase level and on sub-phrase level
Articulation	Staccato	Duration contrast articulation rule
Time deviations	Sharp duration contrasts	Duration contrast rule
	Small tempo variability	Punctuation rule

HAPPINESS

Expressive Cue	Juslin	Macro-Rule in DM
Tempo	Fast	Tone IOI is shortened by 15%
Sound level	High	Sound level is increased by 3 dB
Articulation	Staccato	Duration contrast articulation rule
	Large articulation variability	Score articulation rules
Time deviations	Sharp duration contrasts	Duration contrast rule
	Small timing variations	Punctuation rule

SADNESS

Continued on next page

Table 7.2: (continued)

Expressive Cue	Juslin	Macro-Rule in DM
Tempo	Slow	Tone IOI is lengthened by 30%
Sound level	Low	Sound level is decreased by 6 dB
Articulation	Legato	Duration contrast articulation rule
Articulation	Small articulation variability	Score legato articulation rule
Time deviations	Soft duration contrasts	Duration contrast rule
	Large timing variations	Phrase arch rule applied on phrase level and sub-phrase level
		Phrase arch rule applied on sub-phrase level
Final ritardando		Obtained from the Phrase rule with the *next* parameter

7.3 Applications

7.3.1 A fuzzy analyser of emotional expression in music and gestures

An overview of the analysis of emotional expression is given in Chapter 5. We will here[3] focus on one of such analysis systems aimed at real time applications. As mentioned, for basic emotions such as happiness, sadness or anger, there is a rather simple relationship between the emotional description and the cue

[3] This section is a modification and shortening of the paper by Friberg (2005)

values (i.e. measured parameters such as tempo, sound level or articulation). Since we are aiming at real-time playing applications we will focus here on performance cues such as tempo and dynamics. The emotional expression in body gestures has also been investigated but to a lesser extent than in music. Camurri et al. (2003) analysed and modelled the emotional expression in dancing. Boone and Cunningham (1998) investigated children's movement patterns when they listened to music with different emotional expressions. Dahl and Friberg (2004) investigated movement patterns of a musician playing a piece with different emotional expressions. These studies all suggested particular movement cues related to the emotional expression, similar to how we decode the musical expression. We follow the idea that musical expression is intimately coupled to expression in body gestures and biological motion in general (see Friberg and Sundberg, 1999; Juslin et al., 2002). Therefore, we try to apply similar analysis approaches to both domains. Table 7.3 presents typical results from previous studies in terms of qualitative descriptions of cue values. As seen in the Table, there are several commonalities in terms of cue descriptions between motion and music performance. For example, anger is characterised by both fast gestures and fast tempo. The research regarding emotional expression yielding the qualitative descriptions as given in Table 7.3 was the starting point for the development of current algorithms.

The first prototype that included an early version of the fuzzy analyser was a system that allowed a dancer to control the music by changing dancing style. It was called The Groove Machine and was presented in a performance at Kulturhuset, Stockholm 2002. Three motion cues were used, QoM, maximum velocity of gestures in the horizontal plane, and the time between gestures in the horizontal plane, thus slightly different from the description above. The emotions analysed were (as in all applications here) anger, happiness, and sadness. The mixing of three corresponding audio loops was directly controlled by the fuzzy analyser output (for a more detailed description see Lindstrom et al., 2005).

Emotion	Motion cues	Music performance cues
Anger	Large	Loud
	Fast	Fast
	Uneven	Staccato
	Jerky	Sharp timbre
Sadness	Small	Soft
	Slow	Slow
	Even soft	Legato
Happiness	Large	Loud
	Rather fast	Fast
		Staccato
		Small tempo variability

Table 7.3: A characterisation of different emotional expressions in terms of cue values for body motion and music performance. Data taken from Dahl and Friberg (2004) and Juslin (2001).

7.3.2 Real-time visualisation of expression in music performance

The ExpressiBall, developed by Roberto Bresin, is a way to visualise a music performance in terms of a ball on a computer screen (Friberg et al., 2002). A microphone is connected to the computer and the output of the fuzzy analyser as well as the basic cue values are used for controlling the appearance of the ball. The position of the ball is controlled by tempo, sound level and a combination of attack velocity and spectral energy, the shape of the ball is controlled by the articulation (rounded-legato, polygon-staccato) and the color of the ball is controlled by the emotion analysis (red-angry, blue-sad, yellow-happy), see Figure 7.2. The choice of color mapping was motivated by recent studies relating color to musical expression (Bresin, 2005). The ExpressiBall can be used as a pedagogical tool for music students or the general public. It may give an enhanced feedback helping to understand the musical expression.

Greta Music is another application for visualizing music expression.

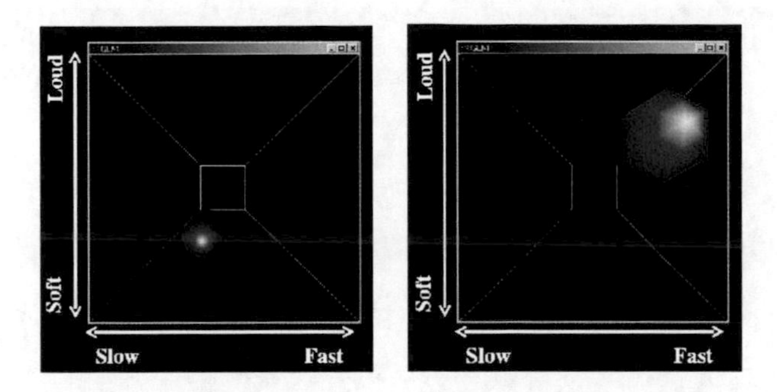

Figure 7.2: Two different examples of the Expressiball giving visual feedback of musical performance. Dimensions used in the interface are: X = tempo, Y = sound pressure level, Z = spectrum (attack time and spectrum energy), Shape = articulation, Colour = emotion. The left figure shows the feedback for a sad performance. The right figure shows the feedback for an angry performance.

In Greta Music the ball metaphor was replaced by the expressive face of the Greta [4] Embodied Conversational Agent (ECA) (Mancini et al., 2007). Here the high-level descriptors, i.e. the emotion labels, are mapped into the emotional expression of the ECA. The values of the extracted acoustical parameters are mapped into movement controls of Greta, e.g. tempo in the musical performance is mapped into the movement speed of Greta, and sound level into the spatial extension of her head movements.

7.3.3 The "Ghost in the Cave" game

Another application that makes use of the fuzzy analyser is the collaborative game Ghost in the Cave (Rinman et al., 2004). It uses as its main input control either body motion or voice. One of the tasks of the game is to express different emotions either with the body or the voice; thus, both modalities are analysed using the fuzzy analyser described above. The game is played in two teams each with a main player, see Figure 7.3. The task for each team is to control a fish avatar in an underwater environment and to go to three different caves. In the caves there is a ghost appearing expressing different emotions. Now the main players have to express the same emotion, causing their fish to

[4]http://www.speech.kth.se/music/projects/gretamusic/

Figure 7.3: Picture from the first realisation of the game Ghost in the Cave. Motion player to the left (in white) and voice player to the right (in front of the microphones).

change accordingly. Points are given for the fastest navigation and the fastest expression of emotions in each subtask. The whole team controls the speed of the fish as well as the music by their motion activity. The body motion and the voice of the main players are measured with a video camera and a microphone, respectively, connected to two computers running two different fuzzy analysers described above. The team motion is estimated by small video cameras (webcams) measuring the Quantity of Motion (QoM). QoM for the team motion was categorised in three levels (high, medium, low) using fuzzy set functions. The music consisted of pre-composed audio sequences, all with the same tempo and key, corresponding to the three motion levels. The sequences were faded in and out directly by control of the fuzzy set functions. One team controlled the drums and one team controlled the accompaniment. The Game has been set up five times since the first realisation at the Stockholm Music Acoustics Conference 2003, including the Stockholm Art and Science festival, Konserthuset, Stockholm, 2004, and Oslo University, 2004.

7.3.4 pDM – Real-time control of the KTH rule system

pDM contains a set of mappers that translate high-level expression descriptions into rule parameters. We have mainly used emotion descriptions (happy, sad, angry, tender) but also other descriptions such as hard, light, heavy or soft have been implemented. The emotion descriptions have the advantages that there has been substantial research made describing the relation between emotions and musical parameters (Sloboda and Juslin, 2001; Bresin and Friberg, 2000). Also, these basic emotions are easily understood by laymen. Typically, these kinds of mappers have to be adapted to the intended application as well as considering the function of the controller being another computer algorithm or a gesture interface. Usually there is a need for interpolation between the descriptions. One option implemented in pDM is to use a 2D plane in which each corner is specified in terms of a set of rule weightings corresponding to a certain description. When moving in the plane the rule weightings are interpolated in a semi-linear fashion. This 2D interface can easily be controlled directly with the mouse. In this way, the well-known Activity-Valence space for describing emotional expression can be implemented (Juslin, 2001). Activity is related to high or low energy and Valence is related to positive or negative emotions. The quadrants of the space can be characterised as happy (high activity, positive valence), angry (high activity, negative valence), tender (low activity, positive valence), and sad (low activity, negative valence). An installation using pDM in which the user can change the emotional expression of the music while it is playing is currently part of the exhibition "Se Hjärnan" (Swedish for "See the Brain") touring Sweden for two years.

7.3.5 A home conducting system

Typically the conductor expresses by gestures overall aspects of the performance and the musician interprets these gestures and fills in the musical details. However, previous conducting systems have often been restricted to the control of tempo and dynamics. This means that the finer details will be static and out of control. An example would be the control of articulation. The articulation is important for setting the gestural and motion quality of

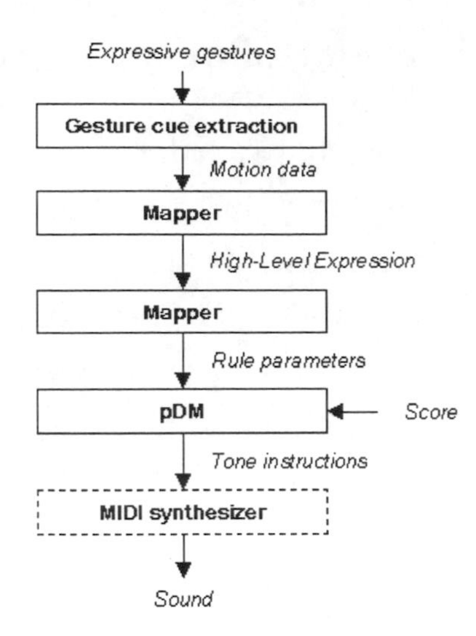

Figure 7.4: Overall schematic view of a home conducting system.

the performance but cannot be applied on an average basis. Amount of articulation (*staccato*) is set on a note-by-note basis dependent on melodic line and grouping, as reported by Bresin and Battel (2000) and Bresin and Widmer (2000). This makes it too difficult for a conductor to control it directly. By using the KTH rule system with pDM described above, these finer details of the performance can be controlled on a higher level without the necessity to shape each individual note. Still the rule system is quite complex with a large number of parameters. Therefore, the important issue when making such a conducting system is the mapping of gesture parameters to music parameters. Tools and models for doing gesture analysis in terms of semantic descriptions of expression have recently been developed (see Chapter 6). Thus, by connecting such a gesture analyser to pDM we have a complete system for controlling the overall expressive features of a score. An overview of the general system is given in Figure 7.4.

Recognition of emotional expression in music has been shown to be an easy task for most listeners including children from about 6 years of age even without any musical training (Peretz, 2001). Therefore, by using simple high-level emotions descriptions such as (happy, sad, angry) the system have the potential of being intuitive and easily understood by most users including children. Thus, we envision a system that can be used by the listeners in their homes rather than a system used for the performers on the stage. Our main design goals have been a system that is (1) easy and fun to use for novices as well as experts, (2) realised on standard equipment using modest computer power. In the following we will describe the system in more detail, starting with the gesture analysis followed by different mapping strategies.

Gesture cue extraction We use a small video camera (webcam) as input device. The video signal is analysed with the EyesWeb tools for gesture recognition (Camurri et al., 2000). The first step is to compute the difference signal between video frames. This is a simple and convenient way of removing all background (static) information in the picture. Thus, there is no need to worry about special lightning, clothes or background content. For simplicity, we have been using a limited set of tools within EyesWeb such as the overall quantity of motion (QoM), x y position of the overall motion, size and velocity of horizontal and vertical gestures.

Mapping gesture cues to rule parameters Depending on the desired application and user ability the mapping strategies can be divided in three categories:

Level 1 (listener level) The musical expression is controlled in terms of basic emotions (happy, sad, angry). This creates an intuitive and simple music feedback comprehensible without the need for any particular musical knowledge.

Level 2 (simple conductor level) Basic overall musical features are controlled using for example the energy-kinematics space previously found relevant for describing the musical expression (Canazza et al., 2003).

Level 3 (advanced conductor level) Overall expressive musical features or emotional expressions in level 1 and 2 are combined with the explicit control of each beat similar to the Radio-Baton system.

Using several interaction levels makes the system suitable both for novices, children and expert users. Contrary to traditional instruments, this system may "sound good" even for a beginner when using a lower interaction level. It can also challenge the user to practice in order to master higher levels similar to the challenge provided in computer games.

Bibliography

R.T. Boone and J.G. Cunningham. Children's decoding of emotion in expressive body movement: The development of cue attunement. *Developmental Psychology*, 34(5):1007–1016, 1998.

J. Borchers, E. Lee, and W. Samminger. Personal orchestra: a real-time audio/video system for interactive conducting. *Multimedia Systems*, 9(5):458–465, 2004.

R. Bresin. What color is that music performance? In *International Computer Music Conference - ICMC 2005*, Barcelona, 2005.

R. Bresin and G. U. Battel. Articulation strategies in expressive piano performance. Analysis of legato, staccato, and repeated notes in performances of the andante movement of Mozart's sonata in G major (K 545). *Journal of New Music Research*, 29(3):211–224, 2000.

R. Bresin and A. Friberg. Emotional coloring of computer-controlled music performances. *Computer Music Journal*, 24(4):44–63, 2000.

R. Bresin and G. Widmer. Production of staccato articulation in Mozart sonatas played on a grand piano. Preliminary results. *TMH-QPSR, Speech Music and Hearing Quarterly Progress and Status Report*, 2000(4):1–6, 2000.

R. Bresin, A. Friberg, and J. Sundberg. Director musices: The KTH performance rules system. In *SIGMUS-46*, pages 43–48, Kyoto, 2002.

A. Camurri, S. Hashimoto, M. Ricchetti, R. Trocca, K. Suzuki, and G. Volpe. EyesWeb: Toward gesture and affect recognition in interactive dance and music systems. *Computer Music Journal*, 24(1):941–952, Spring 2000.

A. Camurri, I. Lagerlöf, and G. Volpe. Recognizing emotion from dance movement: Comparison of spectator recognition and automated techniques. *International Journal of Human-Computer Studies*, 59(1):213–225, July 2003.

S. Canazza, G. De Poli, A. Rodà, and A. Vidolin. An abstract control space for communication of sensory expressive intentions in music performance. *Journal of New Music Research*, 32(3):281–294, 2003.

S. Dahl and A. Friberg. Expressiveness of musician's body movements in performances on marimba. In A. Camurri and G. Volpe, editors, *Gesture-based Communication in Human-Computer Interaction, LNAI 2915*. Springer Verlag, February 2004.

A. Friberg. Generative rules for music performance: A formal description of a rule system. *Computer Music Journal*, 15(2):56–71, 1991.

A. Friberg. A fuzzy analyzer of emotional expression in music performance and body motion. In J. Sundberg and B. Brunson, editors, *Proceedings of Music and Music Science, October 28-30, 2004*, Stockholm: Royal College of Music, 2005.

A. Friberg and G. U. Battel. Structural communication. In R. Parncutt and G. E. McPherson, editors, *The Science and Psychology of Music Performance: Creative Strategies for Teaching and Learning*, pages 199–218. Oxford University Press, New York and Oxford, 2002.

A. Friberg and J. Sundberg. Does music performance allude to locomotion? A model of final ritardandi derived from measurements of stopping runners. *Journal of the Acoustical Society of America*, 105(3):1469–1484, 1999.

A. Friberg, R. Bresin, L. Frydén, and J. Sundberg. Musical punctuation on the microlevel: Automatic identification and performance of small melodic units. *Journal of New Music Research*, 27(3):271–292, 1998.

A. Friberg, V. Colombo, L. Frydén, and J. Sundberg. Generating musical performances with Director Musices. *Computer Music Journal*, 24(3):23–29, 2000.

A. Friberg, E. Schoonderwaldt, P. N. Juslin, and R. Bresin. Automatic real-time extraction of musical expression. In *International Computer Music Conference - ICMC 2002*, pages 365–367, Göteborg, 2002.

T. Ilmonen. The virtual orchestra performance. In *Proceedings of the CHI 2000 Conference on Human Factors in Computing Systems, Haag, Netherlands*, pages 203–204. Springer Verlag, 2000.

P. N. Juslin. Communicating emotion in music performance: A review and a theoretical framework. In P. N. Juslin and J. A. Sloboda, editors, *Music and emotion: Theory and research*, pages 305–333. Oxford University Press, New York, 2001.

P. N. Juslin, A. Friberg, and R. Bresin. Toward a computational model of expression in performance: The GERM model. *Musicae Scientiae*, Special issue 2001-2002:63–122, 2002.

E. Lee, T.M. Nakra, and J. Borchers. You're the conductor: A realistic interactive conducting system for children. In *Proc. of NIME 2004*, pages 68–73, 2004.

E. Lindstrom, A. Camurri, A. Friberg, G. Volpe, and M. L. Rinman. Affect, attitude and evaluation of multisensory performances. *Journal of New Music Research*, 34(1):69Â—86, 2005.

M. Mancini, R. Bresin, and C. Pelachaud. A virtual head driven by music expressivity. *IEEE Transactions on Audio, Speech and Language Processing*, 15 (6):1833–1841, 2007.

T. Marrin Nakra. *Inside the conductor's jacket: analysis, interpretation and musical synthesis of expressive gesture*. PhD thesis, MIT, 2000.

M. V. Mathews. The conductor program and the mechanical baton. In M. Mathews and J. Pierce, editors, *Current Directions in Computer Music Research*, pages 263–282. The MIT Press, Cambridge, Mass, 1989.

I. Peretz. Listen to the brain: a biological perspective on musical emotions. In P. N. Juslin and J. A. Sloboda, editors, *Music and emotion: Theory and research*, pages 105–134. Oxford University Press, New York, 2001.

M.-L. Rinman, A. Friberg, B. Bendiksen, D. Cirotteau, S. Dahl, I. Kjellmo, B. Mazzarino, and A. Camurri. Ghost in the cave - an interactive collaborative game using non-verbal communication. In A. Camurri and G. Volpe, editors, *Gesture-based Communication in Human-Computer Interaction, LNAI 2915*, volume LNAI 2915, pages 549–556, Berlin Heidelberg, 2004. Springer-Verlag.

J.A. Sloboda and P.N. Juslin, editors. *Music and Emotion: Theory and Research*, 2001. Oxford University Press.

J. Sundberg. How can music be expressive? *Speech Communication*, 13:239–253, 1993.

J. Sundberg, A. Askenfelt, and L. Frydén. Musical performance: A synthesis-by-rule approach. *Computer Music Journal*, 7:37–43, 1983.

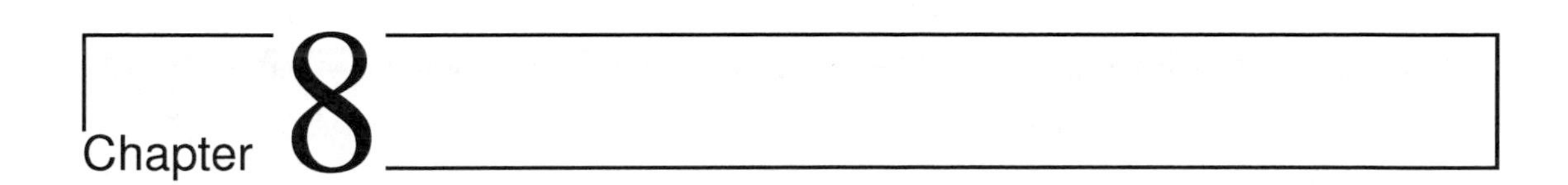

Physics-Based Sound Synthesis

Cumhur Erkut, Vesa Välimäki, Matti Karjalainen, and Henri Penttinen

Laboratory of Acoustics and Audio Signal Processing,
Helsinki University of Technology

About this chapter

This chapter provides the current status and open problems in the field of physics-based sound synthesis. Important concepts and methods of the field are discussed, the state of the art in each technique is presented. The focus is then shifted towards the current directions of the field. The future paths are derived and problems that deserve detailed collaborative research are indicated.

8.1 Introduction

Physics-based sound synthesis focuses on developing efficient digital audio processing algorithms built upon the essential physical behaviour of various sound production mechanisms. The model-based representation of audio can be used in many digital audio applications, including digital sound synthesis, structural analysis of sounds, automatic transcription of musical signals, and parametric audio coding.

Physics-based sound synthesis is currently one of the most active research areas in audio signal processing (Välimäki et al., 2007); many refinements to existing algorithms, as well as several novel techniques are emerging. The aim of this chapter is to provide the current status in physics-based sound synthesis by summarizing various approaches and methodologies within the field, capture the current directions, and indicate open problems that deserve further research. A recent comprehensive review of physics-based sound synthesis methods is given by Välimäki et al. (2006), who also provide pointers to other reviews and tutorials in the field. Our aim is not to duplicate that effort; we rather focus on selective aspects related to each method. Section 8.2 presents background information about these aspects. An important point is that we structurally classify the physics-based sound synthesis methods into two main groups according to the variables used in computation.

In Section 8.3, without going into technical details (the reader is referred to Välimäki et al., 2006, for a detailed discussion of each method), we briefly outline the basics, indicate recent research, and enlist available implementations. We then consider some current directions in physics-based sound synthesis in Section 8.3.3, including the discussion on recent systematic efforts to combine the two structural groups of physics-based sound synthesis.

A unified modular modelling framework, in our opinion, is one of the most important open problems in the field of physics-based sound synthesis. There are, however, other problems, which provide the content of Section 8.4.

8.2 General concepts

A number of physical and signal processing concepts are of paramount importance in physics-based sound synthesis. The background provided in this section is crucial for understanding problem definition in Section 8.2.3, as well as the state of the art and the open problems discussed in the subsequent sections.

8.2.1 Different flavours of modelling tasks

Physical mechanisms are generally complex, and those related to sound production are no exceptions. A useful approach for dealing with complexity is to use a *model*, which typically is based on an abstraction that suppresses the non-essential details of the original problem and allows selective examination with the essential aspects[1]. Yet, an abstraction is task-dependent and it is used for a particular purpose, which in turn determines what is important and what can be left out.

One level of abstraction allows us to derive *mathematical models* (e.g. differential equations) of physical phenomena. Differential equations summarise larger-scale temporal or spatio-temporal relationships of the original phenomena on an infinitesimally small basis. *Musical acoustics*, a branch of physics, relies on simplified mathematical models for a better understanding of the sound production in musical instruments (Fletcher and Rossing, 1998). Similar models are used to study the biological sound sources by Fletcher (1992).

Computational models have been for long a standard tool in various disciplines. At this level, the differential equations of the mathematical models are discretised and solved by computers, one small step at a time. Computational models inherit the abstractions of mathematical models, and add one more level of abstraction by imposing an *algorithm* for solving them (Press et al., 2002). Among many possible choices, *digital signal processing* (DSP) provides

[1]As in Einstein's famous dictum: everything should be made as simple as possible, but no simpler.

an advanced theory and tools that emphasise computational issues, particularly maximal efficiency.

Computational models are the core of physics-based sound synthesis. In addition, physics-based sound synthesis inherits constraints from the task of sound synthesis (Välimäki et al., 2007), i.e. representing huge amount of audio data preferably by a small number of meaningful parameters. Among a wide variety of synthesis and processing techniques, physically-based methods have several advantages with respect to their parameters, control, efficiency, implementation, and sound quality (Jaffe, 1995; Välimäki et al., 2007). These advantages have also been more recently exploited in interactive sound design and sonification (see Hermann and Hunt, 2005; Hermann and Ritter, 2005 and, for a general overview, Chapter 10).

8.2.2 Physical domains, systems, variables, and parameters

Physical phenomena occur in different *physical domains*: string instruments operate in *mechanical*, wind instruments in *acoustical*, and electro-acoustic instruments (such as the analog synthesisers) operate in *electrical* domains. The domains may interact, as in the electro-mechanical Fender Rhodes, or they can be used as *analogies* (equivalent models) of each other. Analogies make unfamiliar phenomena familiar to us. It is therefore not surprising to find many electrical circuits as analogies to describe phenomena of other physical domains in a musical acoustics textbook (Fletcher and Rossing, 1998).

A physical system is a collection of objects united by some form of interaction or interdependence. A mathematical model of a physical system is obtained through rules (typically differential equations) relating measurable quantities that come in pairs of *variables*, such as force and velocity in the mechanical domain, pressure and volume velocity in the acoustical domain, or voltage and current in the electrical domain. If there is a linear relationship between the dual variables, this relation can be expressed as a *parameter*, such as impedance $Z = U/I$ being the ratio of voltage U and current I, or by its inverse, admittance $Y = I/U$. An example from the mechanical domain is mobility (mechanical admittance) as the ratio of velocity and force. When using such parameters, only one of the dual variables is needed explicitly, because the other one is achieved through the constraint rule.

The physics-based sound synthesis methods use two types of variables for computation, *K-variables* and *wave variables* (Välimäki et al., 2006; Rabenstein et al., 2007). K-variables refer to the Kirchhoff continuity rules of dual quantities mentioned above, in contrast to wave components of physical variables. Instead of pairs of K-variables, the wave variables come in pairs of *incident* and *reflected* wave components. The decomposition into wave components is clear in such wave propagation phenomena, where opposite-travelling waves add up to the actual observable K-quantities. A wave quantity is directly observable only when there is no other counterpart. It is, however, a highly useful abstraction to apply wave components to any physical cases, since this helps in solving computability (causality) problems in discrete-time modelling.

8.2.3 Problem definition and schemes

Important concepts in physics-based sound synthesis come into the play in the structure, design, implementation, and execution of the physics-based sound synthesis techniques. These concepts are enlisted in Table 8.1. In general, the properties in the first column are easier to handle compared to those in the second. Thus, the properties in the second column readily point out open research problems. We will elaborate these problems in Section 8.4.

For the purposes of this chapter, the main problem of physics-based sound synthesis is to derive efficient, causal, and explicit computational models for high-quality, natural-sounding synthetic audio, which are optimally balancing accuracy, efficiency, and ease of control in interactive contexts. These models should operate in the widest range of physical domains and handle the nonlinearities and parameter updates in a robust and predictable manner. In this respect, a DSP-based formulation and stability guarantee are desirable features. Port-based formulations and modular schemes have certain advantages when attempting to design a general, unified framework for physics-based sound synthesis. And finally, there is a challenging *inverse problem*: identifica-

Common	Challenge
causal	non-causal
explicit	implicit
lumped	distributed
linear	nonlinear
time-invariant	time-varying
terminals	ports
passive	active
no stability guarantee	stability guarantee
monolithic	modular

Table 8.1: Important concepts in physics-based sound synthesis

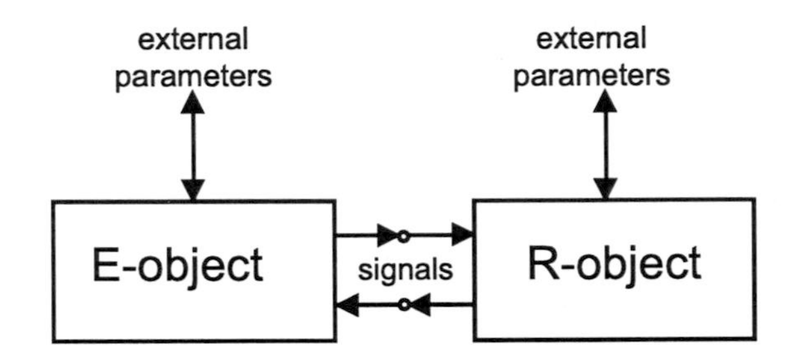

Figure 8.1: Excitation plus resonator paradigm of physics-based sound synthesis.

tion of model parameters starting from sound recordings.

Based on these concepts and the problem definition, two general schemes for physics-based sound synthesis emerge. One way of decomposing a physics-based sound synthesis system is highlighting the functional elements *exciter* and *resonator* (Borin et al., 1992) (abbreviated in Figure 8.1 as E-object and R-object, respectively). In this generic scheme, the exciter and the resonator are connected through ports. The exciter is usually nonlinear, whereas resonator is usually linear, and can be decomposed in sub-models. The interaction between the objects is usually handled implicitly within the system.

Alternatively, a modular system with explicit local interactions is schemat-

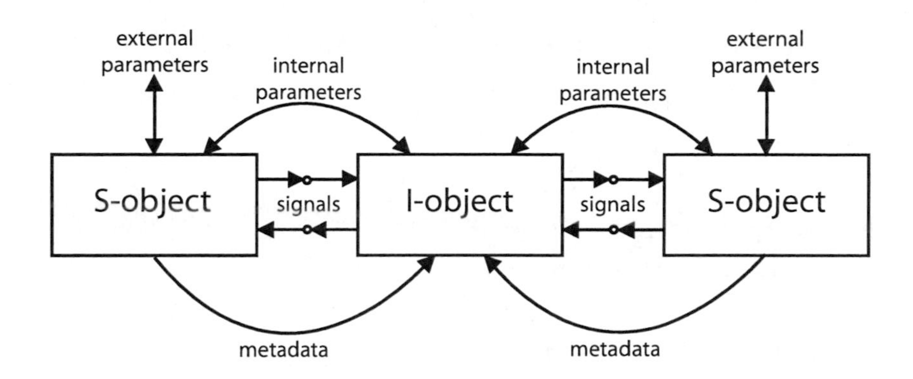

Figure 8.2: Modular interaction diagram.

ically illustrated in Figure 8.2. This scheme was first proposed by Borin et al. (1992), but only recently it is being used for implementing physics-based sound synthesis systems (Rocchesso and Fontana, 2003; Rabenstein et al., 2007).

In Figure 8.2, an *S-object* represents a synthesis module that can correspond to both the exciter and the resonator of Figure 8.1. An *I-object* is an explicit interconnection object (connector); it can generically model different physical interactions (impact, friction, plucking, etc) between the S-objects. Each synthesis module has *internal* and *external* parameters, with a reference of their accessibility from the connector. Internal parameters of a synthesis module (such as port admittances) are used by a connector for distributing the outgoing signals; they are only meaningful if the objects are linked. The external parameters are specific attributes of a synthesis module, such as its resonance characteristics. These attributes can be modified by the user asynchronously. Each I-object may be connected to other I-objects, or accept control information provided by the user or an algorithm. Finally, metadata contains descriptors such as the domain or the type of the synthesis module.

Note that locality implies that only neighbouring synthesis modules are connected to a connector. However, the modular scheme of Figure 8.2 may be considered as a building block for block-based descriptions of sound sources, allowing a constructivist approach to sound modelling (Rabenstein et al., 2007).

A reader who is already familiar with the concepts mentioned so far may want to proceed to Section 8.3, and read how the available methods relate to the general schemes presented here, and what is the current status in each of them. For others, these concepts are explained in the rest of this section.

8.2.4 Important concepts explained

Physical structure and interaction

Physical phenomena are observed as structures and processes in space and time. As a universal property in physics, the interaction of entities in space always propagates with a finite velocity. *Causality* is a fundamental physical property that follows from the finite velocity of interaction from a cause to the corresponding effect. The requirement of causality introduces special computability problems in discrete-time simulation, because two-way interaction with no delay leads to the *delay-free loop problem*. An evident solution is to insert a unit delay into the delay-free loop. However, this arbitrary delay has serious side effects (Borin et al., 2000; Avanzini, 2001). The use of wave variables is advantageous, since the incident and reflected waves have a causal relationship.

Taking the finite propagation speed into account requires using a spatially *distributed* model. Depending on the case at hand, this can be a full three dimensional (3-D) model such as that used for room acoustics, a 2-D model such as for a drum membrane, or a 1-D model such as for a vibrating string. If the object to be modelled behaves homogeneously as a whole, for example due to its small size compared to the wavelength of wave propagation, it can be considered as a *lumped* system that does not need spatial dimensions.

Signals, signal processing, and discrete-time modelling

The word *signal* typically means the value of a measurable or observable quantity as a function of time and possibly as a function of place. In signal processing, signal relationships typically represent one-directional cause-effect chains. Modification of signals can be achieved technically by active electronic components in analog signal processing or by numeric computation in digital signal processing. This simplifies the design of circuits and algorithms compared to two-way interaction that is common in (passive) physical systems, for example in systems where the reciprocity principle is valid. In true physics-based modelling, the two-way interactions must be taken into account. This means that, from the signal processing viewpoint, such models are full of feedback loops, which further implicates that the concepts of computability (causality) and stability become crucial, as will be discussed later.

We favour the discrete-time signal processing approach to physics-based modelling whenever possible. The motivation for this is that digital signal processing is an advanced theory and tool that emphasises computational issues, particularly maximal efficiency. This efficiency is crucial for real-time simulation and sound synthesis. Signal flow diagrams are also a good graphical means to illustrate the algorithms underlying the simulations.

The sampling rate and the spatial sampling resolution need more focus in this context. According to the sampling theorem (Shannon, 1948), signals must be sampled so that at least two samples must be taken per period or wavelength for sinusoidal signal components or their combinations, in order to make the perfect reconstruction of a continuous-time signal possible. This limit frequency, one half of the sampling rate f_s, is called the *Nyquist frequency* or the *Nyquist limit*. If a signal component higher in frequency f_x is sampled by rate f_s, it will be *aliased*, i.e., mirrored by the Nyquist frequency back to the *base band* by $f_a = f_s - f_x$. In audio signals, this will be perceived as very disturbing distortion, and should be avoided. In linear systems, if the inputs are bandlimited properly, the aliasing is not a problem because no new frequency components are created, but in nonlinear systems aliasing is problematic. In modelling physical systems, it is also important to remember that *spatial aliasing* can be a problem if the spatial sampling grid is not dense enough.

Linearity and time invariance

Linearity of a system means that the superposition principle is valid, i.e. quantities and signals in a system behave additively "without disturbing" each other. Mathematically, this is expressed so that if the responses $\{y_1(t), y_2(t)\}$ of the system to two arbitrary input signals $\{x_1(t), x_2(t)\}$, respectively, are $x_1(t) \rightarrow y_1(t)$ and $x_2(t) \rightarrow y_2(t)$, then the response to $Ax_1(t) + Bx_2(t) \rightarrow Ay_1(t) + By_2(t)$ is the same as the sum of the responses to $Ax_1(t)$ and $Bx_2(t)$, for any constants A and B.

A linear system cannot create any signal components with new frequencies. If a system is nonlinear, it typically creates harmonic (integer multiples) or intermodulation (sums and differences) frequency components. This is particularly problematic in discrete-time computation because of the aliasing of new signal frequencies beyond the Nyquist frequency.

If a system is both *linear and time invariant* (LTI), there are constant-valued parameters that effectively characterise its behaviour. We may think that in a time-varying system its characteristics (parameter values) change according to some external influence, while in a nonlinear system the characteristics change according to the signal values in the system.

Linear systems or models have many desirable properties. In digital signal processing, LTI systems are not only easier to design but also are typically more efficient computationally. A linear system can be mapped to transform domains where the behaviour can be analyzed by algebraic equations (Oppenheim et al., 1996). For continuous-time systems, the Laplace and Fourier transforms can be applied to map between the time and frequency domains, and the Sturm-Liouville transform (Trautmann and Rabenstein, 2003) applies similarly to the spatial dimension[2]. For discrete-time systems, the Z-transform and the discrete Fourier transform (DFT and its fast algorithm, FFT) are used.

For nonlinear systems, there is no such elegant theory as for the linear ones; rather, there are many forms of nonlinearity, which require different methods, depending on which effect is desired. In discrete-time modelling,

[2]A technical detail: unlike the Laplace transform, the Sturm-Liouville transform utilises a non-unique kernel that depends on the boundary conditions.

nonlinearities bring problems that are difficult to solve. In addition to aliasing, the delay-free loop problem and stability problems can become worse than they are in linear systems. If the nonlinearities in a system to be modelled are spatially distributed, the modelling task is even more difficult than with a localised nonlinearity.

Energetic behaviour and stability

The product of dual variables such as voltage and current gives power, which, when integrated in time, yields energy. Conservation of energy in a closed system is a fundamental law of physics that should also be obeyed in physics-based modelling.

A physical system can be considered *passive* in the energetic sense if it does not produce energy, i.e. if it preserves its energy or dissipates it into another energy form, such as thermal energy. In musical instruments, the resonators are typically passive, while excitation (plucking, bowing, blowing, etc.) is an *active* process that injects energy to the passive resonators.

The *stability* of a physical system is closely related to its energetic behaviour. Stability can be defined so that the energy of the system remains finite for finite-energy excitations. In this sense, a passive system always remains stable. From the signal processing viewpoint, stability may also be meaningful if it is defined so that the variables, such as voltages, remain within a linear operating range for possible inputs in order to avoid signal clipping and distortion.

Recently, Bilbao (2007) has applied the principles of conservation and dissipation of physical energy to time-domain stability analysis of certain numerical methods that contain, as an added feature, the conservation of an exact numerical counterpart to physical energy that bounds numerical dynamics. His method extends well to strongly nonlinear systems and allows for a convenient analysis of boundary conditions.

For system transfer functions, stability is typically defined so that the system poles (roots of the denominator polynomial) in a Laplace transform remain in the left half plane, or that the poles in a Z-transform in a discrete-time system remain inside the unit circle (Oppenheim et al., 1996). This guarantees that there are no responses growing without bounds for finite excitations.

In signal processing systems with one-directional interaction between stable sub-blocks, an instability can appear only if there are feedback loops. In general, it is impossible to analyze the stability of such a system without knowing its whole feedback structure. Contrary to this, in models with physical two-way interaction the passivity rule is a sufficient condition of stability, i.e. if each element is passive, then any arbitrary network of such elements remains stable.

Modularity and locality of computation

For a computational realisation, it is desirable to decompose a model systematically into blocks and their interconnections. Such an object-oriented approach helps to manage complex models through the use of the modularity principle. The basic modules can be formulated to correspond to elementary objects or functions in the physical domain at hand. Abstractions of new macro-blocks on the basis of more elementary ones helps hiding details when building excessively complex models.

For one-directional interactions in signal processing, it is enough to have input and output terminals for connecting the blocks. For physical interaction, the connections need to be done through ports, with each port having a pair of K- or wave variables depending on the modelling method used. This follows the mathematical principles used for electrical networks (Rabenstein et al., 2007).

Locality of interaction is a desirable modelling feature, which is also related to the concept of causality. In a physical system with a single propagation speed of waves, it is enough that a block interacts only with its nearest neighbours; it does not need global connections to perform its task. If the properties of one block in such a localised model vary, the effect automatically propagates throughout the system. On the other hand, if some effects propagate for example at the speed of light but others with the speed of sound in air, the light waves are practically simultaneously everywhere. If the sampling rate in a discrete-time model is tuned to audio bandwidth (typically 44.1 or 48 kHz sample rate), the unit delay between samples is too long to represent light wave propagation between blocks. Two-way interaction with zero delay means a delay-free loop, the problem that we often face in physics-based sound synthesis. Technically it is possible to realise fractional delays (Laakso et al., 1996), but delays shorter than the unit delay contain a delay-free component, so the problem is hard to avoid. There are ways to make such systems computable (Borin et al., 2000), but they may impose additional constraints in real-time processing in terms of the cost in time (or accuracy).

Types of complexity in physics-based modelling

Models are always just approximations of real physical phenomena. Therefore, they reduce the complexity of the target system. This may be desired for a number of reasons, such as keeping the computational cost manageable, or more generally forcing some cost function below an allowed limit. These constraints are particularly important in real-time sound synthesis and simulation.

The complexity of a model is often the result of the fact that the target system is conceptually over-complex for a scientist or engineer developing the model, and thus cannot be improved by the competence or effort available. An over-complex system may be deterministic and modellable in principle but not in practice: it may be stochastic due to noise-like signal components, or it may be chaotic so that infinitesimally small disturbances lead to unpredictable states.

A particularly important form of complexity is perceptual complexity. For example, in sound synthesis there may be no need to make the model more precise, because listeners cannot hear the difference. Phenomena that are physically prominent but do not have any audible effect can be excluded in such cases.

8.3 State-of-the-art

This section starts with an overview of physics-based *methods and techniques* for modelling and synthesizing musical instruments with an emphasis on the state of the art in each technique. The methods are grouped according to their variables. Wherever possible, we indicate their relation to the concepts and general schemes discussed in Section 8.2; Section 8.3.1 thus focuses on the K-models (finite difference models, mass-spring networks, modal synthesis, and source-filter models) whereas Section 8.3.2 discusses wave models (wave digital filters and digital waveguides).

The second part of this section is devoted to a discussion of the current status of the *field*; we discuss block-based interconnection strategies, modelling and control of musical instruments and everyday sounds, perceptual evaluation of physics-based models, and finally outline emerging applications such as physics-based audio restoration. This part helps us to extrapolate the current trends into the future paths and indicate the open problems of the field in Section 8.4.

Our discussion so far has (indirectly) pointed out many fields related to the physics-based sound synthesis, including physics (especially musical acoustics), mathematics, computer science, electrical engineering, digital signal processing, computer music, perception, human-computer interaction, and control. A novel result in these fields surely affects our field. However in order to emphasise our *methodology*, to keep our focus directed and the size of this chapter manageable, we have excluded these fields in our discussion, with the expense of shifting the balance between sound and sense towards sound. We hope that complementary chapters in this book will altogether provide a more balanced overview.

8.3.1 K-models

Finite difference models

A finite difference scheme is a generic tool for numerically integrating differential equations (Strikwerda, 1989; Bilbao, 2007). In this technique, the mathematical model, which is typically distributed on a bounded spatio-temporal domain, corresponds to the excitation plus resonator paradigm (see Figure 8.1). This mathematical model is discretised with the help of grid functions and difference operators. The numerical model can be *explicit* or *implicit* (in this case, iteration may be needed, see Strikwerda, 1989). In either case, the operations are local. Typically one physical K-variable is directly observable, and the other is hidden in the states of the system.

Early examples of using finite differences in physics-based sound synthesis can be found in the works by Hiller and Ruiz (1971a), Hiller and Ruiz (1971b) and Chaigne (1992). Since then, finite differences have been applied successfully to multi-dimensional structures; Chaigne (2002) systematically extends this line of research.

The finite difference model parameters are typically derived from the physical material properties, although the loss terms in most cases are simplified due to the lack of a general theory. Therefore, some difference schemes may be preferred over the others based on their numerical properties, as done by Bensa et al. (2003), who have showed that a mixed derivative term has superior numerical properties for modelling frequency dependent losses compared to higher-order temporal differences. A promising direction concerning the correlation of perception and the model parameters is the work of McAdams et al. (2004).

Standard DSP tools for analysis and design cannot be facilitated for finite difference models, as they do not follow regular DSP formulations. Recent DSP-oriented (re)formulations attempt to fill this gap (Smith, 2007, 2004a; Pakarinen, 2004; Karjalainen and Erkut, 2004).

The DSP tools aside, *von Neumann* analysis provides a standard technique for investigating the stability of an LTI finite-difference structure (Strikwerda, 1989; Press et al., 2002; Savioja, 1999; Bilbao, 2004). Finite difference schemes have provisions for modelling nonlinear and time-varying systems, but it has been difficult to analyze their stability and passivity. A recent work by Bilbao (2007) is a remedy in this sense.

Although the locality of the finite difference structures have been used for parallel processing in general applications, in sound synthesis a parallel implementation has been rarely addressed, except a parallel hardware implementation of a 2-D plate equation by Motuk et al. (2005). Despite the large number of publications in the field, available sound synthesis software consists of a few Matlab toolboxes that focus on 1-D structures (Kurz and Feiten, 1996; Kurz, 1995; Karjalainen and Erkut, 2004). The DSP-oriented finite difference structures have been implemented in BlockCompiler[3] by Karjalainen (2003) (see also Rabenstein et al., 2007).

Mass-spring networks

This group of techniques (also referred to as *mass-interaction, cellular* or *particle* systems) decompose the original physical system in its structural atoms (Cadoz et al., 1983). These structural atoms are masses, springs, and dash-pots in the mechanical domain. The interactions between the atoms are managed via explicit interconnection elements that handle the transfer of the K-variables between the synthesis objects. By imposing a constraint on the causality of action and reaction, and by using finite-difference formalism, modularity is also achieved (Cadoz et al., 1983). Thus, it is possible to construct complex modular cellular networks that are in full compliance with the diagram in Figure 8.2. Mass-spring systems typically include special interaction objects for implementing time-varying or nonlinear interactions (Florens and Cadoz, 1991). However, the energetic behaviour and stability analysis of the resulting network is hard to estimate, since the existing analysis tools apply only to LTI cases.

The principles of mass-spring networks for physics-based sound synthesis were introduced by Cadoz et al. (1993) within their system CORDIS-

[3]http://www.acoustics.hut.fi/software/BlockCompiler/

ANIMA, which is a comprehensive audio-visual-haptic system. In this paradigm, sound synthesis becomes part of a more general constructive, physics-based approach to multisensory interaction and stimulation. Moreover, by regarding sound synthesis as a part of a broader approach including the micro-structural creation of sound and the macro-temporal construction of music, the same paradigm can be used both for synthesis of sound and composition of music (Cadoz, 2002). The developments, achievements, and results along this line in a large time-span are outlined in a review article by Cadoz et al. (2003).

Despite the successful results, constructing a detailed mass-spring network is still a hard task, since the synthesis objects and their interaction topology require a large number of parameters. To address this issue, Cadoz and his coworkers developed helper systems for support, authoring, analysis, and parameter estimation of mass-spring networks (Castagné and Cadoz, 2002; Cadoz et al., 2003; Szilas and Cadoz, 1998).

A renewed interest (probably due to the intuitiveness of the mass-spring metaphor) in cellular networks resulted in other systems and implementations, which are built upon the basic idea of the modular interactions but placing additional constraints on computation, sound generation, or control. An example implementation is PMPD[4] by Henry (2004).

PMPD closely follows the CORDIS-ANIMA formulation for visualisation of mass-spring networks within the pd-GEM environment (Puckette, 1997), and defines higher-level aggregate geometrical objects such as squares and circles in 2-D or cubes or spheres in 3-D. Although the package only generates slowly-varying control-rate signals, it is a very valuable tool for understanding the basic principles of mass-spring networks and physics-based control. Other systems that are based on similar principles are TAO by Pearson (1995) and CYMATIC by Howard and Rimell (2004).

[4]PMPD has multi-platform support and it is released as a free software under the GNU Public License (GPL). It can be downloaded from `http://drpichon.free.fr/pmpd/`.

Modal synthesis

Linear resonators can also be described in terms of their vibrational modes in the frequency domain. This representation is particularly useful for sound sources that have a small number of relatively sharp resonances (such as the xylophone or the marimba), and may be obtained by experimental modal analysis (Ewins, 2000; Bissinger, 2003).

The modal description, essentially a frequency-domain concept, was successfully applied to discrete-time sound synthesis by Adrien (1989, 1991). In his formulation, the linear resonators (implemented as a parallel filter-bank) are described in terms of their modal characteristics (frequency, damping factor, and mode shape for each mode), whereas connections (representing all non-linear aspects) describe the mode of interaction between objects (e.g. strike, pluck, or bow). These ideas were implemented in MOSAIC software platform (Morrison and Adrien, 1993), which was later ported and extended to Modalys[5]. Note that, although the basic idea of the modal synthesis resembles the excitation plus resonator scheme in Figure 8.1, Modalys and more recent implementations are modular and fully support the bidirectional interaction scheme of Figure 8.2 (usually by iteration).

Modal synthesis is best suited for mechanical domain and uses K-variables. The resonator filterbank is essentially a lumped model, however a matrix block brings back the spatial characteristics of a distributed system by transforming the input force to modal coordinates for weighting the individual resonances. An excellent DSP formulation of modal synthesis, based on the state-space formalism, is given by Smith (2007).

If the modal density of a sound source is high (such as a string instrument body), or if there are many particles contained in a model (such as a maracas) modal synthesis becomes computationally demanding. If the accuracy is not of paramount importance, instead of a detailed bookkeeping of each mode or particle, using stochastic methods significantly reduces the computational cost without sacrificing the perceived sound quality (Cook, 1997). The basic building blocks of modal synthesis, as well as stochastic extensions

[5]Proprietary software of IRCAM, see `http://www.ircam.fr/logiciels.html`

are included in STK (Cook, 2002; Cook and Scavone, 1999)[6].

Two linear modal resonators linked by an interaction element as in Figure 8.2 has been reported by Rocchesso and Fontana (2003). The interaction element simulates nonlinear impact or friction iteratively and provides energy to the modal resonators. These impact and friction models are implemented as pd plugins[7].

The functional transform method (FTM) is a recent development closely related to the modal synthesis (Trautmann and Rabenstein, 2003). In FTM, the modal description of a resonator is obtained directly from the governing PDEs by applying two consecutive integral transforms (Laplace and Sturm-Liouville) to remove the temporal and spatial partial derivatives, respectively. The advantage of this approach is that while traditional modal synthesis parameters are bound to the measured modal patterns of complex resonators, FTM can more densely explore the parameter space, if the problem geometry is simple enough and physical parameters are available. More recently, nonlinear and modular extensions of the method, as well as multirate implementations to reduce the computational load have been reported (Trautmann and Rabenstein, 2004; Petrausch and Rabenstein, 2005).

Source-filter models

When an exciter in Figure 8.1 is represented by a signal generator, a resonator by a time-varying filter, and the bidirectional signal exchange between them is reduced to unidirectional signal flow from the exciter towards the resonator, we obtain a *source-filter model*. In some cases, these reductions can be physically justified, however in general they are mere simplifications, especially when the source is extremely complex, such as in the human voice production (Sundberg, 1991; Kob, 2004; Titze, 2004; Arroabarren and Carlosena, 2004).

Since the signal flow is strictly unidirectional, this technique does not

[6]STK has multiplatform support and it is released as open source without any specific license. It can be downloaded from `http://ccrma.stanford.edu/software/stk/`

[7]Available from `http://www.soundobject.org/software.html`; pd externals are provided under the GPL license.

provide good means for interactions. However, the resonators may be decomposed to arbitrary number of sub-blocks, and outputs of several exciters may be added. Thus, to a certain degree, the modularity is provided. The exciter is usually implemented as a switching wavetable, and resonators are simple time-varying filters. An advantage here is that these components are included in every computer music and audio signal processing platform. Thus, source-filter models can be used as early prototypes of more advanced physical models.

This is, for example, the case in *virtual analog synthesis*. This term became popular when the Nord Lead 1 synthesiser was introduced into the market as "an analog-sounding digital synthesiser that uses no sampled sounds[8]". Instead, a source-filter based technique was used. Stilson and Smith (1996) have introduced more physically-oriented sound synthesis models of analog electric circuits (Stilson, 2006), whereas Välimäki and Huovilainen (2006) report useful oscillator and filter algorithms for virtual analog synthesis.

The final reason of our focus on the source-filter models here is the *commuted synthesis* technique (Smith, 1993; Karjalainen et al., 1993), as it converts a port-based physical model into a source signal (usually an inverse-filtered recorded tone) and a terminal-based filter under the assumptions of linearity and time-invariance. This method works very well for plucked (Laurson et al., 2001; Välimäki et al., 2004) or struck (Välimäki et al., 2003) string instruments, and it is especially advantageous for parameter estimation. These advantages have been utilised by Laurson et al. (2005) in their PWGLSynth for high-quality physics-based sound synthesis and control. Välimäki et al. (2006) and Smith (2007) provide comprehensive discussions on commuted synthesis.

8.3.2 Wave models

Wave digital filters

The *wave digital filter* (WDF) theory is originally formulated for conversion of analog filters into digital filters by Fettweis (1986). In physics-based sound

[8] `http://www.clavia.com/`

synthesis, a physical system is first converted to an equivalent electrical circuit using the domain analogies, then each circuit element is discretised (usually by the bilinear transform). Each object is assigned a port resistance and the energy transfer between objects is carried out by explicit interconnection objects (*adaptors*), which implement Kirchhoff laws and eliminate the delay-free loops. Lumped WDF models are mostly used as exciters, as in the pioneering work of Van Duyne et al. (1994), but are also applicable to resonators. Modern descriptions of the wave digital principle as a tool in physics-based sound synthesis are given by Bilbao (2004) and Välimäki et al. (2006).

The WDF technique is fully compliant with, and actually an inspiration to the modular interaction scheme in Figure 8.2. However, in its original form, its disadvantage is that the class of compatible blocks is restricted to those that communicate via wave variables, i.e. only blocks with wave ports can be connected to the WDF adaptors. Rabenstein et al. (2007) present two different directions to open the wave-based interconnection strategy to K-type models.

Digital waveguides

Digital waveguides (DWGs) formulated by Smith (2007) are the most popular physics-based method for 1-D structures, such as strings and wind instruments. The reason for this is their extreme computational efficiency. They have been also used in generalised networks (Rocchesso and Smith, 2003), as well as in 2-D and 3-D modelling in the form of *digital waveguide mesh* (Van Duyne and Smith, 1993, 1995; Savioja et al., 1995; Murphy et al., 2007).

A DWG is a bi-directional delay line pair with an assigned port admittance Y and it accommodates the wave variables of any physical domain. The change in Y across a junction of the waveguide sections causes *scattering*, and the scattering junctions of interconnected ports have to be formulated. Since DWGs are based on the wave components, this is not a difficult task, as the reflected waves can be causally formulated as a function of incoming waves. DWGs are mostly compatible with wave digital filters, but in order to be compatible with K-modelling techniques, special conversion algorithms must be applied to construct hybrid models. The exciting development of the DWG theory and its applications per instrument family can be found in the works by Smith (2007) and Välimäki et al. (2006); here, after discussing a related technique, we will consider its software implementations.

Essl et al. (2004a) discuss the theory of banded waveguides, a new synthesis method for sounds made by solid objects and an alternative method for treating two- and three-dimensional objects. It is a departure from waveguides in the sense that it divides the excitation signal into frequency bands. Each band contains primarily one resonant mode. For each band, digital waveguides are used to model the dynamics of the travelling wave and the resonant frequency of the mode. Amongst many things, the method enables to create non-linear and highly inharmonic sounds, but it is essentially an efficient technique to model complex resonators with few modes. A companion article by Essl et al. (2004b) presents musical applications of this synthesis method, including bar percussion instruments, the musical saw, bowed glasses, bowed bowls, a bowed cymbal, and a tabla.

As in the case of modal synthesis, the basic synthesis blocks of digital and banded waveguides are included in STK as specific classes. STK also offers prototype waveguide models of many instrument families, but excludes the interactors (or I-objects, as depicted in Figure 8.2). In contrast, both the synthesis objects and the interactors are fully supported in the two software environments presented by Rabenstein et al. (2007), namely the *Binary Connection Tree* (BCT) and *BlockCompiler* (BC). Based on the suitable interconnection strategies between the wave and K-based blocks, BCT and BC allow to build complex physics-based sound synthesis systems without burdening the user with problems of block compatibility. We next discuss these suitable interconnection strategies.

8.3.3 Current directions in physics-based sound synthesis

Block-based interconnection strategies

Using hybrid approaches in sound synthesis to maximise strengths and minimise weaknesses of each technique, has been previously addressed by Jaffe

(1995). It has been pointed out that hybridisation typically shows up after a technique has been around for some time and its characteristics have been extensively explored.

A basic question, with increasing research interest and practical application scenarios, is to understand how different discrete-time modelling paradigms are interrelated and can be combined, whereby K-models and wave models can be understood in the same theoretical framework. Rabenstein et al. (2007) and Murphy et al. (2007) indicate recent results in interconnection of wave and K-blocks by using specific interconnector elements (*KW-converters*), both in the form of theoretical discussions and by examples.

Modelling and control of Western and ethnic instruments

Around the mid-1990s, we reached the point where most Western orchestral instruments could be synthesised based on a physical model, and commercial products were introduced to the market place. Comprehensive summaries of physics-based sound synthesis of orchestral instrument families are provided by Smith (2004b) and Välimäki et al. (2006).

More recently, many authors have focused on acoustical analysis and model-based sound synthesis of ethnic and historical musical instruments. An example is the work of Penttinen et al. (2006), many others (e.g. the Finnish kantele, various lutes, ancient Chinese and African flutes, the Tibetan praying bowl, and the Japanese sho) are listed by Välimäki et al. (2006). Note that this trend is not limited to physics-based sound synthesis methods, as discussed by Whalley (2005) in a special issue devoted to the process of combining digital technology and non-Western musical instruments in the creation of new works.

The work of Penttinen et al. (2006) is also important in the sense that it tackles the joint synthesis-control problem by combining a high-level, computer-assisted composition environment with real-time sound synthesis. Here, the control data is provided by a notation module in a deterministic, expressive, repeatable, and precise fashion (Laurson et al., 2005)[9]. The results were found of excellent quality by listeners in informal settings, which indicates that sound

[9]`http://www.siba.fi/PWGL`

quality, expressiveness, and control are an inseparable whole, a single entity in sound generation and modelling.

Modelling and control of everyday sounds

We have also recently seen applications of physical modelling techniques to non-musical sound sources. A summary of this research line is provided by Rocchesso (2004); many other examples (e.g. bird song, wind chimes, footsteps, and beach balls) are enlisted by Välimäki et al. (2006).

An interesting aspect in this line of research, especially in the works by Cook (1997) and by Rocchesso and Fontana (2003) is the stochastic higher-level control blocks that govern the dynamics of simplistic ("cartoonified") low-level resonator structures.

Perceptual evaluation

Another direction is the subjective evaluation of perceptual features and parameter changes in physics-based synthesis, as witnessed by the works of McAdams et al. (2004), Rocchesso and Scalcon (1999), Lakatos et al. (2000), Järveläinen et al. (2001), Järveläinen and Tolonen (2001) and Järveläinen and Karjalainen (2006).

This line of research provides musically relevant information on the relation of timbre and the properties of human hearing. These results help in reducing the complexity of synthesis models, because details that are inaudible need not be modelled. Besides the gaps in our understanding in acoustics of musical instruments and everyday sounds, these perceptual issues are of paramount importance in sound generation and modelling.

Physics-based audio restoration

The first attempts at audio restoration based on physical models were conducted by Esquef et al. (2002). While this can be successful for single tones, the practical application of such methods for recordings including a mix of several instruments is a challenge for future research. The main problem is high-quality source separation, which is required before this kind of restoration process. Sophisticated algorithms have been devised for this task, but generally speaking, separation of a musical signal into individual source signals is still a difficult research problem (Klapuri, 2003; Virtanen, 2007).

8.4 Open problems and future paths

8.4.1 Sound source models and parameter estimation

There are many musical instruments yet to be studied acoustically. An acoustical study may be combined with the physics-based sound synthesis in order to verify the acoustical characteristics of the instrument in focus. Moreover, there is a vast amount of performance characteristics to be explored. Ideally, these characteristics should be extracted from recordings rather than isolated experiments.

Physical parameter extraction techniques need to be extended. For best sound quality, computational methods that automatically calibrate all the parameter values of a physical model according to the sound of a good instrument should exist. An ideal model should not only be well suited to represent the original process, but it should be flexible enough to offer simple ways to perform the processing. The effectiveness of a model can hence be measured in terms of how accurate the reconstruction of the modelled process is, or in terms of its parametric control properties. To date, physical models of sound and voice have been appreciated for their desirable properties in terms of synthesis, control and expressiveness. However, it is also widely recognised that they suffer from being inadequate to be fitted to real observed data, due to the high number of parameters involved, to the fact that control parameters are not related to the produced sound signal in a trivial way, and to the severe nonlinearities in the numerical schemes in some cases.

All these cues make the parametric identification of physics-based models a formidable problem. In the last decade several approaches to the problem of model inversion have been investigated for the identification of structural and of control parameters. Among these, the use of nonlinear optimisation algorithms; the combination of physical models, sound analysis, and approximation theory; the use of techniques from nonlinear time series analysis and dynamical reconstruction theory. Future research in voice and sound physical modelling should certainly take into account the importance of model fitting to real data, both in terms of system structure design and in terms of parametric identification. Joint design of numerical structures and identification procedures may also be a possible path to complexity reduction. It is also desirable that from the audio-based physical modelling paradigm, new model structures emerge which will be general enough to capture the main sound features of broad families of sounds, e.g. sustained tones from wind and string instruments, percussive sounds, etc., and to be trained to reproduce the peculiarities of a given instrument from recorded data.

Physical virtual analog synthesis is an important future path. Building an analog circuit and comparing the measured physical variables with the synthetic ones may improve the tuning of the virtual analog model parameters, and thus the quality of the audio output. Since many analog electrical, mechanical, and acoustical systems can be decomposed into elementary components, it is desirable to build a library of such components. The theory of wave digital filters (Fettweis, 1986) and block-based physical modelling (Rabenstein et al., 2007) may be facilitated for this purpose.

Important future directions in hybrid modelling include analysis of the dynamic behaviour of parametrically varying hybrid models, as well as benchmark tests for computational costs of the proposed structures.

8.4.2 Directivity and sound radiation modelling

Motivated by the immersive and virtual reality applications (Lokki et al., 2002; Savioja et al., 1999), the directivity and distributed radiation research and modelling are expected to be major challenging problems in physics-based sound synthesis in the next decade.

Being distributed sound sources, musical instruments typically exhibit complex sound radiation patterns (Fletcher and Rossing, 1998). The directional masking and reflections caused by the player and other objects in an environment further shape these patterns (Cook and Trueman, 1998). Furthermore, sound radiation involves diffraction effects that are very perceptible and important. In an offline simulation, Derveaux et al. (2003) have demonstrated the evolution of the directivity pattern in the near field of a classical guitar, especially during the attack. However, these phenomena have been largely ignored in real-time applications due to their high computational cost.

The attempts to efficiently incorporate the directional radiation characteristics of sound sources into *direction-dependent physical modelling* have a long history (Huopaniemi et al., 1994; Karjalainen et al., 1995; Huopaniemi et al., 1995). Karjalainen et al. (1995) report three different methods for retaining the directional radiation information during spatial sound synthesis: 1) directional filtering, 2) a set of elementary sources, and 3) a direction-dependent excitation. The first two methods have been used for creating interactive virtual auditory environments, as reported by Lokki et al. (2002) and Savioja et al. (1999). Nevertheless, a significant part of the effort in virtual acoustic simulations has been devoted to simulating the acoustics of room and concert halls, assuming that the sound is provided, for instance, by a recording (see the overview of related work in the paper by James et al., 2006).

James et al. (2006) recently proposed a novel algorithm for real-time audio-visual synthesis of realistic sound radiation from rigid objects, based on the modal synthesis paradigm. They pre-compute the linear vibration modes of an object and relate each mode to its sound pressure field, or acoustic transfer function, using standard methods from numerical acoustics. Each transfer function is then approximated to a specified accuracy using low-order multi-pole sources placed near the object.

Compared to the geometrically vibration sources that James et al. (2006) consider, musical instruments typically exhibit more structured radiation characteristics. For instance, Hill et al. (2004) show how the input admittance of a classical guitar and its sound-pressure response can be characterised and reconstructed using only a small set of acoustical parameters. We therefore expect improvements, generalisations, and a wide-spread use of direction-dependent physical modelling of musical instruments in the near future.

8.4.3 Control

In addition to the stochastic control blocks (Cook, 1997; Rocchesso and Fontana, 2003), nonlinear dynamics (Strogatz, 1994) could also help to control the simplistic low-level sound source models. This potential is surprisingly under-researched (the most mature applications are mentioned by Cadoz et al., 2003 and Rodet and Vergez, 1999). If carefully designed, the discrete-time nonlinear blocks can successfully modify the characteristics of simplistic synthesis objects in a dynamical fashion. This way, coherent and plausible sonic behaviour (including synchronisation and flocking, Strogatz, 2003) of a large group of animate/inanimate objects may be efficiently modelled.

Going back to less exotic sound sources, the user control (or "playing") of physical models of musical instruments is another problem area where general solutions are unavailable. The piano is one of the easiest cases, because the player only controls the fundamental frequency and dynamic level of tones. In the cases of string and wind instruments, the control issue requires clever technical solutions. The control of virtual musical instruments is currently a lively research field (Paradiso, 1997; Cook, 1992; Howard and Rimell, 2004; Karjalainen et al., 2006). Yet, a remaining challenge is how to make controllability and interactivity central design principles in physics-based sound synthesis.

8.4.4 Applications

An ultimate dream of physical modelling researchers and instrument builders is virtual prototyping of musical instruments. This application will preeminently require physical models with excellent precision in the simulation of sound production. A musical instrument designer should have the possibility to modify a computer model of a musical instrument and then play it to verify that the design is successful. Only after this would the designed instrument be manufactured. Naturally, fine details affecting the timbre of the instrument should be faithfully simulated, since otherwise this chain of events would be fruitless. Current research is still far away from this goal and more research work is required.

The concept of Structured Audio (Vercoe et al., 1998), which is introduced as part of the MPEG-4 international multimedia standard and aims for parametric description and transmission of sound, opens up a new paradigm that, together with other MPEG4 ideas such as the Audio BIFS (Binary Format for Scenes), remains to be explored.

In addition to synthesizing musical sounds, in the future, the physical modelling techniques are expected to be applied to numerous everyday sound sources for human-computer interfaces, computer games, electronic toys, sound effects for films and animations, and virtual reality applications.

Despite a long development history, significant recent advances, and premises such as control, efficiency, and sound quality, the physics-based sound synthesis still lacks a wide-spread use in music as a compositional tool for a composer/user (as opposed to a performance tool), although several case studies have been reported by Chafe (2004). We believe that the most important factor behind this is the lack of a unified modular modelling framework in full compliance with the scheme in Figure 8.2. Such a framework should optimally balance accuracy, efficiency, and ease of control, and operate in the widest range of physical domains. It should also handle the parameter updates in a robust and predictable manner in real-time. Useful tools and metaphors should minimise the time devoted to instrument making and maximise the time devoted to music making. Designing such a framework may require a holistic approach spanning the domain from sound to sense, and bringing the expertise in audio, control, and music together. In this respect, promising results have been obtained, but some challenges still remain (Rabenstein et al., 2007), and there is a vast amount of opportunities for further research and development.

Bibliography

J.-M. Adrien. Dynamic modeling of vibrating structures for sound synthesis, modal synthesis. In *Proc. AES 7th Int. Conf.*, pages 291–300, Toronto, Canada, May 1989.

J.-M. Adrien. The missing link: modal synthesis. In G. De Poli, A. Piccialli, and C. Roads, editors, *Representations of Musical Signals*, pages 269–297. The MIT Press, Cambridge, MA, 1991.

I. Arroabarren and A. Carlosena. Vibrato in singing voice: The link between source-filter and sinusoidal models. *EURASIP Journal on Applied Signal Processing*, 2004(7):1007–1020, July 2004. Special issue on model-based sound synthesis.

F. Avanzini. *Computational Issues in Physically-based Sound Models*. PhD thesis, Dept. of Computer Science and Electronics, University of Padova, Italy, 2001.

J. Bensa, S. Bilbao, R. Kronland-Martinet, and J. O. Smith. The simulation of piano string vibration: From physical models to finite difference schemes and digital waveguides. *J. Acoust. Soc. Am.*, 114(2):1095–1107, Aug. 2003.

S. Bilbao. *Wave and Scattering Methods for Numerical Simulation*. John Wiley and Sons, Chichester, UK, 2004. ISBN 0-470-87017-6.

S. Bilbao. Robust physical modeling sound synthesis for nonlinear systems. *Signal Processing Magazine, IEEE*, 24(2):32–41, 2007.

G. Bissinger. Modal analysis of a violin octet. *J. Acoust. Soc. Am.*, 113(4): 2105–2113, Apr. 2003.

G. Borin, G. De Poli, and A. Sarti. Algorithms and structures for synthesis using physical models. *Computer Music J.*, 16(4):30–42, 1992.

G. Borin, G. De Poli, and D. Rocchesso. Elimination of delay-free loops in discrete-time models of nonlinear acoustic systems. *IEEE Trans. Speech and Audio Processing*, 8(5):597–605, Sep. 2000.

C. Cadoz. The physical model as a metaphor for musical creation "pico..TERA", a piece entirely generated by physical model. In *Proc. Int. Computer Music Conf.*, pages 305–312, Gothenburg, Sweden, 2002.

C. Cadoz, A. Luciani, and J.-L. Florens. Responsive input devices and sound synthesis by simulation of instrumental mechanisms: The CORDIS system. *Computer Music J.*, 8(3):60–73, 1983.

C. Cadoz, A. Luciani, and J.-L. Florens. CORDIS-ANIMA. a modeling and simulation system for sound and image synthesis. The general formalism. *Computer Music J.*, 17(1):19–29, Spring 1993.

C. Cadoz, A. Luciani, and J.-L. Florens. Artistic creation and computer interactive multisensory simulation force feedback gesture transducers. In *Proc. Conf. New Interfaces for Musical Expression NIME*, pages 235–246, Montreal, Canada, May 2003.

N. Castagné and C. Cadoz. Creating music by means of "physical thinking": The musician oriented Genesis environment. In *Proc. COST-G6 Conf. Digital Audio Effects*, pages 169–174, Hamburg, Germany, Sep. 2002.

C. Chafe. Case studies of physical models in music composition. In *Proc. 18th International Congress on Acoustics (ICA)*, pages 297 – 300, Kyoto, Japan, April 2004.

A. Chaigne. Numerical simulations of stringed instruments – today's situation and trends for the future. *Catgut Acoustical Society Journal*, 4(5):12–20, May 2002.

A. Chaigne. On the use of finite differences for musical synthesis. Application to plucked stringed instruments. *J. Acoustique*, 5(2):181–211, Apr. 1992.

P. Cook and D. Trueman. A database of measured musical instrument body radiation impulse responses, and computer applications for exploring and utilizing the measured filter functions. In *Proc. Intl. Symp. Musical Acoustics*, pages 303–308, Leavenworth, WA, USA, 1998.

P. R. Cook. *Real sound synthesis for interactive applications*. A. K. Peters, Natick, MA, USA, 2002.

P. R. Cook. A meta-wind-instrument physical model, and a meta-controller for real-time performance control. In *Proc. Int. Computer Music Conf.*, pages 273–276, San Jose, California, October 14-18 1992.

P. R. Cook. Physically informed sonic modeling (PhISM): Synthesis of percussive sounds. *Computer Music J.*, 21(3):38–49, 1997.

P. R. Cook and G. P. Scavone. The Synthesis ToolKit STK. In *Proc. Int. Computer Music Conf.*, Beijing, China, Oct. 1999. For an updated version of STK see `http://ccrma-www.stanford.edu/software/stk/`.

G. Derveaux, A. Chaigne, P. Joly, and E. Bécache. Time-domain simulation of a guitar: Model and method. *J. Acoust. Soc. Am.*, 114:3368–3383, 2003.

P. Esquef, V. Välimäki, and M. Karjalainen. Restoration and enhancement of solo guitar recordings based on sound source modeling. *J. Audio Eng. Soc.*, 50(4):227–236, Apr. 2002.

G. Essl, S. Serafin, P. R. Cook, and J. O. Smith. Theory of banded waveguides. *Computer Music J.*, 28(1):37–50, 2004a.

G. Essl, S. Serafin, P. R. Cook, and J. O. Smith. Musical applications of banded waveguides. *Computer Music J.*, 28(1):51–63, 2004b.

D. J. Ewins. *Modal Testing: Theory, Practice and Application*. Taylor & Francis Group, second edition, 2000.

A. Fettweis. Wave digital filters: Theory and practice. *Proc. IEEE*, 74(2):270–327, Feb. 1986.

N. H. Fletcher. *Acoustic systems in biology*. Oxford University Press, New York, USA, 1992.

N. H. Fletcher and T. D. Rossing. *The Physics of Musical Instruments*. Springer-Verlag, New York, NY, USA, 2nd edition, 1998.

J.-L. Florens and C. Cadoz. The physical model: Modeling and simulating the instrumental universe. In G. De Poli, A. Piccialli, and C. Roads, editors, *Representations of Musical Signals*, pages 227–268. The MIT Press, Cambridge, Massachusetts, 1991.

C. Henry. PMPD: Physical modelling for pure data. In *Proc. Int. Computer Music Conf.*, Coral Gables, Florida, USA, Nov. 2004.

T. Hermann and A. Hunt. Guest editors' introduction: An introduction to interactive sonification. *Multimedia, IEEE*, 12(2):20–24, 2005.

T. Hermann and H. Ritter. Model-based sonification revisited – authors' comments on Hermann and Ritter, ICAD 2002. *ACM Trans. Appl. Percept.*, 2(4): 559–563, October 2005.

T. J. W. Hill, B. E. Richardson, and S. J. Richardson. Acoustical parameters for the characterisation of the classical guitar. *Acta Acustica united with Acustica*, 90(2):335–348, 2004.

L. Hiller and P. Ruiz. Synthesizing musical sounds by solving the wave equation for vibrating objects: Part 1. *J. Audio Eng. Soc.*, 19(6):462–470, June 1971a.

L. Hiller and P. Ruiz. Synthesizing musical sounds by solving the wave equation for vibrating objects: Part 2. *J. Audio Eng. Soc.*, 19(7):542–551, 1971b.

D. M. Howard and S. Rimell. Real-time gesture-controlled physical modelling music synthesis with tactile feedback. *EURASIP Journal on Applied Signal Processing*, 2004(7):1001–1006, July 2004. Special issue on model-based sound synthesis.

J. Huopaniemi, M. Karjalainen, V. Välimäki, and T. Huotilainen. Virtual instruments in virtual rooms–a real-time binaural room simulation environment for physical models of musical instruments. In *Proc. Intl. Computer Music Conf.*, pages 455–462, Aarhus, Denmark, 1994.

J. Huopaniemi, M. Karjalainen, and V. Välimäki. Physical models of musical instruments in real-time binaural room simulation. In *Proc. Int. Congr. Acoustics (ICA'95)*, volume 3, pages 447–450, Trondheim, Norway, 1995.

D. A. Jaffe. Ten criteria for evaluating synthesis techniques. *Computer Music J.*, 19(1):76–87, 1995.

D. L. James, J. Barbic, and D. K. Pai. Precomputed acoustic transfer: output-sensitive, accurate sound generation for geometrically complex vibration sources. In *Proc. ACM SIGGRAPH '06*, pages 987–995, New York, NY, USA, 2006.

H. Järveläinen and T. Tolonen. Perceptual tolerances for decay parameters in plucked string synthesis. *J. Audio Eng. Soc.*, 49(11):1049–1059, Nov. 2001.

H. Järveläinen, V. Välimäki, and M. Karjalainen. Audibility of the timbral effects of inharmonicity in stringed instrument tones. *Acoustics Research Letters Online*, 2(3):79–84, July 2001.

H. Järveläinen and M. Karjalainen. Perceptibility of inharmonicity in the acoustic guitar. *Acta Acustica united with Acustica*, 92(5):842–847, October 2006.

M. Karjalainen. BlockCompiler: A research tool for physical modeling and DSP. In *Proc. COST-G6 Conf. Digital Audio Effects*, pages 264–269, London, UK, Sep. 2003.

M. Karjalainen and C. Erkut. Digital waveguides versus finite difference structures: Equivalence and mixed modeling. *EURASIP Journal on Applied Signal Processing*, 2004(7):978–989, July 2004.

M. Karjalainen, V. Välimäki, and Z. Jánosy. Towards high-quality sound synthesis of the guitar and string instruments. In *Proc. Int. Computer Music Conf.*, pages 56–63, Tokyo, Japan, Sep. 1993.

M. Karjalainen, J. Huopaniemi, and V. Välimäki. Direction-dependent physical modeling of musical instruments. In *Proc. Int. Congr. Acoustics (ICA'95)*, volume 3, pages 451–454, Trondheim, Norway, 1995.

M. Karjalainen, T. Mäki-Patola, A. Kanerva, and A. Huovilainen. Virtual air guitar. *J. Audio Eng. Soc.*, 54(10):964–980, October 2006.

A. Klapuri. Multiple fundamental frequency estimation based on harmonicity and spectral smoothness. *IEEE Trans. Speech and Audio Processing*, 11(6): 804–816, Nov. 2003.

M. Kob. Singing voice modeling as we know it today. *Acta Acustica united with Acustica*, 90(4):649–661, July/Aug. 2004. Selection of papers presented at SMAC 2003.

M. Kurz. Klangsynthese mittels physicalisher Modellierung einer schwingenden Saite durch numerische Integration der Differentialgleichung. Master's thesis, Technical University Berlin, 1995.

M. Kurz and B. Feiten. Physical modelling of a stiff string by numerical integration. In *Proc. Int. Computer Music Conf.*, pages 361–364, Hong Kong, Aug. 1996.

T. I. Laakso, V. Välimäki, M. Karjalainen, and U. K. Laine. Splitting the unit delay — Tools for fractional delay filter design. *IEEE Signal Processing Mag.*, 13(1):30–60, Jan. 1996.

S. Lakatos, P. R. Cook, and G. P. Scavone. Selective attention to the parameters of a physically informed sonic model. *Acoustics Research Letters Online*, Mar. 2000. Published in J. Acoust. Soc. Am. 107, L31-L36.

M. Laurson, C. Erkut, V. Välimäki, and M. Kuuskankare. Methods for modeling realistic playing in acoustic guitar synthesis. *Computer Music J.*, 25(3):38–49, 2001.

M. Laurson, V. Norilo, and M. Kuuskankare. PWGLSynth: A visual synthesis language for virtual instrument design and control. *Computer Music J.*, 29 (3):29–41, 2005.

T. Lokki, L. Savioja, R. Väänänen, J. Huopaniemi, and T. Takala. Creating interactive virtual auditory environments. *Computer Graphics and Applications, IEEE*, 22(4):49–57, 2002.

S. McAdams, A. Chaigne, and V. Roussarie. The psychomechanics of simulated sound sources: Material properties of impacted bars. *J. Acoust. Soc. Am.*, 115 (3):1306–1320, Mar. 2004.

J. Morrison and J. M. Adrien. MOSAIC: A framework for modal synthesis. *Computer Music Journal*, 17(1):45–56, 1993.

E. Motuk, R. Woods, and S. Bilbao. Parallel implementation of finite difference schemes for the plate equation on a FPGA-based multi-processor array. In *Proc. EUSIPCO*, Antalya, Turkey, Sep. 2005. Special session on modal-based sound synthesis.

D. Murphy, A. Kelloniemi, J. Mullen, and S. Shelley. Acoustic modeling using the digital waveguide mesh. *Signal Processing Magazine, IEEE*, 24(2):55–66, 2007.

A. V. Oppenheim, A. S. Willsky, and S. H. Navab. *Signals and Systems*. Prentice-Hall, Englewood Cliffs, NJ, second edition, 1996.

J. Pakarinen. Spatially distributed computational modeling of a nonlinear vibrating string. Master's thesis, Helsinki University of Technology, Espoo, Finland, June 2004.

J. Paradiso. Electronic music interfaces: New ways to play. *IEEE Spectrum*, 34 (12):18–30, December 1997.

M. D. Pearson. Tao: a physical modelling system and related issues. *Organised Sound*, 1(1):43–50, Apr. 1995.

H. Penttinen, J. Pakarinen, V. Välimäki, Mikael Laurson, Henbing Li, and Marc Leman. Model-based sound synthesis of the guqin. *J. Acoust. Soc. America*, 120(6):4052–4063, 2006.

S. Petrausch and R. Rabenstein. Tension modulated nonlinear 2D models for digital sound synthesis with the functional transformation method. In *Proc.*

EUSIPCO, Antalya, Turkey, Sep. 2005. Special session on modal-based sound synthesis. CD-R Proceedings.

W. H. Press, S. A. Teukolsky, W. A. Vetterling, and B. P. Flannery. *Numerical recipes in C++*. Cambridge Univestiy Press, Cambridge, UK, 2nd edition, 2002.

M. Puckette. Pure data. In *Proc. Int. Computer Music Conf.*, pages 224–227, Thessaloniki, Greece, Sep. 1997.

R. Rabenstein, S. Petrausch, A. Sarti, G. De Sanctis, C. Erkut, and M. Karjalainen. Block-based physical modeling for digital sound synthesis. *Signal Processing Magazine, IEEE*, 24(2):42–54, 2007.

D. Rocchesso. Physically-based sounding objects, as we develop them today. *J. New Music Research*, 33(3):305–313, September 2004.

D. Rocchesso and F. Fontana, editors. *The Sounding Object*. Edizioni di Mondo Estremo, Firenze, Italy, 2003.

D. Rocchesso and F. Scalcon. Bandwidth of perceived inharmonicity for musical modeling of dispersive strings. *IEEE Trans. Speech and Audio Processing*, 7(5):597–601, Sep. 1999.

D. Rocchesso and J. O. Smith. Generalized digital waveguide networks. *IEEE Trans. Speech and Audio Processing*, 11(3):242–254, May 2003.

X. Rodet and C. Vergez. Nonlinear dynamics in physical models: From basic models to true musical-instrument models. *Computer Music J.*, 23(3):35–49, September 1999.

L. Savioja. *Modeling Techniques for Virtual Acoustics*. PhD thesis, Helsinki University of Technology, Espoo, Finland, 1999.

L. Savioja, J. Backman, A. Järvinen, and T. Takala. Waveguide mesh method for low-frequency simulation of room acoustics. In *Proc. 15th Int. Congr. Acoust. (ICA'95)*, volume 2, pages 637–640, Trondheim, Norway, June 1995.

L. Savioja, J. Huopaniemi, T. Lokki, and R. Väänänen. Creating interactive virtual acoustic environments. *J. Audio Eng. Soc.*, 47(9):675–705, 1999.

C. E. Shannon. A mathematical theory of communication. *Bell System Technical Journal*, 27:379–423 and 623–656, July and October 1948.

J. O. Smith. Efficient synthesis of stringed musical instruments. In *Proc. Int. Computer Music Conf.*, pages 64–71, Tokyo, Japan, Sep. 1993.

J. O. Smith. *Physical Audio Signal Processing, August 2007 Edition*. `http://ccrma.stanford.edu/~jos/pasp/`, August 2007.

J. O. Smith. On the equivalence of the digital waveguide and finite difference time domain schemes. `http://www.citebase.org/abstract?id=oai:arXiv.org:physics/0407032`, July 2004a.

J. O. Smith. Virtual acoustic musical instruments: Review and update. *Journal of New Music Research*, 33(3):283–304, Sep. 2004b.

T. S. Stilson. *Efficiently-variable Non-oversampled Algorithms in Virtual-analog Music Synthesis: a Root-locus Perspective*. PhD thesis, Stanford, CA, USA, 2006.

T. S. Stilson and J. O. Smith. Analyzing the Moog VCF with considerations for digital implementation. In *Proc. International Computer Music Conference*, pages 398–401, Hong Kong, China, August 1996.

J. C. Strikwerda. *Finite difference schemes and partial differential equations*. Wadsworth, Brooks & Cole, California, 1989.

S. Strogatz. *Sync*. Allen Lane, The Penguin Press, London, UK, 2003.

S. Strogatz. *Nonlinear Dynamics and Chaos*. Studies in nonlinearity. Westview Press, 1994.

J. Sundberg. Synthesizing singing. In G. De Poli, A. Piccialli, and C. Roads, editors, *Representations of Musical Signals*, pages 299–320. The MIT Press, Cambridge, MA, USA, 1991.

N. Szilas and C. Cadoz. Analysis techniques for physical modeling networks. *Computer Music J.*, 22(3):33–48, 1998.

I. R. Titze. Theory of glottal airflow and source-filter interaction in speaking and singing. *Acta Acustica united with Acustica*, 90(4):641–648, July/Aug. 2004.

L. Trautmann and R. Rabenstein. *Digital Sound Synthesis by Physical Modeling Using the Functional Transformation Method.* Kluwer Academic/Plenum Publishers, New York, NY, 2003.

L. Trautmann and R. Rabenstein. Multirate simulations of string vibrations including nonlinear fret-string interactions using the functional transformation method. *EURASIP Journal on Applied Signal Processing*, 2004(7):949–963, June 2004.

S. A. Van Duyne and J. O. Smith. Physical modeling with the 2-d digital waveguide mesh. In *Proc. Int. Computer Music Conf.*, pages 40–47, Tokyo, Japan, 1993.

S. A. Van Duyne and J. O. Smith. The tetrahedral digital waveguide mesh. In *Proc. IEEE Workshop on Applications of Signal Processing to Audio and Acoustics*, pages 463–466, New Paltz, NY, 1995.

S. A. Van Duyne, J. R. Pierce, and J. O. Smith. Traveling wave implementation of a lossless mode-coupling filter and the wave digital hammer. In *Proc. Intl. Computer Music Conf. (ICMC)*, pages 411–418, Aarhus, Denmark, 1994.

B. L. Vercoe, W. G. Gardner, and E. D. Scheirer. Structured audio: creation, transmission, and rendering of parametric sound representations. *Proc. IEEE*, 86(3):922–940, Nov. 1998.

T. Virtanen. Monaural sound source separation by nonnegative matrix factorization with temporal continuity and sparseness criteria. *IEEE Trans. Audio, Speech and Language Proc.*, 15(3):1066–1074, 2007.

V. Välimäki and A. Huovilainen. Oscillator and filter algorithms for virtual analog synthesis. *Comput. Music J.*, 30(2):19–31, June 2006.

V. Välimäki, M. Laurson, and C. Erkut. Commuted waveguide synthesis of the clavichord. *Computer Music J.*, 27(1):71–82, 2003.

V. Välimäki, H. Penttinen, J. Knif, M. Laurson, and C. Erkut. Sound synthesis of the harpsichord using a computationally efficient physical model. *EURASIP Journal on Applied Signal Processing*, 2004(7):934–948, July 2004. Special issue on model-based sound synthesis.

V. Välimäki, J. Pakarinen, C. Erkut, and M. Karjalainen. Discrete-time modelling of musical instruments. *Rep. Prog. Phys.*, 69(1):1–78, January 2006.

V. Välimäki, R. Rabenstein, D. Rocchesso, X. Serra, and J. O. Smith. Signal processing for sound synthesis: Computer-generated sounds and music for all from the Guest Editors. *Signal Processing Magazine, IEEE*, 24(2):9–11, 2007.

I. Whalley. New technology, non-Western instruments and composition. *Organised Sound*, 10(01):1 – 2, Apr. 2005.

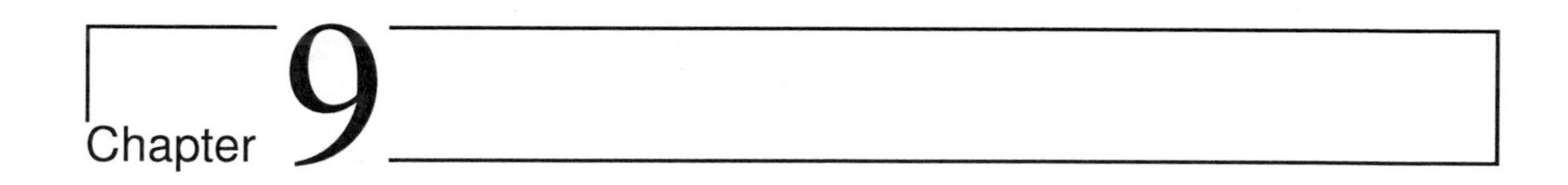

Interactive Sound

Federico Avanzini

Department of Information Engineering, University of Padua

About this chapter

This chapter tries to trace a route that, starting from studies in ecological perception and action-perception loop theories, goes down to sound modelling and design techniques for interactive computer animation and virtual reality applications.

We do not intend to provide an in-depth discussion about different theories of perception. We rather review a number of studies from experimental psychology that we consider to be relevant for research in multimodal virtual environments and interfaces, and we argue that such research needs to become more aware of studies in ecological perception and multimodal perception.

The chapter starts with an analysis of relevant literature in perception, while sound modelling techniques and applications to multimodal interfaces and VR are addressed in the last part of the chapter. The technically inclined reader may turn the chapter upside-down and start reading the last sections, referring to the initial material when needed. Where necessary, we will make use of the notions about physics-based sound synthesis techniques reviewed in Chapter 8.

9.1 Introduction

Most of Virtual Reality (VR) applications make use of visual displays, haptic devices, and spatialised sound displays. Multisensory information is essential for designing immersive virtual worlds, as an individual's perceptual experience is influenced by interactions among sensory modalities. As an example, in real environments visual information can alter the haptic perception of object size, orientation, and shape (Welch and Warren, 1986). Similarly, being able to hear sounds of objects in an environment, while touching and manipulating them, provides a sense of immersion in the environment not obtainable otherwise (Hahn et al., 1998). Properly designed and synchronised haptic and auditory displays are likely to provide much greater immersion in a virtual environment than a high-fidelity visual display alone. Moreover, by skewing the relationship between the haptic and visual and/or auditory displays, the range of object properties that can be effectively conveyed to the user can be significantly enhanced.

The importance of multimodal feedback in computer graphics and interaction has been recognised for a long time (Hahn et al., 1998) and is motivated by our daily interaction with the world. Streams of information coming from different channels complement and integrate each other, with some modality possibly dominating over the remaining ones, depending on the task (Welch and Warren, 1986; Ernst and Bülthoff, 2004). Research in ecological acoustics (Gaver, 1993a,b) demonstrates that auditory feedback in particular can effectively convey information about a number of attributes of vibrating objects, such as material, shape, size, and so on (see also Chapter 10).

Recent literature has shown that sound synthesis techniques based on physical models of sound generation mechanisms allow for high quality synthesis and interactivity, since the physical parameters of the sound models can be naturally controlled by user gestures and actions. Sounds generated by solid objects in contact are especially interesting since auditory feedback is known in this case to provide relevant information about the scene (e.g. object material, shape, size). Sound models for impulsive and continuous contact have been proposed for example in the papers by van den Doel and Pai (1998) and by Avanzini et al. (2003). Physically-based sound models of contact have been applied by DiFilippo and Pai (2000) to the development of an audio-haptic interface for contact interactions.

A particularly interesting research direction is concerned with bimodal (auditory and haptic) perception in contact interaction. Starting from a classic work by Lederman (1979), many studies have focused on continuous contact (i.e. scraping or sliding) and have investigated the relative contributions of tactile and auditory information to judgments of roughness of both real surfaces (Lederman, 1979; Lederman et al., 2002; Guest et al., 2002) and synthetic haptic and auditory textures (McGee et al., 2002). Impulsive contact interactions (i.e. impact) are apparently less investigated. A few studies have investigated the effect of auditory feedback on haptic stiffness perception (DiFranco et al., 1997; Avanzini and Crosato, 2006). Again, results from ecological acoustics (Freed, 1990; Giordano, 2006) provide useful indications about which auditory cues are relevant to stiffness/hardness perception, and can be exploited in the design of synthetic sound feedback.

The chapter is organised as follows. In Section 9.2 we provide a concise overview of the ecological approach to perception, and we focus on the literature on ecological acoustics. Section 9.3 addresses the topic of multisensory perception and interaction, and introduces some powerful concepts like sensory combination/integration, embodiment and enaction, sensory substitution. Finally, Section 9.4 discusses recent literature on interactive computer animation and virtual reality applications with a focus on multimodal feedback and especially auditory feedback. We will emphasise the relevance of studies in ecological acoustics and multimodal perception in aiding the design of multimodal interfaces and virtual environments.

9.2 Ecological acoustics

The ecological approach to perception, originated in the work of Gibson, refers to a particular idea of how perception works and how it should be studied. General introductions to the ecological approach to perception are provided by Gibson (1986) and Michaels and Carello (1981). Carello and Turvey (2002) also provide a synthetic overview of the main concepts of the ecological approach.

The label "ecological" reflects two main themes that distinguish this approach from more established views. First, perception is an achievement of animal-environment systems, not simply animals (or their brains). What makes up the environment of a particular animal is part of this theory of perception. Second, the main purpose of perception is to guide action, so a theory of perception cannot ignore what animals do. The kinds of activities that a particular animal does, e.g. how it eats and moves, are part of this theory of perception.

9.2.1 The ecological approach to perception

Direct versus indirect perception

The ecological approach is considered controversial because of one central claim: perception is direct. To understand the claim we can contrast it with the more traditional view.

Roughly speaking, the classical theory of perception states that perception and motor control depend upon internal referents, such as the retina for vision and cochlea for audition. These internal, psychological referents for the description and control of motion are known as sensory reference frames. Sensory reference frames are necessary if sensory stimulation is ambiguous (i.e. impoverished) with respect to external reality; in this case, our position and motion relative to the physical world cannot be perceived directly, but can only be derived indirectly from motion relative to sensory reference frames. Motion relative to sensory reference frames often differs from motion relative to physical reference frames (e.g. if the eye is moving relative to the external environment). For this reason, sensory reference frames provide only an indirect relation to physical reference frames. For example, when objects in the world reflect light, the pattern of light that reaches the back of the eye (the retina) has lost and distorted a lot of detail. The role of perception is then fixing the input and adding meaningful interpretations to it so that the brain can make an inference about what caused that input in the first place. This means that accuracy depends on the perceiver's ability to "fill in the gaps" between motion defined relative to sensory reference frames and motion defined relative to physical reference frames, and this process requires inferential cognitive processing.

A theory of direct perception, in contrast, argues that sensory stimulation is determined in such a way that there exists a 1:1 correspondence between patterns of sensory stimulation and the underlying aspects of physical reality (Gibson, 1986). This is a very strong assumption, since it basically says that reality is fully specified in the available sensory stimulation. Gibson (1986) provides the following example in the domain of visual perception, which supports, in his opinion, the direct perception theory. If one assumes that objects are isolated points in otherwise empty space, then their distances on a line projecting to the eye cannot be discriminated, as they stimulate the same retinal location. Under this assumption it is correct to state that distance is not perceivable by eye alone. However Gibson argues that this formulation is inappropriate for describing how we see. Instead he emphasises that the presence of a continuous background surface provides rich visual structure.

Including the environment and activity into the theory of perception allows a better description of the input, a description that shows the input to be richly structured by the environment and the animal's own activities. According to Gibson, this realisation opens up the new possibility that perception might be veridical. A relevant consequence of the direct perception approach is that sensory reference frames are unnecessary: if perception is direct, then anything that can be perceived can also be measured in the physical world.

Energy flows and invariants

Consider the following problem in visual perception: how can a perceiver distinguish the motion of an object from his/her own motion? Gibson (1986) provides an ecological solution to this problem, from which some general concepts can be introduced. The solution goes as follows: since the retinal input is ambiguous, it must be compared with other input. A first example of additional input is the information on whether any muscle commands had been issued to move the eyes or the head or the legs. If no counter-acting motor command is detected, then object motion can be concluded; on the contrary, if such motor commands are present then this will allow the alternative conclusion of self-motion. When the observer is moved passively (e.g. in a train), other input must be taken into account: an overall (global) change in the pattern of light indicates self-motion, while a local change against a stationary background indicates object motion.

This argument opened a new field of research devoted to the study of the structure in changing patterns of light at a given point of observation: the optic flow. The goal of this research is to discover particular patterns, called invariants, which are relevant to perception and hence to action of an animal immersed in an environment. Perceivers exploit invariants in the optic flow, in order to effectively guide their activities. Carello and Turvey (2002) provide the following instructive example: a waiter, who rushes towards the swinging door of the restaurant kitchen, adjusts his motion in order to control the collision with the door: he maintains enough speed to push through the door, and at the same time he is slow enough not to hurt himself. In order for his motion to be effective he must know when a collision will happen and how hard the collision will be. One can identify structures in the optic flow that are relevant to these facts: these are examples of quantitative invariants.

The above considerations apply not only to visual perception but also to other senses, including audition (see Section 9.2.2). Moreover, recent research has introduced the concept of global array (Stoffregen and Bardy, 2001). According to this concept, individual forms of energy (such as optic or acoustic flows) are subordinate components of a higher-order entity, the global array, which consists of spatio-temporal structure that extends across many dimensions of energy. The general claim underlying this concept is that observers are not separately sensitive to structures in the optic and acoustic flows but, rather, observers are directly sensitive to patterns that extend across these flows, that is, to patterns in the global array.

Stoffregen and Bardy (2001) exemplify this concept by examining the well known McGurk effect (McGurk and MacDonald, 1976), which is widely interpreted as reflecting general principles of intersensory interaction. Studies of this effect use audio-visual recordings in which the visual portion shows a speaker saying one syllable, while the audio track contains a different syllable. Observers are instructed to report the syllable that they hear, and perceptual reports are strongly influenced by the nominally ignored visible speaker. One of the most consistent and dramatic findings is that perceptual reports frequently are not consistent with either the visible or the audible event. Rather, observers often report "a syllable that has not been presented to either modality and that represents a combination of both". The wide interest in the McGurk effect arises in part from the need to explain why and how the final percept differs from the patterns in both the optic and acoustic arrays. In particular, Stoffregen and Bardy (2001) claim that the McGurk effect is consistent with the general idea that perceptual systems do not function independently, but work in a cooperative manner to pick up higher-order patterns in the global array. If speech perception is based on information in the global array then it is unnatural (or at least uncommon), for observers who can both see and hear the speaker, to report only what they hear. The global array provides information about what is being said, rather than about what is visible or what is audible: multiple perceptual systems are stimulated simultaneously and the stimulation has a single source (i.e. a speaker). In research on the McGurk effect the discrepancy between the visible and audible consequences of speech is commonly interpreted as a conflict between the two modalities, but it could also be interpreted as creating information in the global array that specifies the experimental manipulation, that is, the global array may specify that what is seen and what is heard arise from two different speech acts.

Affordances

The most radical contribution of Gibson's theory is probably the notion of affordance. Gibson (1986, p. 127) uses the term affordance as the noun form of the verb "to afford". The environment of a given animal affords things for that animal. What kinds of things are afforded? The answer is that behaviours are afforded. A stair with a certain proportion of a person's leg length affords climbing (is climbable); a surface which is rigid relative to the weight of an animal affords stance and traversal (is traversable); a ball which is falling with a certain velocity, relative to the speed that a person can generate in running toward it, affords catching (is catchable), and so on. Therefore, affordances are the possibilities for action of a particular animal-environment setting; they are usually described as "-ables", as in the examples above. What is important is that affordances are not determined by absolute properties of objects and environment, but depend on how these relate to the characteristics of a particular animal, e.g. size, agility, style of locomotion, and so on (Stoffregen, 2000).

The variety of affordances constitute ecological reformulations of the traditional problems of size, distance, and shape perception. Note that affordances and events are not identical and, moreover, that they differ from one another in a qualitative manner (Stoffregen, 2000). Events are defined without respect to the animal, and they do not refer to behaviour. Instead, affordances are defined relative to the animal and refer to behaviour (i.e. they are animal-environment relations that afford some behaviour). The concept of affordance thus emphasises the relevance of activity to defining the environment to be perceived.

9.2.2 Everyday sounds and the acoustic array

Ecological psychology has traditionally concentrated on visual perception. There is now interest in auditory perception and in the study of the acoustic array, the auditory equivalent of the optic array.

The majority of the studies in this field deal with the perception of properties of environment, objects, surfaces, and their changing relations, which is a major thread in the development of ecological psychology in general. In all of this research, there is an assumption that properties of objects, surfaces, and events are perceived as such. Therefore studies in audition investigate the identification of sound source properties, such as material, size, shape, and so on.

Two companion papers by Gaver (1993a,b) have greatly contributed to the build-up of a solid framework for ecological acoustics. Specifically, Gaver (1993a) deals with foundational issues, addresses such concepts as the acoustic array and acoustic invariants, and proposes a sort of "ecological taxonomy" of sounds.

Musical listening versus everyday listening

Gaver (1993a) introduces the concept of everyday listening, as opposed to musical listening. When a listener hears a sound, she/he might concentrate on attributes like pitch, loudness, and timbre, and their variations over time. Or she/he might notice its masking effect on other sounds. Gaver refers to these as examples of musical listening, meaning that the considered perceptual dimensions and attributes have to do with the sound itself, and are those used in the creation of music.

On the other hand, the listener might concentrate on the characteristics of the sound source. As an example, if the sound is emitted by a car engine the listener might notice that the engine is powerful, that the car is approaching quickly from behind, or even that the road is a narrow alley with echoing walls on each side. Gaver refers to this as an example of everyday listening, the experience of listening to events rather than sounds. In this case the perceptual dimensions and attributes have to do with the sound-producing event and its environment, rather than the sound itself.

Everyday listening is not well understood by traditional approaches to audition, although it forms most of our experience of hearing the day-to-day world. Descriptions of sound in traditional psychoacoustics are typically based on Fourier analysis and include frequency, amplitude, phase, and duration. Traditional psychoacoustics takes these "primitive" parameters as the main dimensions of sound and tries to map them into corresponding "elemental" sensations (e.g. the correspondence between sound amplitude and perceived loudness, or between frequency and perceived pitch). This kind of approach does not consider higher-level structures that are informative about events.

Everyday listening needs a different theoretical framework, in order to understand listening and manipulate sounds along source-related dimensions instead of sound-related dimensions. Such a framework must answer two fundamental questions. First, it has to develop an account of ecologically relevant perceptual attributes, i.e. the features of events that are conveyed through listening. Thus the first question asked by Gaver (1993a) is: "What do we hear?". Second, it has to develop an ecological acoustics, that describes which acoustic properties of sounds are related to information about the sound sources. Thus the second question asked by Gaver (1993b) is: "How do we hear it?"

Acoustic flow and acoustic invariants

Any source of sound involves an interaction of materials. Let us go back to the above example of hearing an approaching car: part of the energy produced in the engine produces vibrations in the car, instead of contributing to its motion. Mechanical vibrations, in turn, produce waves of acoustic pressure in the air surrounding the car, where the waveforms follows the movement of the car's surfaces (within limits determined by the frequency-dependent coupling of the surface's vibrations to the medium). These pressure waves then contain information about the vibrations that caused them, and result in a sound signal from which a listener might obtain such information. More in general, the patterns of vibration produced by contacting materials depend both on contact forces, duration of contact, and time-variations of the interaction, as well as sizes, shapes, materials, and textures of the objects.

Sound also conveys information about the environment in which the event have occurred. In everyday conditions, a listener's ear is reached not only by the direct sound but also by the reflections of sound over various other objects in the environment, resulting in a coloration of the spectrum.

In addition, the transmitting medium also has an influence on sound signals: dissipation of energy, especially at high-frequency, increases with the path travelled by the sound waves and thus carries information about the distance of the source. Another example is the Doppler effect, which is produced when sound sources and listeners are in relative motion, and results in a shift of the frequencies. Changes in loudness caused by changes in distance from a moving sound source may provide information about time-to-contact in a fashion analogous to changes in visual texture. The result is an acoustic array, analogous to the optical array described previously.

Several acoustic invariants can be associated to sound events: for instance, several attributes of a vibrating solid, including its size, shape, and density, determine the frequencies of the sound it produces. It is quite obvious that a single physical parameters can influence simultaneously many different sound parameters. As an example, changing the size of an object will scale the sound spectrum, i.e. will change the frequencies of the sound but not their pattern. On the other hand, changing the object shape results in a change of both the frequencies and their relationships. Gaver argues that these complex patterns of change may serve as information for distinguishing the responsible physical parameters: ecological acoustics focuses on discovering this kind of acoustic invariants.

Maps of everyday sounds

As already mentioned, Gaver has proposed an ecological categorisation of everyday sounds.

A first category includes sounds generated by solid objects. The pattern of vibrations of a given solid is structured by a number of its physical attributes. Properties can be grouped in terms of attributes of the interaction that has produced the vibration, those of the material of the vibrating objects, and those of the geometry and configuration of the objects.

Aerodynamic sounds are caused by the direct introduction and modification of atmospheric pressure differences from some source. The simplest aerodynamic sound is exemplified by an exploding balloon. Other aerodynamic sounds, e.g. the noise of a fan, are caused by more continuous events. Another sort of aerodynamic event involves situations in which changes in pressure themselves transmit energy to objects and set them into vibration (for example, when wind passes through a wire).

Sound-producing events involving liquids (e.g. dripping and splashing) are similar to those of vibrating solids: they depend on an initial deformation that is counter-acted by restoring forces in the material. The difference is that no audible sound is produced by the vibrations of the liquid. Instead, the resulting sounds are created by the resonant cavities (bubbles) that form and oscillate in the surface of the liquid. As an example, a solid object that hits a liquid pushes it aside and forms a cavity that resonates to a characteristic frequency, amplifying and modifying the pressure wave formed by the impact itself.

Although all sound-producing events involve any of the above categories (vibrating solids, aerodynamic, or liquid interactions), many also depend on complex patterns of simpler events. As an example, footsteps are temporal patterns of impact sounds. The perception of these patterned sounds is also related to the timing of successive events, (e.g. successive footstep sounds must occur within a range of rates and regularities in order to be perceived as walking). A slightly more complex example is a door slam, which involves the squeak of scraping hinges and the impact of the door on its frame. This kind of compound sounds involve mutual constraints on the objects that participate in related events: concatenating the creak of a heavy door closing slowly with the slap of a slammed light door would probably not sound natural.

Starting from these considerations, Gaver derived a tentative map of everyday sounds, which is shown in figure 9.1 and discussed in the following.

- Basic Level Sources: consider, for example, the region describing sounds made by vibrating solids. Four different sources of vibration in solids are indicated as basic level events: deformation, impacts, scraping and rolling.

- Patterned Sources involve temporal patterning of basic events. For instance, walking, as described above, but also breaking, spilling, and so on,

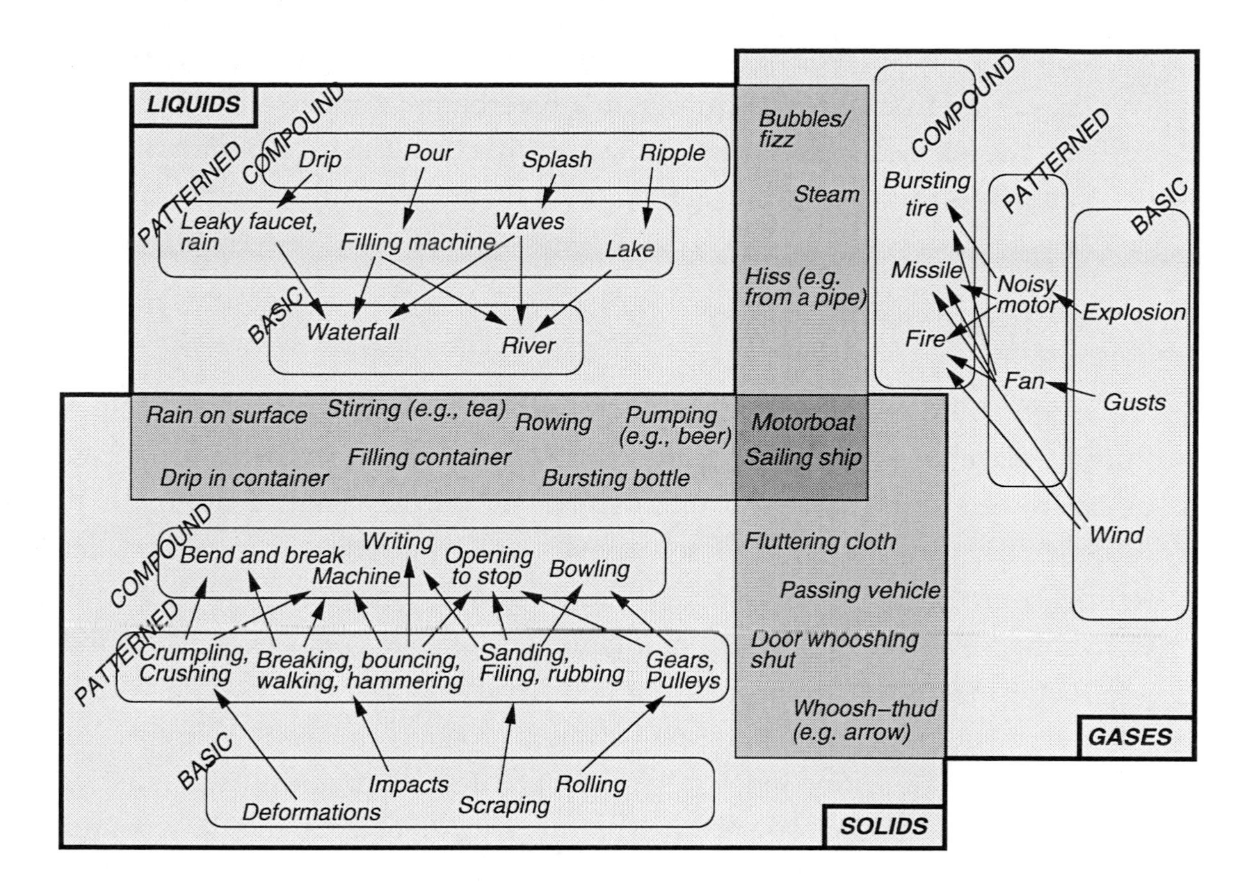

Figure 9.1: A map of everyday sounds. Complexity increases towards the center. Figure based on Gaver (1993a).

are all complex events involving patterns of simpler impacts. Similarly, crumpling or crushing are examples of patterned deformation sounds. In addition, other sorts of information are made available by their temporal complexity. For example, the regularity of a bouncing sound provides information about the symmetry of the bouncing object.

- Compound events involve more than one type of basic level event. An example is the slamming door discussed above. Other examples are the sounds made by writing, which involve a complex series of impacts and scrapes over time, while those made by bowling involve rolling followed by impact sounds.

- Hybrid events involve yet another level of complexity in which more than one basic type of material is involved. As an example, the sounds resulting from water dripping on a reverberant surface are caused both by the surface vibrations and the quickly-changing reverberant cavities, and thus involve attributes both of liquid and vibrating solid sounds.

9.2.3 Relevant studies

Although still quite "young", the literature on ecological acoustics has produced a number of relevant results in the last 20 years. In the following we briefly review some of the most influential studies, classified according to the categorisation by Gaver discussed above: basic, patterned, compound, and hybrid sources. It has to be noted that most of these studies are concerned with sound events produced from interactions of solids objects, while sound-producing events that involve liquids and aerodynamic interactions have been addressed less frequently. A reason for this is probably that sounds from solids are especially interesting when talking about interaction: auditory feedback is frequently generated when we touch or interact with objects, and these sounds often convey potentially useful information regarding the nature of the objects with which we are interacting.

Basic level sources

Many studies have investigated the perception of object material from impact sounds. Wildes and Richards (1988) tried to find an acoustical parameter that could characterise material type independently from variations in other features (e.g. size or shape). Materials can be characterised using a coefficient of internal friction, which measures anelasticity (in ascending order of anelasticity we have steel, glass, wood and rubber). This coefficient is measurable using both the quality factor Q and the decay time t_e of vibration, the latter being the time required for amplitude to decrease to $1/e$ of its initial value. Decreasing anelasticity results in increasing Q and t_e.

Lutfi and Oh (1997) performed a study on material discrimination in synthetic struck clamped bar sounds. Stimuli were synthesised by varying elasticity and density of the bars, with values taken in the ranges of various metals, glass, and crystal. Perturbations on parameter values were applied either to all the frequency components together (lawful covariation) or independently to each of them (independent perturbation). Listeners were presented with a pair of stimuli, were given a target material (either iron or glass), and had to tell which of two presented stimuli was produced by the target materials. Participants performance was analyzed in terms of the weights given to three different acoustical parameters: frequency, decay, and amplitude. Data revealed that discrimination was mainly based on frequency in all conditions, with amplitude and decay rate being of secondary importance.

Klatzky et al. (2000) also investigated material discrimination in stimuli with variable frequency and internal friction. In a first experimental setup subjects were presented with pairs of stimuli and had to judge on a continuous scale the perceived difference in the materials. In another experiment they were presented with one stimulus and had to categorise the material using four response alternatives: rubber, wood, glass and steel. Results indicated that judgments of material difference were significantly influenced by both the friction coefficient and the fundamental frequency. An effect of both these variables was found in a categorisation task: for lower decay factors steel and glass were chosen over rubber and plexiglass. Glass and wood were chosen for higher frequencies than steel and plexiglass.

Besides material, another relevant ecological dimension of impact sounds is the hardness of collision. Freed (1990) tried to relate hardness to some attack-related timbral dimensions. His stimuli were generated by percussing four cooking pans, with variable diameter, by means of six mallets of variable hardness. Mallet hardness ratings were found to be independent of the size of the pans, thus revealing the ability to judge properties of the percussor independently of properties of the sounding object. The analysis of results showed that the useful information for mallet hardness rating was contained in the first 300 ms of the signals. Four acoustical indices were measured in this sound attack portion: average spectral level, spectral level slope (i.e. rate of change in spectral level, a measure of damping), average spectral centroid, and spectral centroid TWA (time weighted average). These acoustical indices were used as predictors in a multiple regression analysis and were found to account for 75% of the variance of the ratings.

When we consider continuous contact (e.g. scraping) instead of impulsive contact, a relevant ecological dimension is surface roughness. In a classic study, Lederman (1979) compared the effectiveness of tactile and auditory information in judging the roughness of real surfaces. Roughness of aluminum plates was manipulated by varying the distance between adjacent grooves of fixed width, or by varying the width of the grooves. Subjects were given the task to rate numerically the roughness. In one condition participants only listened to the sounds generated by the experimenter by moving his fingertips on the plate. In a second condition subjects were asked to move their fingertips onto the plate while wearing cotton plugs and earphones. In a third condition they were able to hear the sounds they generated when touching the plate. Results showed that when both tactile and auditory information were present, the tactile one dominated in determining experimental performance. Roughness estimates were shown to increase as both the distance between grooves and the width of the grooves decreased. More recent research by Lederman and coworkers has focused on roughness perception when the surface is explored using a rigid probe rather than with the bare skin: as the probe provides a rigid link between the skin and the surface, vibratory roughness perception occurs in this case. Lederman et al. (2002) investigated relative contributions of haptic and auditory information to roughness judgments. Stimuli were obtained by asking subjects to explore with a probe a set of plates with periodic textures of varying inter-element spacings. Three conditions were used: touch-only, audition-only, and touch+audition. Results showed that, although dominance of haptic information was still found, sound played a more relevant role than in the case of direct contact with fingers. The authors argue that this may be due not only to the different interaction conditions, but also to the fact that the amplitude of the produced sounds is considerably greater for probe-based exploration than for bare skin contact.

The auditory perception of geometric properties of interacting objects has also been investigated. Carello et al. (1998) studied the recognition of the length of wood rods dropped on the floor. In their experiments subjects judged the perceived length by adjusting the distance of a visible surface in front of them. Subjects were found to be able to scale length of the rods consistently and physical length was found to correlate strongly with estimated length. Analysis of the relationship between the acoustical and perceptual levels was carried on using three acoustical features: signal duration, amplitude and spectral centroid. None of the considered acoustical variables predicted length estimates better than actual length. Length estimates were then explained by means of an analysis of the moments of inertia of a falling rod. Results of these latter analysis show potential analogies between the auditory and the tactile domain.

Patterned and compound sources

According to figure 9.1, patterned sound sources include bouncing, breaking, walking, and so on. Many of these everyday sounds have been investigated in the literature. Warren and Verbrugge (1988) studied acoustic invariants in bouncing and breaking events, and distinguished between two classes of invariants: structural invariants that specify the properties of the objects, and transformational invariants that specify their interactions and changes. Warren and Verbrugge investigated the nature of the transformational invariants that allow identification of breaking and bouncing events. On the basis of a physical analysis the authors hypothesised that the nature of these invariants was essentially temporal, static spectral properties having little or no role. Experimental stimuli were generated by dropping one of three different glass objects on the floor from different heights, so that for each of the objects a bouncing event and a breaking one were recorded. Once the ability of participants to correctly identify these two types of events was assessed with the original stimuli, two further experiments were conducted using synthetic stimuli. The bouncing event was synthesised by superimposing four trains of damped quasi-periodic pulses, each one generated from a recorded frame of a bouncing glass sound, all with the same damping. The breaking event was synthesised by superimposing the same sequences, but using different damping coefficients for each of them. Identification performance was extremely accurate in all cases, despite the strong simplifications of the spectral and temporal profile of the acoustical signal. The transformational invariants for bouncing was then identified as a single damped quasi-periodic sequence of pulses, while that for breaking was identified as a multiple damped quasi-periodic sequence of pulses.

Repp (1987) reports a study on auditory perception of another patterned sound composed of impact events: hands clapping. In particular he hypothesised that subjects are able to recognise size and configuration of clapping hands from the auditory information. Recognition of hands size was also related to recognition of the gender of the clapper, given that males have in general bigger hands than females. In a first experiment, clapper gender and hand size recognition from recorded clapping sounds were investigated. Overall clapper recognition was not good, although listeners performance in the identification of their own claps was much better. Gender recognition was barely above chance. Gender identification appeared to be guided by misconceptions: faster, higher-pitched and fainter claps were judged to be produced by females and vice-versa. In a second experiment subjects had to recognise the configuration of clapping hands. In this case performance was quite good: although hands configuration was a determinant of the clapping sound spectrum, the best predictor of performance was found to be clapping rate, spectral variables having only a secondary role.

A study on gender recognition in walking sounds is reported by Li et al. (1991). Subjects were asked to categorise the gender of the walker on the basis of four recorded walking sequences. Results show that recognition levels are well above chance. Several anthropometric measures were collected on the walkers (height, weight and shoe size). Duration analysis on the recorded walking excerpts indicated that female and male walkers differed with respect to the relative duration of stance and swing phases, but not with respect to walking speed. Nonetheless judged maleness was significantly correlated with this latter variable, and not with the former. Several spectral measures were derived from the experimental stimuli (spectral centroid, skewness, and kurtosis, spectral mode, average spectral level, and low and high spectral slopes). Two components were then derived from principal components analysis, and were then used as predictors for both physical and judged gender. Overall male walkers were characterised by lower spectral centroid, mode and high frequency energy, and by higher skewness, kurtosis and low-frequency slope. These results were then tested in a further experiment. Stimuli were generated by manipulating the spectral mode of the two most ambiguous walking excerpts. Consistently with previous analyses, the probability of choosing the response "male" was found to decrease as spectral mode increased. A final experiment showed that judged gender could be altered by having a walker wear shoes of the opposite gender.

Unlike the previous studies, the work of Gygi et al. (2004) did not focus on a specific event or feature. Instead the authors use for their experiments a large (70) and varied catalogue of sounds, which covers "nonverbal human sounds, animal vocalisations, machine sounds, the sounds of various weather conditions, and sounds generated by human activities". Patterned, compound, and hybrid sounds (according to the terminology used by Gaver) are included, e.g. beer can opening, bowling, bubbling, toilet flushing, etc. The experiments applied to non-verbal sound an approach adopted in speech perception studies, namely the use of low-, high-, and bandpass filtered speech to assess the importance of various frequency regions for speech identification. The third experiment is perhaps the most interesting one. The authors seem to follow an approach already suggested by Gaver (1993b): "[. . .] if one supposes that the temporal features of a sound are responsible for the perception of some event, but that its frequency makeup is irrelevant, one might use the amplitude contour from the original sound to modify a noise burst.". Results from this experiment show that identifiability is heavily affected by experience and has a strong variability between sounds. The authors tried to quantify the relevance of temporal structures through a selection of time- and frequency-domain parameters, including statistics of the envelope (a measure of the envelope "roughness"), autocorrelation statistics (to reveal periodicities in the waveform), and moments of the long term spectrum (to see if some spectral characteristics were preserved when the spectral information was drastically reduced). Correlation of these parameters with the identification results showed that three variables were mainly used by listeners: number of autocorrelation peaks, ratio of burst duration to total duration, cross-channel correlation. These are all temporal features, reflecting periodicity, amount of silence, and coherence of envelope across channels.

9.3 Multimodal perception and interaction

9.3.1 Combining and integrating auditory information

Humans achieve robust perception through the combination and integration of information from multiple sensory modalities. According to some authors, multisensory perception emerges gradually during the first months of life, and experience significantly shapes multisensory functions. By contrast, a different line of thinking assumes that sensory systems are fused at birth, and the single senses differentiate later. Empirical findings in newborns and young children have provided evidence for both views. In general experience seems to be necessary to fully develop multisensory functions.

Sensory combination and integration

Looking at how multisensory information is combined, two general strategies can be identified (Ernst and Bülthoff, 2004): the first is to maximise information delivered from the different sensory modalities (sensory combination). The second strategy is to reduce the variance in the sensory estimate to increase its reliability (sensory integration).

Sensory combination describes interactions between sensory signals that are not redundant: they may be in different units, coordinate systems, or about complementary aspects of the same environmental property. Disambiguation and cooperation are examples for this kind of interactions: if a single modality is not enough to provide a robust estimate, information from several modalities can be combined. As an example, object recognition is achieved through different modalities that complement each other and increase the information content.

By contrast, sensory integration describes interactions between redundant signals. Ernst and Bülthoff (2004) illustrate this concept with an example:

when knocking on wood at least three sensory estimates about the location of the knocking event can be derived: visual, auditory and proprioceptive. In order for these three location signals to be integrated, they first have to be transformed into the same coordinates and units. For this, the visual and auditory signals have to be combined with the proprioceptive neck-muscle signals to be transformed into body coordinates. The process of sensory combination might be non-linear. At a later stage the three signals are then integrated to form a coherent percept of the location of the knocking event.

There are a number of studies that show that vision dominates the integrated percept in many tasks, while other modalities (in particular audition and touch) have a less marked influence. This phenomenon of visual dominance is often termed visual capture. As an example, it is known that in the spatial domain vision can bias the perceived location of sounds whereas sounds rarely influence visual localisation. One key reason for this asymmetry seems to be that vision provides more accurate location information.

In general, however, the amount of cross-modal integration depends on the features to be evaluated or the tasks to be accomplished. The modality precision or modality appropriateness hypothesis by Welch and Warren (1986) is often cited when trying to explain which modality dominates under what circumstances. These hypotheses state that discrepancies are always resolved in favour of the more precise or more appropriate modality. As an example, the visual modality usually dominates in spatial tasks, because it is the most precise at determining spatial information. For temporal judgments however the situation is reversed and audition, being the more appropriate modality, usually dominates over vision. In texture perception tasks, haptics dominates on other modalities, and so on. With regard to this concept, Ernst and Bülthoff (2004) note that the terminology modality precision and modality appropriateness can be misleading because it is not the modality itself or the stimulus that dominates: the dominance is determined by the estimate and how reliably it can be derived within a specific modality from a given stimulus. Therefore, the term estimate precision would probably be more appropriate. The authors also list a series of questions for future research, among which one can find "What are the temporal aspects of sensory integration?". This is a particularly interesting question in the context of this chapter since, as already noted, temporal aspects are especially related to audition.

Auditory capture and illusions

Psychology has a long history of studying intermodal conflict and illusions in order to understand mechanisms of multisensory integration. Much of the literature on multisensory perception has focused on spatial interactions: an example is the ventriloquist effect, in which the perceived location of a sound shifts towards a visual stimulus presented at a different position. Identity interactions are also studied: an example is the already mentioned McGurk effect (McGurk and MacDonald, 1976), in which what is being heard is influenced by what is being seen (for example, when hearing /ba/ but seeing the speaker say /ga/ the final perception may be /da/).

As already noted, the visual modality does not always win in cross-modal tasks. In particular, the senses can interact in time, i.e they interact in determining not what is being perceived or where it is being perceived, but when it is being perceived. The temporal relationships between inputs from the different senses play an important role in multisensory integration. Indeed, a window of synchrony between auditory and visual events is crucial even in the spatial ventriloquist effect, which disappears when the audio-visual asynchrony exceeds approximately 300 ms. This is also the case in the McGurk effect, which fails to occur when the audio-visual asynchrony exceeds 200 – 300 ms.

There is a variety of cross-modal effects that demonstrate that, outside the spatial domain, audition can bias vision. In a recent study, Shams et al. (2002) presented subjects with a briefly flashed visual stimulus that was accompanied by one, two or more auditory beeps. There was a clear influence of the number of auditory beeps on the perceived number of visual flashes. That is, if there were two beeps subjects frequently reported seeing two flashes when only one was presented. Maintaining the terminology above, this effect may be called auditory capture.

Another recent study by Morein-Zamir et al. (2003) has tested a related hypothesis: that auditory events can alter the perceived timing of target lights. Specifically, four experiments reported by the authors investigated whether irrelevant sounds can influence the perception of lights in a visual temporal order judgment task, where participants judged which of two lights appeared first. The results show that presenting one sound before the first light and another one after the second light improves performance relative to baseline (sounds appearing simultaneously with the lights), as if the sounds pulled the perception of lights further apart in time. More precisely, the performance improvement results from the second sound trailing the second light. On the other hand, two sounds intervening between the two lights lead to a decline in performance, as if the sounds pulled the lights closer together. These results demonstrate a temporal analogue of the spatial ventriloquist effect.

These capture effects, or broadly speaking, these integration effects, are of course not only limited to vision and audition. In principle they can occur between any modalities (even within modalities). In particular some authors have investigated whether audition can influence tactile perception similarly to what Shams et al. (2002) have done for vision and audition. Hötting and Röder (2004) report upon a series of experiments where a single tactile stimulus was delivered to the right index finger of subjects, accompanied by one to four task-irrelevant tones. Participants (both sighted and congenitally blind) had to judge the number of tactile stimuli. As a test of whether possible differences between sighted and blind people were due to the availability of visual input during the experiment, half of the sighted participants were run with eyes open (sighted seeing) and the other half were blindfolded (sighted blindfolded). The first tone always preceded the first tactile stimulus by 25 ms and the time between the onsets of consecutive tones was 100 ms. Participants were presented with trials made of a single tactile stimulus accompanied by no, one, two, three or four tones. All participants reported significantly more tactile stimuli when two tones were presented than when no or only one tone was presented. Sighted participants showed a reliable illusion for three and four tones as well, while blind participants reported a lower number of perceived tactile stimuli than sighted seeing or sighted blindfolded participants. These results extend the finding of the auditory-visual illusion established by Shams et al. (2002) to the auditory-tactile domain. Moreover, the results (especially the discrepancies between sighted and congenitally blind participants) suggest that interference by a task-irrelevant modality is reduced if processing accuracy of the task-relevant modality is high.

Bresciani et al. (2005) conducted a very similar study, and investigated whether the perception of tactile sequences of two to four taps delivered to the index fingertip can be modulated by simultaneously presented sequences of auditory beeps when the number of beeps differs (less or more) from the number of taps. This design allowed to systematically test whether task-irrelevant auditory signals can really modulate (influence in both directions) the perception of tactile taps, or whether the results of Hötting and Röder (2004) merely reflected an original but very specific illusion. In a first experiment, the auditory and tactile sequences were always presented simultaneously. Results showed that tactile tap perception can be systematically modulated by task-irrelevant auditory inputs. Another interesting point is the fact that subjects responses were significantly less variable when redundant tactile and auditory signals were presented rather than tactile signals alone. This suggests that even though auditory signals were irrelevant to the task, tactile and auditory signals were probably integrated. In a second experiment, the authors investigate how sensory integration is affected by manipulation of the timing between auditory and tactile sequences. Results showed that the auditory modulation of tactile perception was weaker when the auditory stimuli were presented immediately before the onset or after the end of the tactile sequences. This modulation completely vanished with a 200 ms gap between the auditory and tactile sequences. Shams et al. (2002) found that the temporal window in which audition can bias the perceived number of visual flashes is about 100 ms. These results suggest that the temporal window of auditory-tactile integration might be wider than for auditory-visual integration.

The studies discussed above provide evidence of the fact that the more salient (or reliable) a signal is, the less susceptible to bias this signal should be. In the same way, the more reliable a biasing signal is, the more bias it should induce. Therefore, the fact that auditory signals can bias both visual and tactile perception probably indicates that, when counting the number of events presented in a sequence, auditory signals are more reliable than both visual and tactile signals. When compared to the studies by Shams et al. (2002), the effects observed on tactile perception are relatively small. This difference in the magnitude of the auditory-evoked effects likely reflects a higher saliency of tactile than visual signals in this kind of non-spatial task.

Other authors have studied auditory-tactile integration in surface texture perception. Lederman and coworkers, already mentioned in Section 9.2.3, have shown that audition had little influence on texture perception when participants touched the stimulus with their fingers (Lederman, 1979). However, when the contact was made via a rigid probe, with a consequent increase of touch-related sound and a degradation of tactile information, auditory and tactile cues were integrated (Lederman et al., 2002). These results suggest that although touch is mostly dominant in texture perception, the degree of auditory-tactile integration can be modulated by the reliability of the single-modality information

A related study was conducted by Guest et al. (2002). In their experimental setup, participants had to make forced-choice discrimination responses regarding the roughness of abrasive surfaces which they touched briefly. Texture sounds were captured by a microphone located close to the manipulated surface and subsequently presented through headphones to the participants in three different conditions: veridical (no processing), amplified (12dB boost on the 2 – 20kHz band), and attenuated (12dB attenuation in the same band). The authors investigated two different perceptual scales: smooth-rough, and moist-dry. Analysis of discrimination errors verified that attenuating high frequencies led to a bias towards an increased perception of tactile smoothness (or moistness), and conversely the boosted sounds led to a bias towards an increased perception of tactile roughness (or dryness). This work is particularly interesting from a sound-design perspective, since it investigates the effects of a non-veridical auditory feedback (not only the spectral envelope is manipulated, but sounds are picked up in the vicinity of the surface and are therefore much louder than in natural listening conditions).

9.3.2 Perception is action

Embodiment and enaction

According to traditional mainstream views of perception and action, perception is a process in the brain where the perceptual system constructs an internal representation of the world, and eventually action follows as a subordinate function. This view of the relation between perception and action makes then two assumptions. First, the causal flow between perception and action is primarily one-way: perception is input from world to mind, action is output from mind to world, and thought (cognition) is the mediating process. Second, perception and action are merely instrumentally related to each other, so that each is a means to the other. If this kind of "input-output" picture is right, then it must be possible, at least in principle, to disassociate capacities for perception, action, and thought.

Although everyone agrees that perception depends on processes taking place in the brain, and that internal representations are very likely produced in the brain, more recent theories have questioned such a modular decomposition in which cognition interfaces between perception and action. The ecological approach discussed in Section 9.2 rejects the one-way assumption, but not the instrumental aspect of the traditional view, so that perception and action are seen as instrumentally interdependent. Others argue that a better alternative is to reject both assumptions: the main claim of these theories is that it is not possible to disassociate perception and action schematically, and that every kind of perception is intrinsically active and thoughtful: perception is not a process in the brain, but a kind of skillful activity on the part of the animal as a whole. As stated by Noë (2005), only a creature with certain kinds of bodily skills (e.g. a basic familiarity with the sensory effects of eye or hand movements, etc.) could be a perceiver.

One of the most influential contributions in this direction is due to Varela et al. (1991) (see O'Regan and Noë, 2001, for a detailed review of other works based on similar ideas). They presented an "enactive conception" of experience, which is not regarded as something that occurs inside the animal, but rather as something that the animal enacts as it explores the environment in

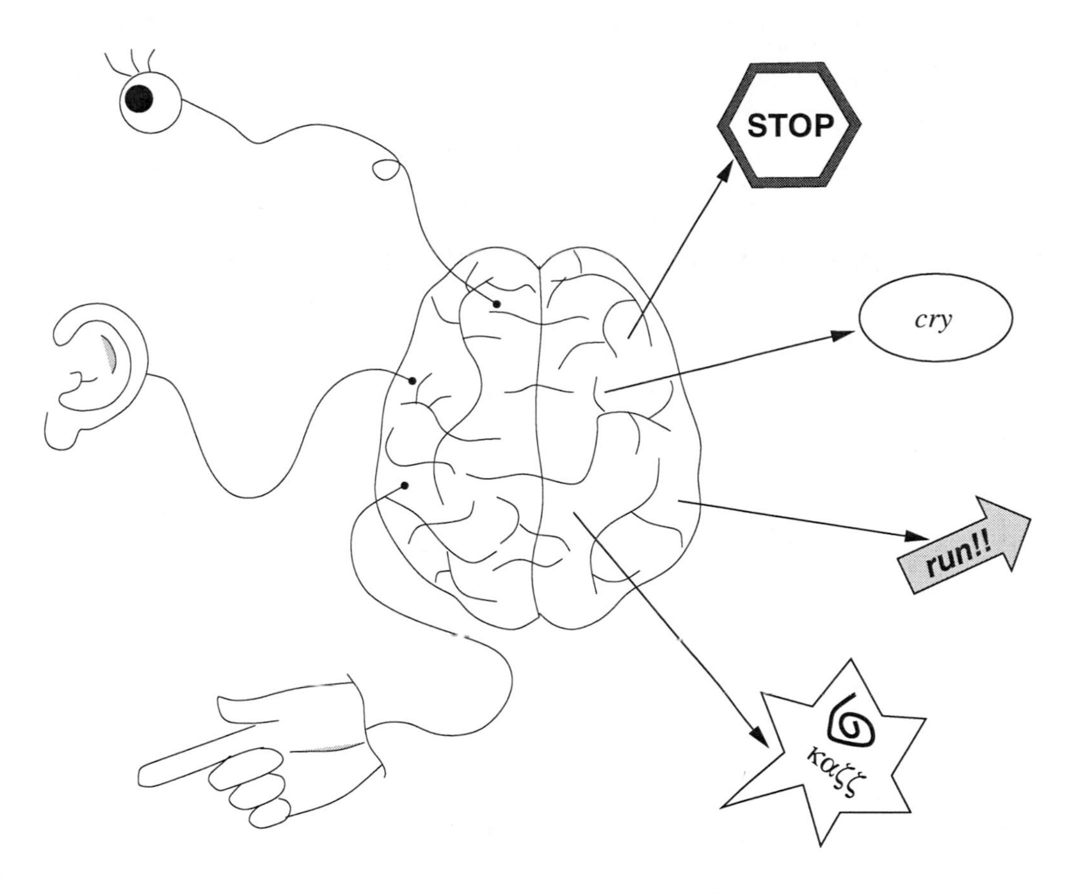

Figure 9.2: A cartoon representation of traditional views of the perception-action functions as a causal one-way flow.

which it is situated. In this view, the subject of mental states is the embodied, environmentally-situated animal. The animal and the environment form a pair in which the two parts are coupled and reciprocally determining. Perception is thought of in terms of activity on the part of the animal. The term "embodied" is used by the authors as a mean to highlight two points: first, cognition depends upon the kinds of experience that are generated from specific sensorimotor capacities. Second, these individual sensorimotor capacities are themselves embedded in a biological, psychological, and cultural context. Sensory and motor processes, perception and action, are fundamentally inseparable in cognition.

O'Regan and Noë (2001) have introduced closely related concepts, according to which perception consists in exercising an exploratory skill. The authors illustrate their approach with an example: the sensation of softness that one might experience in holding a sponge consists in being aware that one can exercise certain practical skills with respect to the sponge: one can for example press it, and it will yield under the pressure. The experience of softness of the sponge is characterised by a variety of such possible patterns of interaction with the sponge. The authors term sensorimotor contingencies the laws that describe these interactions. When a perceiver knows, in an implicit, practical way, that at a given moment he is exercising the sensorimotor contingencies associated with softness, then he is in the process of experiencing the sensation of softness.

O'Regan and Noë (2001) then classify sensory inputs according to two criteria, i.e. corporality and alerting capacity. Corporality is the extent to which activation in a neural channel systematically depends on movements of the body. Sensory input from sensory receptors like the retina, the cochlea, and mechanoreceptors in the skin possesses corporality, because any body motion will generally create changes in the way sensory organs are positioned in space, and consequently in the incoming sensory signals (the situation is less clear for the sense of smell, but sniffing, blocking the nose, moving the head, do affect olfactory stimulation). Proprioceptive input from muscles also possesses corporality, because there is proprioceptive input when muscle movements produce body movements. The authors argue that corporality is one important factor that explains the extent to which a sensory experience will appear to an observer as being truly sensory, rather than non-sensory, like a thought, or a memory. The alerting capacity of sensory input is the extent to which that input can cause automatic orienting behaviours that capture cognitive processing resources. According to these definitions, vision, touch, hearing, and smell have not only high corporality but also high alerting capacity. With high corporality and high alerting capacity, vision, touch, hearing and smell have strong phenomenal presence. This is in accordance with the usual assumption that they are the prototypical sensory modalities.

A possible objection to the definitions of perception and action given above is that most sensations can be perceived without any exploratory skill being engaged. For example, having the sensation of red or of a bell ringing does not seem to involve the exercising of skills. An immediate counter-objection is that sensations are never instantaneous, but are always extended over time, and that at least potentially, they always involve some form of activity. O'Regan and Noë (2001) refer to a number of experiments, especially in the domain of visual perception, that support this idea. Experiments on "change blindness" present observers with displays of natural scenes and ask them to detect cyclically repeated changes (e.g. large object shifting, changing colors, and so on). Under normal circumstances a change of this type would create a transient signal in the visual system that would be detected by low-level visual mechanisms and would attract attention to the location of the change. In the "change blindness" experiments, however, conditions were arranged in such a way that these transients were hidden by superimposing a brief global flicker over the whole visual field at the moment of the change. It was shown that in this condition observers have great difficulty seeing changes, even when the changes are extremely large (and are perfectly visible to someone who knows what they are). Such results contrast with the subjective impression of "seeing everything" in an observed scene or picture. The authors regard them as a support to the view that an observer sees the aspects of a scene which he/she is currently "visually manipulating", which makes it reasonable that only a subset of scene elements that share a particular scene location can be perceived at a given moment.

A related example, again in the domain of visual perception, is discussed by Noë (2005) who introduces the concept of "experiential blindness" and reports upon cases where this phenomenon has been observed. According to Nöe there are, broadly speaking, two different kinds of blindness: blindness due to damage or disruption of the sensitive apparatus (caused by e.g. cataracts, retinal injury, and so on), and blindness that is not due to the absence of sensation or sensitivity, but rather to the person's inability to integrate sensory stimulation with patterns of movement and thought. The latter is termed experiential blindness because it occurs despite the presence of normal visual sensation. The author considers attempts to restore sight in congenitally blind individuals whose blindness is due to cataracts impairing the eye's sensitivity by obstructing light on its passage to the retina. The medical literature reports that surgery restores visual sensation, at least to a significant degree, but that it does not restore sight. In the period immediately after the operation, patients suffer blindness despite rich visual sensations. This clearly contrasts with the traditional input-output picture described at the beginning of this section, according to which removing the cataract and letting in the light should enable normal vision. A related phenomenon is that of blindness caused by paralysis. Normally the eyes are in nearly constant motion, engaging in sharp movements several times a second. If the eyes cease moving, they loose their receptive power. A number of studies are reported by Noë (2005), which show that images stabilised on the retina fade from view. This is probably an instance of the more general phenomenon of sensory fatigue thanks to which we do not continuously feel our clothing on our skin, the glasses resting on the bridge of our nose, or a ring on our finger. This suggests that some minimal amount of eye and body movement is necessary for perceptual sensation.

Audition and sensory substitution

According to the theories discussed above, the quality of a sensory modality does not derive from the particular sensory input channel or neural activity involved in that specific modality, but from the laws of sensorimotor skills that are exercised. The difference between "hearing" and "seeing" lies in the fact that, among other things, one is seeing if there is a large change in sensory input when blinking; on the other hand, one is hearing if nothing happens when one blinks but there is a left/right difference when one turns the head, and so on. This line of reasoning implies that it is possible to obtain a visual experience from auditory or tactile input, provided the sensorimotor laws that are being obeyed are the laws of vision.

The phenomenon of sensory substitution is coherent with this view. Perhaps the first studies on sensory substitution are due to Bach-y-Rita who, starting from 1967, has been experimenting with devices to allow blind people to "see" via tactile stimulation provided by a matrix of vibrators connected to a video camera. A comprehensive review of this research stream is provided by Kaczmarek et al. (1991). The tactile visual substitution systems developed by Bach-y-Rita and coworkers use matrices of vibratory or electrical cutaneous stimulators to represent the luminance distribution captured by a camera on a skin area (the back, the abdomen, the forehead, or the fingertip). Due to technical reasons and to bandwidth limitations of tactile acuity, these devices have a rather poor spatial resolution, being generally matrices of not more than 20×20 stimulators. One interesting result from early studies was that blind subjects were generally unsuccessful in trying to identify objects placed in front of a fixed camera. It was only when the observer was allowed to actively manipulate the camera that identification became possible. Although subjects initially located the stimulation on the skin area being stimulated, with practice they started to locate objects in space (although they were still able to feel local tactile sensation). This point supports the idea that the experience associated with a sensory modality is not wired into the neural hardware, but is rather a question of exercising sensorimotor skills: seeing constitutes the ability to actively modify sensory impressions in certain law-obeying ways.

A certain amount of studies investigate sensory substitution phenomena that involve audition. One research stream deals with the use of echolocation devices to provide auditory signals to a user, depending on the direction, distance, size and surface texture of nearby objects. Such devices have been studied as prostheses for the blind. Ifukube et al. (1991) designed an apparatus in which a frequency-modulated ultrasound signal (with carrier and modulating frequencies in a similar range as that produced by bats for echolocation) is emitted from a transmitting array with broad directional characteristics in order to detect obstacles. Reflections from obstacles are picked up by a two-channel receiver and subsequently digitally down-converted by a 50:1 factor, resulting in signals that are in the audible frequency range and can be presented binaurally through earphones. The authors evaluated the device through psychophysical experiments in order to establish whether obstacles may be perceived as localised sound images corresponding to the direction and the size of the obstacles. Results showed that the auditory feedback was successfully used for the recognition of small obstacles, and also for discriminating between several obstacles at the same time without any virtual image.

While such devices cannot provide a truly visual experience, they nevertheless provide users with the clear impression of things being "out in front of them". In this sense, these devices can be thought as variants of the blind person's cane. Blind people using a cane sense the external environment that is being explored through the cane, rather than the cane itself. The tactile sensations provided by the cane are "relocated" onto the environment, and the cane itself is forgotten or ignored. O'Regan and Noë (2001) prefer to say that sensations in themselves are situated nowhere, and that the location of a sensation is an abstraction constructed in order to account for the invariance structure of the available sensorimotor contingencies.

A related research was conducted by Meijer (1992), who developed an experimental system for the conversion of a video stream into sound patterns, and investigated possible applications of such a device as a vision substitution device for the blind. According to the image-to-sound mapping chosen by Meijer, a $N \times M$ pixel image is sampled from the video stream at a given rate, and converted into a spectrogram in which grey level of the image corresponds to partial amplitude. Therefore the device potentially conveys more detailed information than the one developed by Ifukube et al. (1991), since it provides a representation of the entire scene rather than simply detecting obstacles and isolated objects. The approach followed by Mejer resembles closely the work by Bach-y-Rita, except that audition instead of tactile stimulation is used as substitute for vision. Although from a purely mathematical standpoint the chosen image-to-sound mapping ensures the preservation of visual information to a certain extent, it is clear that perceptually such a mapping is highly abstract and a priori completely non-intuitive. Accordingly, Meijer (1992) remarks that the actual perception of these sound representations remains to be evaluated. However, it must also be noted that users of such devices sometimes testify that a transfer of modalities indeed takes place[1]. Again, this finding is consistent with the sensorimotor theories presented above, since the key ingredient is the possibility for the user to actively manipulate the device.

[1]The experience of a visually impaired user, who explicitly described herself as seeing with the visual-to-auditory substitution device, is reported at `http://www.seeingwithsound.com/tucson2002f.ram`

9.4 Sound modelling for multimodal interfaces

In this final section we discuss recent literature on interactive computer animation and virtual reality applications. All of these applications involve direct interaction of an operator with virtual objects and environments and require multimodal feedback in order to enhance the effectiveness of the interaction. We will especially focus on the role of auditory feedback, and will emphasise the relevance of studies in ecological acoustics and multimodal perception, which we have previously discussed, in aiding the design of multimodal interfaces and virtual environments.

The general topic of the use of sound in interfaces is also addressed in Chapter 10.

9.4.1 Interactive computer animation and VR applications

The need for multisensory feedback

Typical current applications of interactive computer animation and VR applications (Srinivasan and Basdogan, 1997) include medicine (surgical simulators for medical training, manipulation of micro and macro robots for minimally invasive surgery, remote diagnosis for telemedicine, aids for the disabled such as haptic interfaces for non-sighted people), entertainment (video games and simulators that enable the user to feel and manipulate virtual solids, fluids, tools, and avatars), education (e.g. interfaces giving students the feel of phenomena at nano, macro, or astronomical scales, "what if" scenarios for non-terrestrial physics, display of complex data sets), industry (e.g. CAD systems in which a designer can manipulate the mechanical components of an assembly in an immersive environment), and arts (virtual art exhibits, concert rooms, museums in which the user can log in remotely, for example to play musical instruments or to touch and feel haptic attributes of the displays, and so on). Most of the virtual environments (VEs) built to date contain complex visual displays, primitive haptic devices such as trackers or gloves to monitor hand position, and spatialised sound displays. However it is being more and more acknowledged that accurate auditory and haptic displays are essential in order to realise the full promise of VEs.

Being able to hear, touch, and manipulate objects in an environment, in addition to seeing them, provides a sense of immersion in the environment that is otherwise not possible. It is quite likely that much greater immersion in a VE can be achieved by synchronizing even simple haptic and auditory displays with the visual one, than by increasing the complexity of the visual display alone. Moreover, by skewing the relationship between the haptic and visual and/or auditory displays, the range of object properties that can be effectively conveyed to the user can be significantly enhanced. Based on these considerations, many authors (see for example Hahn et al., 1998 and Srinivasan and Basdogan, 1997) emphasise the need to make a more concerted effort to bring the three modalities together in VEs.

The problem of generating effective sounds in VEs has been addressed in particular by Hahn et al. (1998), who identify three sub-problems: sound modelling, sound synchronisation, and sound rendering. The first problem has long been studied in the field of computer music (see also Chapter 8). However, the primary consideration in VE applications is the effective parametrisation of sound models so that the parameters associated with motion (changes of geometry in a scene, user's gestures) can be mapped to the sound control parameters, resulting in an effective synchronisation between the visual and auditory displays. Finally, sound rendering refers to the process of generating sound signals from models of objects and their movements within a given environment, which is in principle very much equivalent to the process of generating images from their geometric models: the sound energy being emitted needs to be traced within the environment, and perceptual processing of the sound signal may be needed in order to take into account listener effects (e.g. filtering with Head Related Transfer Functions). The whole process of rendering sounds can be seen as a rendering pipeline analogous to the image rendering pipeline.

Until recently the primary focus for sound generation in VEs has been in spatial localisation of sounds. On the contrary, research about models for sound sources and mappings between object motion/interaction and sound control is far less developed. In Section 9.4.2 we will concentrate on this latter topic.

Learning the lessons from perception studies

Given the needs and the requirements addressed in the previous section, many lessons can be learned from the studies in direct (ecological) perception and in the action-perception loop that we have reviewed in the first part of this chapter.

The concept of "global array" proposed by Stoffregen and Bardy (2001) is a very powerful one: the global array provides information that can optimise perception and performance, and that is not available in any other form of sensory stimulation. Humans may detect informative global array patterns, and they may routinely use this information for perception and control, in both VE and daily life. According to Stoffregen and Bardy (2001), in a sense VE designers do not need to make special efforts to make the global array available to users: the global array is already available to users. Rather than attempting to create the global array, designers need to become aware of the global array that already exists, and begin to understand how multisensory displays structure the global array. The essential aspect is the initial identification of the relevant global array parameters, which makes it possible to construct laboratory situations in which these parameters can be manipulated, and in which their perceptual salience and utility for performance in virtual environments can be evaluated.

For the specific case of auditory information, the description of sound producing events by Gaver (1993b) provides a framework for the design of environmental sounds. Gaver emphasises that, since it is often difficult to identify the acoustic information of events from acoustic analysis alone, it is useful to supplement acoustic analyses with physical analyses of the event itself. Studying the physics of sound-producing events is useful both in suggesting relevant source attributes that might be heard and in indicating the acoustic information for them. Resynthesis, then, can be driven by the resulting physical simulations of the event.

The studies on multimodal perception reviewed in Section 9.3 also provide a number of useful guidelines and even quantitative data. We have seen that streams of information coming from different channels complement and integrate each other, with some modality possibly dominating over the remaining ones depending on the features to be evaluated or the tasks to be accomplished (the modality precision or modality appropriateness hypothesis by Welch and Warren, 1986). In particular, when senses interact in time, a window of synchrony between the feedback of different modalities (e.g. auditory and visual, or auditory and haptic feedbacks) is crucial for multisensory integration. Many of the studies previously discussed (e.g., Shams et al., 2002; Guest et al., 2002; Bresciani et al., 2005) report quantitative results about "integration windows" between modalities. These estimates can be used as constraints for the synchronisation of rendering pipelines in a multimodal architectures.

9.4.2 Sound models

Physics-based approaches

Sound synthesis techniques traditionally developed for computer music applications (e.g. additive, subtractive, frequency modulation, Zölzer, 2002) provide abstract descriptions of sound signals. Although well suited for the representation of musical sounds, these techniques are in general not effective for the generation of non-musical interaction sounds. We have seen in Section 9.2 that research in ecological acoustics points out that the nature of everyday listening is rather different and that auditory perception delivers information which goes beyond attributes of musical listening.

On the other hand, physically-based sound modelling approaches (see Chapter 8) generate sound from computational structures that respond to physical input parameters, and therefore they automatically incorporate complex responsive acoustic behaviours. Moreover, the physical control parameters do not require in principle manual tuning in order to achieve realistic output. Again, results from research in ecological acoustics aid in determining what sound features are perceptually relevant, and can be used to guide the tuning process.

A second advantage of physically-based approaches is interactivity and ease in associating motion to sound control. As an example, the parameters needed to characterise collision sounds, e.g. relative velocity at collision, are computed in the VR physical simulation engine and can be directly mapped into control parameters of a physically-based sound model. The sound feedback consequently responds in a natural way to user gestures and actions. This is not the case with traditional approaches to sound synthesis, where the problem of finding a motion-correlated parametrisation is not a trivial one. Think about the problem of parameterizing real recorded sounds by their attributes such as amplitude and pitch: this corresponds to a sort of "reverse engineering" problem where one tries to determine how the sounds were generated starting from the sounds themselves. Designing effective mappings between user gestures and sound control parameters is important especially in the light of the studies in action-perception loop, that we have addressed in Section 9.3.2.

Finally, physically-based sound models can in principle allow the creation of dynamic virtual environments in which sound rendering attributes are incorporated into data structures that provide multimodal encoding of object properties: shape, material, elasticity, texture, mass, and so on. In this way a unified description of the physical properties of an object can be used to control the visual, haptic, and sound rendering, without requiring the design of separate properties for each thread. This problem has already been studied in the context of joint haptic-visual rendering, and recent haptic-graphic APIs (Technologies, 2002; Sensegraphics, 2006) adopt a unified scene graph that takes care of both haptics and graphics rendering of objects from a single scene description, with obvious advantages in terms of synchronisation and avoidance of data duplication. Physically-based sound models may allow the development of a similar unified scene, that includes description of audio attributes as well.

For all these reasons, it would be desirable to have at disposal sound modelling techniques that incorporate complex responsive acoustic behaviours and can reproduce complex invariants of primitive features: physically-based models offer a viable way to synthesise naturally behaving sounds from computational structures that respond to physical input parameters. Although traditionally developed in the computer music community and mainly applied to the faithful simulation of existing musical instruments, physical models are now gaining popularity for sound rendering in interactive applications (Cook, 2002).

Contact sounds

As already remarked in Section 9.2, an important class of sound events is that of contact sounds between solids, i.e. sounds generated when objects come in contact with each other (collision, rubbing, etc.: see also figure 9.1). Various modelling approaches have been proposed in the literature.

Van den Doel and coworkers (van den Doel and Pai, 1998; van den Doel et al., 2001) proposed modal synthesis (Adrien, 1991) as an efficient yet accurate framework for describing the acoustic properties of objects. Modal synthesis techniques have been already presented in Chapter 8. Here, we recall that if a resonating object is modelled as a network of N masses connected with springs and dampers, then a geometrical transformation can be found that turns the system into a set of decoupled equations. The transformed variables $\{q_n\}_{n=1}^{N}$ are generally referred to as modal displacements, and obey a second-order linear oscillator equation:

$$\ddot{q}_n(t) + g_n\dot{q}_n(t) + \omega_n^2 q_n(t) = \frac{1}{m_n} f(t), \tag{9.1}$$

where q_n is the oscillator displacement and f represents any driving force, while ω_n is the oscillator center frequency. The parameter $1/m_n$ controls the "inertial" properties of the oscillator (m_n has the dimension of a mass), and g_n is the oscillator damping coefficient and relates to the decay properties of the system. Modal displacements q_n are related to physical displacement through an $N \times K$ matrix $\boldsymbol{A}$, whose elements a_{nk} weigh the contribution of the nth mode at a location k. If the force f is an impulse, the response q_n of each mode is a damped sinusoid and the physical displacement at location k is given by

$$x_k(t) = \sum_{n=1}^{N} a_{nk} q_n(t) = \sum_{n=1}^{N} a_{nk} e^{-g_n t/2} \sin(\omega_n t). \quad (9.2)$$

Any pre-computed contact force signal can then be convolved to the impulse response and thus used to drive the modal synthesiser.

The modal representation of a resonating object is naturally linked to many ecological dimensions of the corresponding sounds. The frequencies and the amount of excitation of the modes of a struck object depend on the shape and the geometry of the object. The material determines to a large extent the decay characteristics of the sound. The amplitudes of the frequency components depend on where the object is struck (as an example, a table struck at the edges makes a different sound than when struck at the center). The amplitude of the emitted sound is proportional to the square root of the energy of the impact.

The possibility of linking the physical model parameter to ecological dimensions of the sound has been demonstrated in the paper by Klatzky et al. (2000), already discussed in Section 9.2. In this work, the modal representation proposed by van den Doel and Pai (1998) has been applied to the synthesis of impact sounds with material information.

An analogous modal representation of resonating objects was also adopted by Avanzini et al. (2003). The main difference with the above mentioned works lies in the approach to contact force modelling. While van den Doel and coworkers adopt a feed-forward scheme in which the interacting resonators are set into oscillation with driving forces that are externally computed or recorded, the models proposed by Avanzini et al. (2003) embed direct computation of non-linear contact forces. Despite the complications that arise in the synthesis algorithms, this approach provides some advantages. Better quality is achieved due to accurate audio-rate computation of contact forces: this is especially true for impulsive contact, where contact times are in the order of few ms. Interactivity and responsiveness of sound to user actions is also improved. This is especially true for continuous contact, such as stick-slip friction (Avanzini et al., 2005). Finally, physical parameters of the contact force models provide control over other ecological dimensions of the sound events. As an example, the impact model used by Avanzini et al. (2003), and originally proposed by Hunt and Crossley (1975), describe the non-linear contact force as

$$f(x(t), v(t)) = \begin{cases} kx(t)^\alpha + \lambda x(t)^\alpha \cdot v(t)\,(1 + \mu v(t)) & x > 0, \\ 0 & x \leq 0, \end{cases} \tag{9.3}$$

where x is the interpenetration of the two colliding objects and $v = \dot{x}$. Then force parameters, such as the force stiffness k, can be related to ecological dimensions of the produced sound, such as perceived stiffness of the impact. Similar considerations apply to continuous contact models (Avanzini et al., 2005).

It has been shown that this approach allows for a translation of the map of everyday sounds proposed by Gaver into a hierarchical structure in which "patterned" and "compound" sounds models are built upon low-level, "basic" models of impact and friction (see 9.1). Models for bouncing, breaking, rolling, crumpling sounds are described in the works by Rath and Fontana (2003) and Rath and Rocchesso (2005). See also Chapter 10 for a description of "sounding objects" synthesised with this approach.

A different physically-based approach has been proposed by O'Brien et al. (2001, 2002). Rather than making use of heuristic methods that are specific to particular objects, their approach amounts to employing finite-element simulations for generating both animated video and audio. This task is accomplished by analyzing the surface motions of objects that are animated using a deformable body simulator, and isolating vibrational components that correspond to audible frequencies. The system then determines how these surface motions will generate acoustic pressure waves in the surrounding medium and models the propagation of those waves to the listener. In this way, sounds arising from complex nonlinear phenomena can be simulated, but the heavy computational load prevents real-time sound generation and the use of the method in interactive applications.

Other classes of sounds

The map of everyday sounds developed by Gaver (see figure 9.1) comprises three main classes: solids, liquids, and gases. Research on sound modelling is clearly biased toward the first of these classes, while less has been done for the others.

A physically-based liquid sound synthesis methodology has been developed by van den Doel (2005). The fundamental mechanism for the production of liquid sounds is identified as the acoustic emission of bubbles. After reviewing the physics of vibrating bubbles as it is relevant to audio synthesis, the author has developed a sound model for isolated single bubbles and validated it with a small user study. A stochastic model for the real-time interactive synthesis of complex liquid sounds such as those produced by streams, pouring water, rivers, rain, and breaking waves is based on the synthesis of single bubble sounds. It is shown by van den Doel (2005) how realistic complex high dimensional sound spaces can be synthesised in this manner.

Dobashi et al. (2003) have proposed a method for creating aerodynamic sounds. Examples of aerodynamic sound include sound generated by swinging swords or by wind blowing. A major source of aerodynamic sound is vortices generated in fluids such as air. The authors have proposed a method for creating sound textures for aerodynamic sounds by making use of computational fluid dynamics. Next, they have developed a method using the sound textures for real-time rendering of aerodynamic sound according to the motion of objects or wind velocity.

This brief overview shows that little has been done in the literature about models of everyday sounds in the "liquids" and "gases" categories (we are sticking to the terminology used by Gaver (1993a), and reported in figure 9.1). These are topics that need more research to be carried out in the future.

9.4.3 Applications to multimodal interfaces

Multimodal rendering

An important consequence of using physically-based sound models is that synchronisation with other modalities is in principle straightforward, since the parameters that are needed to characterise the sounds resulting from mechanical contact come directly from the simulation. In other cases where only simple kinematic information like trajectory is present, needed information like velocity and acceleration can be calculated.

A particularly interesting problem is simultaneous audio-haptic rendering. There is a significant amount of literature that deals with the design and the evaluation of interfaces that involve auditory feedback in conjunction with haptic/tactile feedback. In order to be perceived as realistic, auditory and haptic cues have to be synchronised so that they appear simultaneous. They must also be perceptually similar – a rough surface has to both sound and feel rough. Synchronizing the two modalities is more than synchronizing two separate events. Rather than triggering a pre-recorded audio sample or tone, the audio and the haptics change together when the user applies different forces to the object.

Rendering a virtual surface, i.e. simulating the interaction forces that arises when touching a stiff object, is the prototypical haptic task. Properly designed visual (Wu et al., 1999) and/or auditory (DiFranco et al., 1997) feedback can be combined with haptics in order to improve perception of stiffness, or even compensate for physical limitations of haptic devices and enhance the range of perceived stiffness that can be effectively conveyed to the user. Physical limitations (low sampling rates, poor spatial resolution of haptic devices) constrain the values for haptic stiffness rendering to ranges that are often far from typical values for stiff surfaces (Kuchenbecker et al., 2006). Ranges for haptic stiffness are usually estimated by requiring the system to be passive (Colgate and Brown, 1994), thus guaranteeing stability of the interaction, while higher stiffness values can cause the system to become unstable, i.e. to oscillate in an uncontrolled way.

Perceptual experiments on a platform that integrates haptic and sound displays were reported by DiFranco et al. (1997). Prerecorded sounds of contact between several pairs of objects were played to the user through the headphones to stimulate the auditory senses. The authors studied the influence of auditory information on the perception of object stiffness through a haptic interface. In particular, contact sounds influenced the perception of object stiffness during tapping of virtual objects through a haptic interface. These results suggest that, although the range of object stiffness that can be displayed by a haptic interface is limited by the force-bandwidth of the interface, the range perceived by the subject can be effectively increased by the addition of properly designed impact sounds.

While the auditory display adopted by DiFranco et al. (1997) was rather poor (the authors used recorded sounds), a more sophisticated approach amounts to synthesise both auditory and haptic feedback using physically-based models. This approach was taken in the work of DiFilippo and Pai (2000). In this work the modal synthesis techniques described by van den Doel and Pai (1998) were applied to audio-haptic rendering. Contact forces are computed at the rate of the haptic rendering routine (e.g., 1kHz), then the force signals are upsampled at the rate of the audio rendering routine (e.g., 44.1kHz) and filtered in order to remove spurious impulses at contact breaks and high frequency position jitter. The resulting audio force is used to drive the modal sound model. This architecture ensures low latency between haptic and audio rendering (the latency is 1ms if the rate of the haptic rendering routine is 1kHz), which is below the perceptual tolerance for detecting synchronisation between auditory and haptic contact events.

A related study was recently conducted by Avanzini and Crosato (2006). In this paper the sound models proposed by Avanzini et al. (2003, 2005) were integrated into a multimodal rendering architecture, schematically depicted in Fig. 9.3, which extends typical haptic-visual architectures (Salisbury et al., 2004). The sound rendering thread runs at audio rate (e.g. 44.1kHz) in parallel with other threads. Computation of audio contact forces is triggered by collision detection from the haptic rendering thread. Computation of 3D sound can be cascaded to the sound synthesis block. It was shown that the proposed

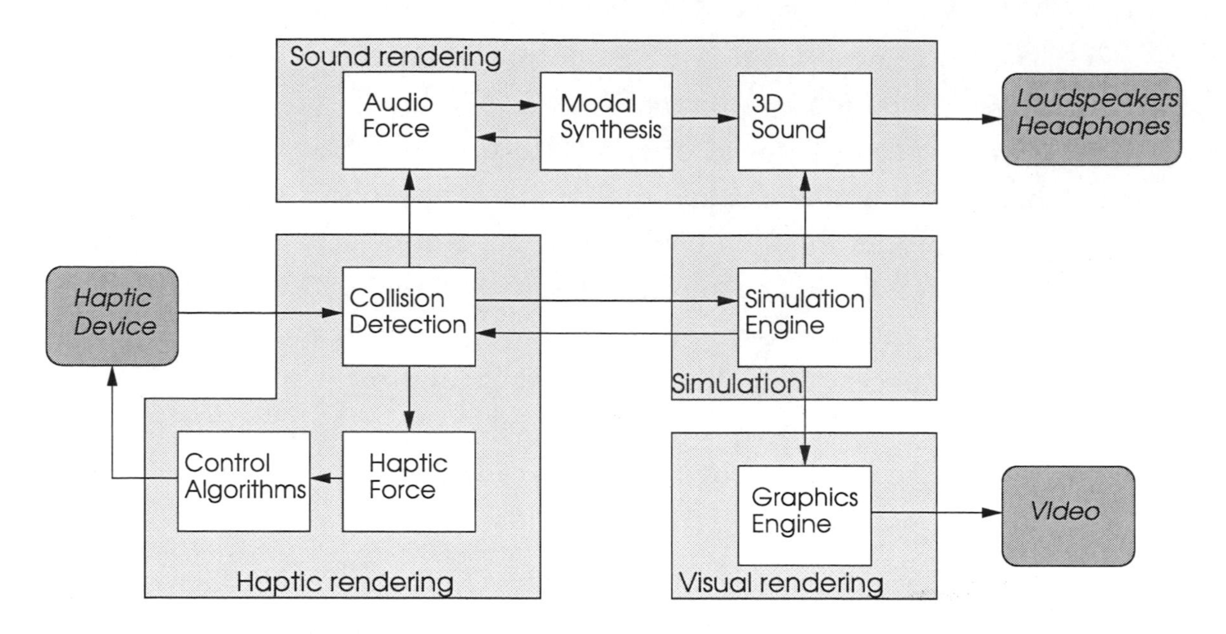

Figure 9.3: An architecture for multimodal rendering of contact interactions. Adapted from Fig. 3 in Salisbury et al. (2004).

rendering scheme allows tight synchronisation of the two modalities, as well as a high degree of interactivity and responsiveness of the sound models to gestures and actions of a user. The setup was used to run an experiment on the relative contributions of haptic and auditory information to bimodal judgments of contact stiffness: experimental results support the effectiveness of auditory feedback in modulating haptic perception of stiffness.

Substituting modalities

In Section 9.3.2 we have already reviewed some studies that address the topic of sensory substitution with applications to the design of interfaces. The focus of such studies (Ifukube et al., 1991; Kaczmarek et al., 1991; Meijer, 1992) is especially substitution systems for visually-impaired users. The very same idea of sensory substitution can be exploited in a different direction: having an interface which is not able to provide feedback of a given modality (e.g. a passive device such as a standard mouse is not able to provide haptic feedback), that modality can effectively substituted with feedback of other modalities, provided that it uses the same sensory-motor skills. We will try to clarify this concept in the remainder of this section. The studies briefly reviewed in the following, although not specifically related with audio but rather with visual and haptic feedback, contain interesting ideas that may be applied to auditory rendering.

Lécuyer et al. (2000) developed interaction techniques for simulating contact without a haptic interface, but with a passive input device combined with the visual feedback of a basic computer screen. The authors exemplify the general idea as follows: assume that a user manipulates a cube in a VE using a passive device like a mouse, and has to insert it inside a narrow duct. As the cube is inserted in the duct, it "visually resists" motion by reducing its speed, and consequently the user increases the pressure on the mouse which results in an increased feedback force by the device. The combined effects of the visual slowing down of the cube and the increased feedback force from the device provides the user with an "illusion" of force feedback, as if a friction force between the cube and the duct was rendered haptically. Lecuyer and coworkers have applied this idea to various interactive tasks, and have shown that properly designed visual feedback can to a certain extent provide a user with "pseudo-haptic" feedback.

Similar ideas have driven the work of van Mensvoort (2002), who developed a cursor interface in which the cursor position is manipulated to give feedback to the user. The user has main control over the cursor movements, but the system is allowed to apply tiny displacements to the cursor position. These displacements are similar to those experienced when using force-feedback systems, but while in force-feedback systems the location of the cursor changes due to the force sent to the haptic display, in this case the cursor location is directly manipulated. These active cursor displacements result in interactive animations that induce haptic sensations like stickiness, stiffness, or mass.

The same approach may be experimented with auditory instead of visual feedback: audition indeed appears to be an ideal candidate modality to support illusion of substance in direct manipulation of virtual objects, while in many applications the visual display does not appear to be the best choice as a replacement of kinesthetic feedback. Touch and vision represent different priorities, with touch being more effective in conveying information about "intensive" properties (material, weight, texture, and so on) and vision emphasizing properties related to geometry and space (size, shape). Moreover, the auditory system tends to dominate in judgments of temporal events, and intensive properties strongly affect the temporal behaviour of objects in motion, thus producing audible effects at different time scales.

Bibliography

J. M. Adrien. The missing link: Modal synthesis. In G. De Poli, A. Piccialli, and C. Roads, editors, *Representations of Musical Signals*, pages 269–297. MIT Press, Cambridge, MA, 1991.

F. Avanzini and P. Crosato. Integrating physically-based sound models in a multimodal rendering architecture. *Comp. Anim. Virtual Worlds*, 17(3-4): 411–419, July 2006.

F. Avanzini, M. Rath, D. Rocchesso, and L. Ottaviani. Low-level sound models: resonators, interactions, surface textures. In D. Rocchesso and F. Fontana, editors, *The Sounding Object*, pages 137–172. Mondo Estremo, Firenze, 2003.

F. Avanzini, S. Serafin, and D. Rocchesso. Interactive simulation of rigid body interaction with friction-induced sound generation. *IEEE Trans. Speech Audio Process.*, 13(6), Nov. 2005.

J. P. Bresciani, M. O. Ernst, K. Drewing, G. Bouyer, V. Maury, and A. Kheddar. Feeling what you hear: auditory signals can modulate tactile tap perception. *Exp. Brain Research*, In press, 2005.

C. Carello and M. T. Turvey. The ecological approach to perception. In Encyclopedia of cognitive science. London, Nature Publishing Group, 2002.

C. Carello, K. L. Anderson, and A. Kunkler-Peck. Perception of object length by sound. *Psychological Science*, 9(3):211–214, May 1998.

J. E. Colgate and J. M. Brown. Factors Affecting the Z-Width of a Haptic Display. In *Proc. IEEE Int. Conf. on Robotics & Automation*, pages 3205–3210, San Diego, May 1994.

P. R. Cook. *Real sound synthesis for interactive applications*. A. K. Peters, Natick, MA, USA, 2002.

D. DiFilippo and D. K. Pai. The AHI: An audio and haptic interface for contact interactions. In *Proc. ACM Symp. on User Interface Software and Technology (UIST'00)*, San Diego, CA, Nov. 2000.

D. E. DiFranco, G. L. Beauregard, and M. A. Srinivasan. The effect of auditory cues on the haptic perception of stiffness in virtual environments. In *Proceedings of the ASME Dynamic Systems and Control Division, Vol.61*, 1997.

Y. Dobashi, T. Yamamoto, and T. Nishita. Real-time rendering of aerodynamic sound using sound textures based on computational fluid dynamics. In *Proc. ACM SIGGRAPH 2003*, pages 732–740, San Diego, CA, July 2003.

M. O. Ernst and H. H. Bülthoff. Merging the senses into a robust percept. *TRENDS in Cognitive Sciences*, 8(4):162–169, Apr. 2004.

D. J. Freed. Auditory correlates of perceived mallet hardness for a set of recorded percussive events. *J. Acoust. Soc. Am.*, 87(1):311–322, Jan. 1990.

W. W. Gaver. What in the world do we hear? an ecological approach to auditory event perception. *Ecological Psychology*, 5(1):1–29, 1993a.

W. W. Gaver. How do we hear in the world? explorations of ecological acoustics. *Ecological Psychology*, 5(4):285–313, 1993b.

J. J. Gibson. *The ecological approach to visual perception*. Lawrence Erlbaum Associates, Mahwah, NJ, 1986.

B. Giordano. *Sound source perception in impact sounds*. PhD thesis, Department of General Psychology, University of Padova, Italy, 2006. URL `http://www.music.mcgill.ca/~bruno/`.

S. Guest, C. Catmur, D. Lloyd, and C. Spence. Audiotactile interactions in roughness perception. *Exp. Brain Research*, 146(2):161–171, Sep. 2002.

B. Gygi, G. R. Kidd, and C. S. Walson. Spectral-temporal factors in the identification of environmental sounds. *J. Acoust. Soc. Am.*, 115(3):1252–1265, Mar. 2004.

J. K. Hahn, H. Fouad, L. Gritz, and J. W. Lee. Integrating sounds in virtual environments. *Presence: Teleoperators and Virtual Environment*, 7(1):67–77, Feb. 1998.

K. Hötting and B. Röder. Hearing Cheats Touch, but Less in Congenitally Blind Than in Sighted Individuals. *Psychological Science*, 15(1):60, Jan. 2004.

K. H. Hunt and F. R. E. Crossley. Coefficient of restitution interpreted as damping in vibroimpact. *ASME J. Applied Mech.*, 42:440–445, June 1975.

T. Ifukube, T. Sasaki, and C. Peng. A blind mobility aid modeled after echolocation of bats. *IEEE Trans. Biomedical Engineering*, 38(5):461–465, May 1991.

K. A. Kaczmarek, J. G. Webster, P. Bach-y-Rita, and W. J. Tompkins. Electrotactile and vibrotactile displays for sensory substitution systems. *IEEE Trans. Biomedical Engineering*, 38(1):1–16, Jan. 1991.

R. L. Klatzky, D. K. Pai, and E. P. Krotkov. Perception of material from contact sounds. *Presence: Teleoperators and Virtual Environment*, 9(4):399–410, Aug. 2000.

K. J. Kuchenbecker, J. Fiene, and G. Niemeyer. Improving Contact Realism through Event-Based Haptic Feedback. *IEEE Trans. on Visualization and Comp. Graphics*, 13(2):219–230, Mar. 2006.

A. Lécuyer, S. Coquillart, and A. Kheddar. Pseudo-haptic feedback: Can isometric input devices simulate force feedback? In *IEEE Int. Conf. on Virtual Reality*, pages 83–90, New Brunswick, 2000.

S. J. Lederman. Auditory texture perception. *Perception*, 8(1):93–103, Jan. 1979.

S. J. Lederman, R. L. Klatzki, T. Morgan, and C. Hamilton. Integrating multimodal information about surface texture via a probe: Relative contribution of haptic and touch-produced sound sources. In *Proc. IEEE Symp. Haptic Interfaces for Virtual Environment and Teleoperator Systems (HAPTICS 2002)*, pages 97–104, Orlando, FL, 2002.

X. Li, R. J. Logan, and R. E. Pastore. Perception of acoustic source characteristics: Walking sounds. *J. Acoust. Soc. Am.*, 90(6):3036–3049, Dec. 1991.

R. A. Lutfi and E. L. Oh. Auditory discrimination of material changes in a struck-clamped bar. *J. Acoust. Soc. Am.*, 102(6):3647–3656, Dec. 1997.

M. R. McGee, P. Gray, and S. Brewster. Mixed feelings: Multimodal perception of virtual roughness. In *Proc. Int. Conf EuroHaptics*, pages 47–52, Edinburgh, July 2002.

H. McGurk and J. MacDonald. Hearing lips and seeing voices. *Nature*, 264 (5588):746–748, Dec. 1976.

P. B. L. Meijer. An experimental system for auditory image representations. *IEEE Trans. Biomedical Engineering*, 39(2):112–121, Feb. 1992.

C. F. Michaels and C. Carello. *Direct Perception.* Prentice-Hall, Englewood Cliffs, NJ, 1981.

S. Morein-Zamir, S. Soto-Faraco, and A. Kingstone. Auditory capture of vision: examining temporal ventriloquism. *Cognitive Brain Research*, 17:154–163, 2003.

A. Noë. *Action in perception.* MIT press, Cambridge, Mass., 2005.

J. F. O'Brien, P. R. Cook, and G. Essl. Synthesizing sounds from physically based motion. In *Proc. ACM SIGGRAPH 2001*, pages 529–536, Los Angeles, CA, Aug. 2001.

J. F. O'Brien, C. Shen, and C. M. Gatchalian. Synthesizing sounds from rigid-body simulations. In *Proc. ACM SIGGRAPH 2002*, pages 175–181, San Antonio, TX, July 2002.

J. K. O'Regan and A. Noë. A sensorimotor account of vision and visual consciousness. *Behavioral and Brain Sciences*, 24(5):883–917, 2001.

M. Rath and F. Fontana. High-level models: bouncing, breaking, rolling, crumpling, pouring. In Davide Rocchesso and Federico Fontana, editors, *The Sounding Object*, pages 173–204. Mondo Estremo, Firenze, 2003.

M. Rath and D. Rocchesso. Continuous sonic feedback from a rolling ball. *IEEE Multimedia*, 12(2):60–69, Apr. 2005.

B. H. Repp. The sound of two hands clapping: an exploratory study. *J. Acoust. Soc. Am.*, 81(4):1100–1109, Apr. 1987.

K. Salisbury, F. Conti, and F. Barbagli. Haptic Rendering: introductory concepts. *IEEE Computer Graphics and Applications*, 24(2):24–32, Mar. 2004.

Sensegraphics. Website, 2006. http://www.sensegraphics.se.

L. Shams, Y. Kamitani, and S. Shimojo. Visual illusion induced by sound. *Cognitive Brain Research*, 14(1):147–152, June 2002.

M. A. Srinivasan and C. Basdogan. Haptics in virtual environments: taxonomy, research status, and challenges. *Comput. & Graphics*, 21(4):393–404, July 1997.

T. A. Stoffregen. Affordances and events. *Ecological Psychology*, 12(1):1–28, Winter 2000.

T. A. Stoffregen and B. G. Bardy. On specification and the senses. *Behavioral and Brain Sciences*, 24(2):195–213, Apr. 2001.

Novint Technologies. The interchange of haptic information. In *Proc. Seventh Phantom Users Group Workshop (PUG02)*, Santa Fe, Oct. 2002.

K. van den Doel. Physically based models for liquid sounds. *ACM Trans. Appl. Percept.*, 2(4):534–546, Oct. 2005.

K. van den Doel and D. K. Pai. The sounds of physical shapes. *Presence: Teleoperators and Virtual Environment*, 7(4):382–395, Aug. 1998.

K. van den Doel, P. G. Kry, and D. K. Pai. Foleyautomatic: Physically-based sound effects for interactive simulation and animation. In *Proc. ACM SIGGRAPH 2001*, pages 537–544, Los Angeles, CA, Aug. 2001.

K. van Mensvoort. What you see is what you feel – exploiting the dominance of the visual over the haptic domain to simulate force-feedback with cursor displacements. In *Proc. ACM Conf. on Designing Interactive Systems (DIS2004)*, pages 345–348, London, June 2002.

F. Varela, E. Thompson, and E. Rosch. *The Embodied Mind.* MIT Press, Cambridge, MA, 1991.

W. H. Warren and R. R. Verbrugge. Auditory perception of breaking and bouncing events: Psychophysics. In W. A. Richards, editor, *Natural Computation*, pages 364–375. MIT Press, Cambridge, Mass., 1988.

R. B. Welch and D. H. Warren. Intersensory interactions. In K. R. Boff, L. Kaufman, and J. P. Thomas, editors, *Handbook of Perception and Human Performance – Volume 1: Sensory processes and perception*, pages 1–36. John Wiley & Sons, New York, 1986.

R. P. Wildes and W. A. Richards. Recovering material properties from sound. In W. A. Richards, editor, *Natural Computation*, pages 357–363. MIT Press, Cambridge, Mass., 1988.

W. C. Wu, C. Basdogan, and M. A. Srinivasan. Visual, haptic, and bimodal perception of size and stiffness in virtual environments. In *Proceedings of the ASME Dynamic Systems and Control Division, (DSC-Vol.67)*, 1999.

U. Zölzer, editor. *DAFX – Digital Audio Effects.* John Wiley & Sons, 2002.

Chapter 10

Sound Design and Auditory Displays

Amalia de Götzen[1], Pietro Polotti[2], Davide Rocchesso[3]

[1]Department of Information Engineering, University of Padua
[2]Department of Computer Science, University of Verona
[3]Department of Art and Industrial Design, IUAV University of Venice

About this chapter

The goal of this chapter is to define the state of the art of research in sound design and auditory display. The aim is to provide a wide overview of the extremely different fields, where these relatively new disciplines find application. These fields range from warning design and computer auditory display to architecture and media.

10.1 Introduction

Sounds in human-computer interfaces have always played a minor role as compared to visual and textual components. Research efforts in this segment of Human-Computer Interaction (HCI) have also been relatively little, as testified by the relatively new inclusion of Sound and Music Computing (H.5.5 in the ACM Computing Classification System[2]) as a sub-discipline of Information Interfaces and Presentation (H.5). The words sound or audio do not appear in any other specification of level-one or level-two items of the hierarchy. On the other hand, for instance, computer graphics is a level-two item on its own (I.3), and Image Processing and Computer Vision is another level–two item (I.4).

So, the scarcity of literature, especially the lack of surveys of the field, do not come as a surprise. Indeed, a survey was published by Hereford and Winn (1994), where a deep investigation of the state of the art of sound usage in Human-Computer Interaction was presented. The main important topics of that overview are: Earcons (symbolic and iconic), and sound in data sonification and in virtual reality environments. The literature study follows some important applications, pointing out successes and problems of the interfaces, always pushing the reader to think about lack of knowledge and need of further explorations. The paper ends with useful guidelines for the interface designer who uses sound, trying to stress the need to improve the knowledge about how people interpret auditory messages and about how sound can be used in human-computer interfaces to convey the information contained in some data set. The knowledge about sound perception is not enough to perform good interactions, as the nature of the interface affects the creation of users' mental models of the device.

The rest of this chapter intends to go beyond Hereford's survey, in several ways. We consider a selection of major works that appeared in the field in the last couple of decades. These works have been either very influential for following researches, or have appeared in respected journals thus being likely to affect wide audiences.

The chapter organisation into six main sections reflects the topics that

[2]http://www.acm.org/class/

have been most extensively studied in the literature, i.e. warnings, earcons, auditory icons, mapping, sonification, and sound design. Indeed, there are wide overlaps between these areas (e.g. sound design and mapping can be considered as part of a task of sonification). However, several important works have been found that are in some way specific and representative of each area. The overall progression of the chapter goes from simple and low-level to complex and high-level. This is similar to what one would find in the table of contents of a textbook of the relative, more consolidated, discipline of information visualisation (Ware, 2004): from basic to complex issues in information and perception, towards information spaces and data mapping, and finally to interaction with displays and data. In our chapter, special relevance has been assigned to the concept of mapping, for the importance it had in auditory display, as well as in the different context of computer music programming and performance.

10.2 Warnings, alerts and audio feedback

Auditory warnings are perhaps the only kind of auditory displays that have been thoroughly studied and for whom solid guidelines and best design practices have been formulated. A milestone publication summarizing the multi-faceted contributions to this sub-discipline is the book edited by Stanton and J.Edworthy (1999a). This book opening chapter summarises well the state of the art in human factors for auditory warnings as it was in the late nineties. Often warnings and alerts are designed after anecdotal evidence, and this is also the first step taken by the authors as they mention problems arising in pilot cockpits or central control rooms. Then, auditory displays are confronted against visual displays, to see how and when to use one sensory channel instead of the other. A good observation is that hearing tends to act as a natural warning sense. It is the ears-lead-eyes pattern[1] that should be exploited. The authors identify four areas of applications for auditory warnings: personal devices, transport, military, and control rooms. Perhaps a fifth important area is geographic-scale alerts, as found in a paper by Avanzini et al. (2004).

[1] http://c2.com/cgi/wiki?SonificationDesignPatterns

The scientific approach to auditory warnings is usually divided into the two phases of hearing and understanding, the latter being influenced by training, design, and number of signals in the set. Studies in hearing triggered classic guidelines such as those individuated by Patterson (1990). He stated, for instance, that alarms should be set between 15 and 25 dB above the masked threshold of environment. Patterson faced also the issue of design for understanding, by suggesting a sound coding system that would allow mapping different levels of urgency.

The possibility that using naturalistic sounds could be better for retention is discussed in the introductory chapter of Stanton and J.Edworthy (1999a), especially with reference to the works of Blattner et al. (1989) and Gaver (1994). The problem of the legacy with traditional warnings is also discussed (sirens are usually associated with danger, and horns with mechanical failures). The retention of auditory signals is usually limited to 4 to 7 items that can be acquired quickly, while going beyond is hard. In order to ease the recalls, it is important to design the temporal pattern accurately. Moreover, there is a substantial difference in discriminating signals in absolute or relative terms. In the final part of their introductory chapter, Stanton and Edworthy focus on their own work on the classification of alarm-related behaviours, especially Alarm-Initiated Activities (AIA) in routine events (where ready-made responses are adequate) and critical events (where deductive reasoning is needed). In the end, designing good warnings means balancing between attention-getting quality of sound and impact on routine performance of operators.

In the same book, the chapter "Auditory warning affordances" (Stanton and J.Edworthy, 1999b) is one of the early systematic investigations on the use of "ecological" stimuli as auditory warnings. The expectation is that sounds that are representative of the event to which they are alarming would be more easily learnt and retained. By using evocative sounds, auditory warnings should express a potential for action: for instance, sound from a syringe pump should confer the notion of replacing the drug. Here, a methodology for designing "ecological" auditory warnings is given, and it unrolls through the phases of highlighting a reference function, finding or generating appropriate sounds, ranking the sounds for appropriateness, evaluating properties in terms of learning and confusion, mapping urgency onto sounds. A study aimed at testing the theory of auditory affordances is conducted by means of nomic (heartbeat for ECG monitor), symbolic (nursery chime for infant warmer), or metaphoric (bubbles for syringe pump) sound associations. Some results are that:

- Learned mappings are not easy to override;
- There is a general resistance to radical departures in alarm design practice;
- Suitability of a sound is easily outweighed by lack of identifiability of an alarm function.

However, for affordances that are learnt through long-time practice, performance can still be poor if an abstract sound is chosen. As a final remark for further research, the authors recommend to get the end users involved when designing new alarms. This is a call for more participatory design practices that should apply to auditory interface components in general, and not only to warnings.

If one considers a few decades of research in human-machine interfaces, the cockpit is one of the most extensively studied environments, even from an acoustic viewpoint. It is populated by alarms, speech communications, and it is reached by "natural" sounds, here intended as produced by system processes or events, such as mechanical failures. In the framework of the functional sounds of auditory warning affordances, Ballas (1999) proposes five linguistic functions used to analyze the references to noise in accident briefs: exclamation, deixis (directing attention), simile (interpretation of an unseen process), metaphor (referring to another type of sound-producing event), and onomatopoeia. To see how certain acoustic properties of the sounds affect the identification of brief sound phenomena, an acoustic analysis was performed on a set of 41 everyday sounds. A factor related to perceptual performance turned out to be the union of (i) harmonics in continuous sounds or (ii) similar spectral patterns in bursts of non-continuous sounds. This union is termed H_{st} and it describes a form of spectral/temporal entropy. The author notices that the warning design principles prescribe similar spectral patterns in repeated bursts, a property similar to H_{st}. An innovative point of this paper is that counting the pulses can give a hint for identification performance. Experimental results give some evidence that the repetition of a component improves identification, whereas the aggregation of different components impairs identification. In the last section, the chapter describes the work of F. Guyot, who investigated the relationship between cognition and perception in categorisation of everyday sounds. She suggested three levels for the categorisation (abstraction) process:

1. Type of excitation;

2. Movement producing the acoustic pattern;

3. Event identification.

Her work is related with the work of Schafer (1994), Gaver (1993b), and Ballas (1993). In particular, Ballas' perceptual and cognitive clustering of 41 sounds resulted in the categories:

- Water-related;

- Signalling and danger-related;

- Doors and modulated noises;

- Two or more transient components.

Finally, Ballas' chapter provides a connection with the soundscape studies of ecological acousticians (see also Section 10.7.3).

Special cases of warnings are found where it is necessary to alert many people simultaneously. Sometimes, these people are geographically spread, and new criteria for designing auditory displays come into play. In Avanzini et al. (2004) the authors face the problem of a system alert for the town of Venice, periodically flooded by the so-called "acqua alta", i.e. the high tide that covers most of the town with 10-40 cm of water. For more than three decades, a system of eight electromechanical and omnidirectional sirens have been providing an alert system for the whole historic town.

A study of the distribution of the signal levels throughout the town was first performed. A noise map of the current alert system used in Venice was realised by means of a technique that extracts building and terrain data from digital city maps in ArcView format with reasonable confidence and limited user intervention. Then a sound pressure level map was obtained by importing the ArcView data into SoundPLAN, an integrated software package for noise pollution simulations. This software is mainly based on a ray tracing approach. The result of the analysis was a significantly non-uniform distribution of the SPL throughout the town. One of the goals of this work is, thus, the redefinition and optimisation of the distribution of the loudspeakers. The authors considered a Constraint Logic Programming (CLP) approach to the problem. CLP is particularly effective for solving combinatorial minimisation problems. Various criteria were considered in proposing new emission points. For instance, the aforementioned Patterson's recommendations require that the acoustic stimulus must be about 15 dB above the background noise to be clearly perceived. Also, installation and maintenance costs make it impractical to install more than 12 loudspeakers in the city area. By taking into account all of these factors, a much more effective distribution of the SPL of the alert signals was achieved. The second main issue of this work is the sound design of the alert signals. In this sense the key questions here considered are:

- How to provide information not only about the arrival of the tide but also about the magnitude of the phenomenon;

- How to design an alert sound system that would not need any listening-training, but only verbal/textual instructions.

Venice being a tourist town, this latter point is particularly important. It would mean that any person should intuitively understand what is going on, not only local people. The choices of the authors went towards abstract signals, i.e. earcons, structured as a couple of signals, according to the concept of "attenson" (attention-getting sounds). The two sound stages specify the rising of the tide and the tide level, respectively. Also, the stimulus must be noticeable without being threatening. The criteria for designing sounds providing different urgency levels were the variation of the fundamental frequency, the sound inharmonicity and the temporal patterns.

The validation of the model concludes the paper. The subjects did not receive any training but only verbal instructions. The alert signal proved to be effective, and no difference between Venetians and non-Venetians was detected. In conclusion, a rich alert model for a very specific situation and for a particular purpose was successfully designed and validated. The model takes into account a number of factors ranging from the topography and architecture of Venice, to the need of culturally non-biased alert signal definition, as well as to the definition of articulated signals able to convey the gravity of the event in an intuitive way.

10.3 Earcons

Blattner, Sumikawa and Greenberg introduced the concept of *earcons* (Blattner et al., 1989), defining them as "non-verbal audio messages that are used in computer/user interfaces to provide information to the user about some computer object, operation or interaction". These messages are called *motives*, "brief succession of pitches arranged in such a way as to produce a tonal pattern sufficiently distinct to allow it to function as an individual recognisable entity". Earcons must be learned, since there is no intuitive link between the sound and what it represents: the earcons are abstract/musical signals as opposed to auditory icons (Gaver, 1993a), where natural/everyday sounds are used in order to build auditory interfaces (see Section 10.4).

In 1998, Brewster presented a new structured approach to auditory display defining composing rules and a hierarchical organisation of musical parameters (timbre, rhythm, register, etc.), in order to represent hierarchical organisations of computer files and folders (Brewster, 1998). Typical applications of this work are telephone-based interfaces (TBIs), where navigation is a problem because the visual display is small or absent. As already mentioned, the main idea is to define a set of sound-design/composing rules for very simple "musical atoms", the earcons, with the characteristics of being easily distinguishable one from the other.

The three experiments described in the paper explore different aspects of the earcons. The first one is more "abstract" and aims at defining easily recognisable and distinguishable earcons. The second one addresses the very concrete problem of lo-fi situations, where monophony of signals and a limited bandwidth are strong limitations. In the same experiment the fundamental aspect of "musical memory" is considered: the navigation test was carried out right after the training and repeated after one week. As far as retention is concerned, very good results were achieved: there was no significant difference between the performances right after the training and after one week. On the contrary, in some cases the listeners were even more skilled in recognizing the earcons one week later than immediately after the training. An interesting feedback coming from the experiments was that the listeners developed mnemonic strategies based on the identification of the earcons with something external as, for example, geometric shapes (triangles and so on). This could be a good cue for earcon sound design. The third experiment is a bit more artificial: the idea is to identify a sound (timbre+ register) with numbers and to represent hierarchies in a book-like style (chapter, sections, subsections) by means of "sounding numbers". In general, these experiments show how problematic the design of earcons is, when many hierarchical levels are involved or when many items are present: one needs to think about very articulated or even polyphonic earcons, challenging the listening skills of the user. In any case, situations which do not present very complex navigation requirements (as in the case of TBI applications), can build upon earcons a robust and extensible method for representing hierarchies.

The work presented in a recent paper by McGookin and Brewster (2004) faces the problem of concurrent earcon presentation. Before tackling the problem, the paper provides a very good three-page survey about auditory display, sonification, auditory icons and earcons. The main concepts about auditory scene analysis are then presented, since they provide useful guidelines for designing more robust earcons. Two experiments are illustrated, which are also exemplary for their use of statistical analysis and workload measures. In the first experiment, the goal is to see how recognition of earcons and their parameters gets worse as the number of concurrent earcons is increased. In the second experiment, new design solutions are tested in their ability to increase the earcon robustness against concurrent presentation. It turns out that using multiple timbres or staggering the onsets will improve attribute identification. As a practical final result of the experiments, four guidelines for designing robust concurrent earcons are given.

10.4 Auditory icons

Another concept has been introduced in the nineties by Bill Gaver (1993b,c) as an earcon counterpart: auditory icons. The basic idea is to use natural and everyday sounds to represent actions and sounds within an interface. The two papers by Gaver can be considered as a foundation for later works on everyday listening: the author presents a fundamental aspect of our way of perceiving the surrounding environment by means of the auditory system. Trying to reply to the question "what do we hear in the world?" (Gaver, 1993b), a first and most relevant consideration emerges: a lot of research efforts were and are devoted to the study of musical perception, while our auditory system is first of all a tool for interacting with the outer world in everyday life.

When we consciously listen to or hear more or less unconsciously "something" in our daily experience, we do not really perceive and recognise sounds but rather events and sound sources. This "natural" listening behaviour is denoted by Gaver as "everyday listening" as opposed to "musical listening", where the perceptual attributes are those considered in the traditional research in audition. As an example, Gaver writes: "while listening to a string quartet we might be concerned with the patterns of sensation the sounds evoke (musical listening), or we might listen to the characteristics and identities of the instruments themselves (everyday listening). Conversely, while walking down a city street we are likely to listen to the sources of sounds - the size of an approaching car, how close it is and how quickly it is approaching."

Despite the importance of non-musical and non-speech sounds, the research in this field is scarce. What Gaver writes is true: we do not really know how our senses manage to gather so much information from a situation like the one of the approaching car described above. Traditional research on audition was and is concerned mainly with a Fourier approach, whose parameters are frequency, amplitude, phase and duration. On the contrary, new research on everyday sounds focuses on the study of different features and dimensions, i.e. those concerning the sound source. The new approach to perception is "ecological" (see also Chapter 9). New perceptual dimensions like size and force are introduced. More generally, the fundamental idea is that complex perceptions are related to complex stimuli (Gaver also talks about "perceptual information") and not to the integration of elementary sensations: "For instance, instead of specifying a particular waveform modified by some amplitude envelope, one can request the sound of an 8-inch metal bar struck by a soft mallet". The map of everyday sounds compiled by Gaver is based on the knowledge of how a sound source first and the environment afterwards determine the perceptive structure of an acoustical signal. "Sound provides information about an interaction of materials at a location in an environment". As an evidence of the importance of the arguments, the paper reports the results of a test based on the "simple" question: "what do you hear?" If the source of a sound is identified, people tend to answer in terms of an object and a space-time context, i.e. an event and, possibly, a place in the environment. The answer concerns the perceptual attributes of the sound only if the source is not identified. Gaver tries also to define maps based on a hierarchical organisation of everyday sounds, as described later in Section 10.7.2.

A paper complementary to that discussed above is the one in which Gaver introduces the question "how do we hear in the world?" (Gaver, 1993c). While in the previous paper the relevant perceptual dimensions of the sound-generating events were investigated, here the focus is on the acoustical information through which we gather information about the events. The starting point is once again the difference between the experience of sounds themselves (musical listening) and "the perception of the sound-producing events" (everyday listening). Taking into account the framework developed in the companion article, Gaver proposes a variety of algorithms that allow everyday sounds to be synthesised and controlled along some dimensions of their sources. He proposes to use the analysis and synthesis approach to study everyday sounds: both sounds and events can be analyzed in order to reduce the data, re-synthesised and then compared to the originals. While in synthesizing sounds of traditional musical instruments the goal is to achieve a perceptually identical version of the original, the main difference here is that we just need to convey the same information about a given aspect of the event. In order to suggest relevant source attributes, the acoustical analysis of a sound event must then be supported by a physical analysis of the event itself.

Gaver gives some examples of algorithms, starting from what he identifies as the three basic sound events (impact, scraping, and dripping) and concludes with three examples of temporally-complex events (breaking, bouncing and spilling). An example of informal physical analysis in describing complex machine sounds is given too. All these examples provide a methodology to explore acoustic information using causal listening (i.e. the capacity of identifying the causes of sounds) to guide the development of the algorithms. The final discussion of the paper concerns the methodological issues that are connected to the validation of synthesis models and suggests their application to the creation of auditory icons.

This investigation is developed in another paper written by Gaver (1993a), which can be considered a fundamental one in the history of sound design and auditory icon definition: Gaver defines in a very clear way the goals of auditory icons as vehicle of "useful information about computer events". Being a paper from the beginning of the 90s, it is not surprising that it is still very concerned about lack-of-capability and computational inefficiency of digital sampling and synthesis techniques. Some of these concerns, related to the parametrisation of physics-based synthesis algorithms, are of course still open problems (see Chapter 8), while some other issue seems to belong (fortunately) to the past. According to what are the main computer events and interactions with a computer desktop, the author follows the classification described in the previous paper and analyzes different kinds of interaction-sounds as: a) impact sounds, b) breaking/bouncing and spilling sounds and c) scraping sounds. A final section is devoted to machine sounds. This subdivision of interaction sounds is extremely clear, ranging from a simple impact to groups of impact sounds involving temporal patterns and organisation, and concluding with continuous interaction (scraping). All these classes of sounds are considered in terms of their psychoacoustic attributes, in the perspective of the definition of some physical model or spectral considerations aiming at the synthesis and parametric control of auditory icons. The fundamental statement of Gaver is that the nonlinearity of the relationship between physical parameters and perceptual results should be bypassed through a simplification of the model. The result is what he calls cartoon sounds (see Section 10.7.2).

10.5 Mapping

Auditory display in general, and sonification in particular, are about giving an audible representation to information, events, and processes. These entities can take a variety of forms and can be reduced to space- or time-varying data. In any case, the main task of the sound designer is to find an effective mapping between the data and the auditory objects that are supposed to represent them, in a way that is perceptually and cognitively meaningful.

Kramer (1994) gave a first important contribution describing the role of mediating structures between the data and the listener or, in other words, of mapping. The term audification was proposed to indicate a "direct translation of a data waveform to the audible domain for purposes of monitoring and comprehension". Examples are found in electroencephalography, seismology and, as explained in the introductory chapter of that book, in sonar signal analysis. In sonification, instead, data are used to control a sound generation, and the generation technique is not necessarily in direct relationship to the data. For instance, we can associate pitch, loudness, and rhythm of a percussive sound source with the physical variables being read from sensors in an engine. Audiation is the third term introduced here. As compared to audification and sonification, it had less fortune among the researchers. It is used to indicate all those cases where recall of the sonic experience (or auditory imagery) is necessary.

Kramer gives a nice description of "parameter nesting", a method to codify many data dimensions (multivariate information) into sound signals. He distinguishes between loudness, pitch, and brightness nesting. Nesting is resemblant of the procedure proposed by Patterson (1990) to design auditory warnings. The chapter continues discussing the advantages and drawbacks of realistic vs. abstract sounds. Also, the important issue of parameter overlap and orthogonality is discussed. When the same audible variable is used on different time scales, it is likely that a loss of clarity results. More generally, changing one sound parameter can affect another parameter. This can be advantageous for mapping related variables, otherwise it could be a problem. It is argued that orthogonality between parameters, although desirable in principle, is very difficult if not impossible to achieve. The design of a balanced display can be obtained through a combination of scaling, multiple mappings, and experiments with map sequencing and interpolation.

In designing a sonification, it is important to use beacons, which are points of orientation for data analysis. The concept of beacon is also used in navigation of virtual environments, even with an acoustic sense. In Kramer (1994), conversely, the orientation provided by beacons is not necessarily spatial. Beacons are considered as the cornerstones to build mappings, or routes from data to auditory parameters. Examples are given in process monitoring and data analysis, where the role of emerging gestalts from multivariate auditory streams is recognised. Data are often naturally grouped in families and it is useful to preserve the relationships in the auditory display. A way to do that is by using streams, as defined and researched by Bregman (1990). Data can be mapped to streams through per stream, inter-stream, and global variables. These concepts are exemplified well by an example with different plant species. A problem with streams is that it is not possible to follow more than one of them at a given time, even though ensembles of streams can be perceived as gestalts or, as some other people like to say, as textures. The chapter is concluded by discussing problems related to memory, cognition, and affection. A major problem is how to recall the mappings (see also Ballas, 1999). This can be done via metaphors (e.g. high pitch = up) or feelings (e.g. harsh = bad situation), and the interactions between the two. These aspects are still very hot and open for further research nowadays.

10.5.1 Direct (audification)

The most straightforward kind of mapping is the one that takes the data to feed the digital-to-analog converters directly, thus playing back the data at an audio sampling rate. This can be of some effectiveness only if the data are temporal series, as it is the case in seismology.

The idea of listening to the data produced by seismograms to seek relevant phenomena and improve understanding is quite old, as it is described in two papers of the sixties by Speeth (1961) and Frantti and Leverault (1965). After those exploratory works, however, no seismic audio activities had been recorded in the literature until the presentations made at early ICAD conferences and until the paper by Hayward (1994). Here it is argued that audification (direct transposition of data into sound with a minimal amount of processing) makes sense in a few cases, but seismic data offer one such case because they are produced by physical phenomena (elastic waves) that are similar for the propagation of earthquakes in rocks and for the propagation of acoustic waves in air. So, if the seismic signals are properly conditioned and transposed in frequency, they sound pretty natural to our ears, and we can use our abilities in interpreting noises in everyday conditions. The authors give a clear and brief introduction to the problems of seismology, distinguishing between exploration seismology and planetary seismology, and highlighting the problem of discrimination of the different kinds of explosions. Then, an extensive list of possible applications of auditory display in seismology is given, including education for operators, quality control of measurements, and event recognition.

One of the main motivations for using auditory display is that there are important events that are difficult to detect in visual time-series displays of noisy data, unless using complex spectral analysis. Conversely, these events are easily detected by ear. There are several problems that have to be faced when trying to sonify seismic data, especially related with the huge dynamic range (> 100 dB) and with the frequency bandwidth which, albeit restricted below 40 Hz, spans more than 17 octaves. Many of the mentioned problems cause headaches to visual analysts as well. In order to let relevant events become audible, the recorded signals have to be subject to a certain amount of processing, like gain control, time compression, frequency shift or transposition, annotation, looping, stereo placement. All these techniques are described fairly accurately by Hayward (1994), with special emphasis on frequency doubling: for this effect it is recommended to derive the analytic signal via Hilbert transform and from this calculate the double-frequency signal via straightforward trigonometric formulas. The technique works well for sine waves but it is not free of artifacts for real signals.

The most valuable part of Hayward's chapter is in the audification examples, which are given with reference to the soundfiles enclosed in the companion CD. First, synthetic data from an earth model are sonified, then field data of different types are analyzed. Listening to these examples is the best way to have an idea of the possibilities of audification of seismic data. A remark is given for annotation tones, which are necessary to help orienting the listener. These are similar to beacons in other applications of auditory display, but Hayward recommends to make them "similar to the signal or the listener will perceive two unrelated streams and it will be difficult to relate the timing between the two". This problem of the accurate temporal localisation of diverse auditory events is a relevant phenomenon that should be always considered when designing auditory displays. To conclude, the contribution by Hayward has been very important to launch further studies and experimentations in seismic audification. As the author wrote, the best use of audification will be obtained when these techniques will be integrated with visual displays and given to operators for their routine work.

10.5.2 Naturalistic

In some cases, it is possible to use natural or mechanical sounds to convey information of various kinds. This is especially effective when the information is physically related to the reference sound sources, so that our everyday physical experience can be exploited in interpreting the sounds.

The chapter by Fitch and Kramer (1994) is probably the first rigorous study that tries to compare the auditory and visual sensory channels in a complex monitoring task, where actions have to be taken in response to a variety of configurations of system variables. The system to be monitored is the human body, and the visual display is chosen from standard practice in anesthesiology. The auditory display is designed from scratch as a hybrid of realistic and abstract sounds. The work of Gaver (1994) is explicitly cited as a source of design inspiration and guidelines for realistic (everyday) auditory displays. Moreover, layers of parameters are superimposed on the auditory streams by the principles of parameter nesting (Kramer, 1994). No use of spatial cues in sound is made. This choice is theoretically well founded as it is supported by the evidence that, while space is the principal dimension of vision, time is the principal dimension of audition. This echoes the theory of indispensable attributes by Kubovy and Van Alkenburg (2001). The task under analysis and the data considered here have a temporal structure and are inherently concurrent. The authors make sure that there is informational equivalence between the visual and the auditory display, and this is assessed in the early experimental stages by measuring the accuracy in recognizing different configurations.

The experimental results show that, for this situation, users react faster with the auditory display than with the visual display. Moreover, a mixed audio-visual display does not give any advantage over the pure visual display, thus indicating a visual bias in presence of dual stimuli. The authors emphasise the emergence of gestalts from complex auditory information. In other words, users are capable to process an ensemble of audio streams as a whole and to readily identify salient configurations.

10.5.3 Abstract

A good mapping can be the key to demonstrate the superiority of auditory over other forms of display for certain applications. Indeed, researchers in sonification and auditory display have long been looking for the killer application for their findings and intuitions. This is especially difficult if the data are not immediately associable with sound objects, and abstract mappings have to be devised. Some researchers looked at the massive data generated by stock market exchanges to see if sonification could help enhancing the predictive capabilities of operators. The paper by Nesbitt and Barrass (2004) documents a large-scale effort aimed at providing a multimodal display for the exploration of stock market data, where 3D immersive graphics is combined with 2D manipulated voice sounds. For visuals, the proposed mapping supports focus and context. For sounds, at the schema level a Bid-Ask landscape metaphor is used, where the audio echoes the changing tension between buyers and sellers. At the perceptual level, data are mapped to timbre (bids vs. asks), loudness, and pitch. With a within-subjects test, the prediction capabilities under auditory, visual, or multisensory feedback are tested. Data show that the auditory and multisensory feedbacks perform similarly, thus indicating sensory redundancy for this specific application. Analyzing comments, it emerged that such redundancy turned into increased sense of presence and decreased workload.

In a recent paper, Cifariello Ciardi (2004) illustrates a smart example of sonification based on an abstract mapping. The data to be represented come from the stock market too. The paper contains a concise but effective discussion on the state of the art of sonification over the last 15 years, i.e. starting from the very first attempt based on a direct data audification made by Frysinger (1990). Even if the mapping strategies are not explicitly illustrated, Ciardi writes that two of the main guidelines inspiring his work are centered on the possibility of representing the rate of changes in stock prices, which is often sharp and unexpected, and on "the inner and subtle correlation of stock price variations" that "may be considered similar to the correlation of patterns within a musical composition." The latter point becomes also an argument to state that "art-technology collaborations are often crucial to the successful design of audio displays, where expertise in sonic treatment is necessary." Moreover, as a further reference point, Ciardi considers the schema-based auditory organisation introduced by Bregman in his auditory scene analysis (Bregman, 1990). According to this schema a listener is able to switch between hearing a sequence of sounds as one or two streams by changing the attentional focus. This is fundamental in order to define the mapping strategies of multivariate data. Accordingly, the problem of orthogonality of the mapped parameters is also taken into consideration.

The author also points out how the sonification of highly active markets poses significant problems: it can easily produce "a continuous hubbub texture in which multiple streams are difficult to detect". Once more the contribution of a composer accustomed to balance the sonic mass both in time and in synchronous situations can be decisive in order to avoid a "repetitive and artificial ... audio output" that "can become irritating and distracting". From an acoustical point of view, Ciardi's work sounds extremely effective. It is interesting to notice how the result tends to be comparable to musical excerpts. This can be seen as an a posteriori musical mapping, i.e. a mapping practice that generates a musically organised sound stream. In the next section, we discuss briefly about what we could denote as a priori musical mapping, i.e. an actual mapping of data onto pre-existing musical materials.

10.5.4 Musical

Music has its own laws and organizing principles, but sometimes these can be bent to follow flows of data. The paper by Barra et al. (2002) is an example of the use of music for auditory display of complex time-varying information. The idea is simple: since many people use low-volume FM radio to mask the background noise of their offices, why not using a continuous musical stream that has the additional property of varying according to important system information? In this case, the information comes from accesses to web servers, as these are of interest for webmasters and system administrators. Not much space is dedicated to how the mapping from data to music is actually done, even though it is reported that the authors used "a musical structure that's neutral with respect to the usual and conventional musical themes". Instead, the paper focuses on architectural aspects of the auditory display system. Three software layers collaborate in a pipeline that goes from the HTTP requests to the sound rendering. In the middle, a Collector processes the events provided by the web server and sends requests to the WebPlayer component. The information is of three different kinds: server workload, errors, details on normal behaviour. Within an Apache module, information is processed according to a set of rules and directives describing which data are relevant and how they must be mapped into sounds. Support for visual peripheral display is also provided. [3]

10.6 Sonification

Sonification can be considered as the auditory equivalent of graphic representation in the visual domain. The main goal of sonification is to define a way for representing reality by means of sound. Carla Scaletti (1994) proposed a working definition of sonification as "a mapping of numerically represented relations in some domain under study to relations in an acoustic domain for the purpose of interpreting, understanding, or communicating relations in the domain under study".

Another less restrictive interpretation of sonification is found by transposition of what people currently intend with the word visualisation. For example, Ware's textbook (Ware, 2004) stands on the three pillars of data, perception, and tasks. We agree that the whole area of perceptualisation, defined as a superset of sonification and visualisation, may be considered as standing on those three pillars. In the following sections we analyze two aspects of sonification whose center of gravity stands closer to data and tasks, respectively. These two aspects are the definition of a methodology for representing information by means of sound (Section 10.6.1) and sonification in interactive contexts (Section 10.6.2).

10.6.1 Information Sound Spaces (ISS)

The concept of Information Sound Spaces (ISS) was introduced by Stephen Barrass in his thesis work (Barrass, 1997). Barrass aims at defining a methodology for representing information by means of sonification processes. The initial motivation of his work could be summarised in the following sentence: "The computer-based workplace is unnaturally quiet... and disquietingly un-

[3]Unfortunately, the project web address does not work anymore and no further details are given about musical mapping strategies. However, details on a user study and the musical material used are given in the previous paper by Barra et al. (2001).

natural...". In other words, the starting point of his work was the problem of the development of auditory displays for the computer. The first goal becomes, thus, to solve the contrast between the informative soundscape of the everyday world and the silence of the computer-based workplace. On the other side the danger is that a "noisy/musical" computer could easily become an annoying element. This concern, according to Barrass, highlights the need to design useful but not intruding/obsessive sounds.

More into detail, his thesis addresses the problems pointed out by previous researchers in the field of auditory display, as:

- The definition of a method for evaluating the usefulness of sounds for a specific activity;
- The definition of methods for an effective representation of data relations by means of sounds;
- The achievement of a psychoacoustic control of auditory displays;
- The development of computer aided tools for auditory information design.

Barrass illustrates a set of already existing approaches to auditory display design. A possible classification of these approaches is:

- Syntactic and grammar-based (es. Morse code, Earcons);
- Pragmatic: materials, lexicon and/or palette-based;
- Semantic: the sound is semantically related to what is meant to represent.

  In particular, the semantic relationships can be subdivided, using the terminology of Semiotics, into:

  - Symbolic: the signifier does not resemble the signified;
  - Indexical: the signified is causally related to the signifier (e.g. the sound of a tennis ball);

- Iconical: the signifier resembles the signified.

About the concept of sign in general, Barrass writes: "The concept that the sign stands for is called the "denotation" and additional signifieds are called "connotations". Cultural associations generate connotations by metonym and metaphor. A metonym invokes an idea or object by some detail or part of the whole - a picture of a horseshoe can be a metonym for a horse. A metaphor expresses the unfamiliar in terms of the familiar - a picture of a tree can be a metaphor for a genealogy". This point is very important in the definition of Barrass' methodology for auditory display design. Further issues as identifiability and learnability, as well as potential problems, concerning masking, discriminability and conflicting mappings are also discussed.

Barrass proposes different approaches for auditory display design. Among the others, a Pragmatic approach and a Task-oriented approach are discussed. The Pragmatic approach concerns design principles of warnings and alarms (see also Section 10.2).

A task-oriented approach takes a particular role in the following developments of the thesis, in terms of sound design for information display. Task analysis is a method developed in Human-Computer Interaction (HCI) design to analyze and characterise the information required in order to manipulate events, modes, objects and other aspects of user interfaces. The methodology is based on the so-called Task analysis and Data characterisation (TaDa). According to this analysis, it is possible to define the requirements necessary for some specific information representation on a certain display, addressing a specific kind of user. One of the possible strategies to take into consideration the user from the very first step of the design process is to use a story to describe a problem. As in a general design practice, the tools become storyboards, scenarios, interviews, and case studies. Barrass names the case studies with the term "Earbenders", from a colloquial expression denoting short stories coming out from everyday life and worth listening to (bending one's ear to).

It is then a problem of collecting and selecting good data. In order to collect "stories", Barrass proposes to ask people questions as: "During the next few weeks if you notice yourself using your hearing to help you, say, find a lost pin, or search for hollow spaces in a wall, or notice something wrong with your car, or tell whether the dog is hungry, anything at all, then please email me with a brief story about the occasion". Beside the story and in order to define a methodology, Barrass characterises application scenarios by means of some precise keys: a "critical" question; an answer; a characterizing subject; and a characterizing sound.

The EarBenders stories remind another approach to design in general, becoming more and more popular in the HCI community: the Cultural Probes introduced in the paper by Gaver et al. (1999). In that work, the three authors describe a design process aiming at introducing elder people to new technologies. The main points of their methodology are those of being extremely open to the dialogue with the users, and closer to a conception of their actions as those of designers-artists rather than scientists. The idea is to collect "inspirational data" by means of a personal interaction with the users, based on some cultural provocations or, better said, "probes".

Another important part of Barrass' thesis is the definition of an Information-Sound Space, what he calls a cognitive artefact for auditory design. Barrass starts from the example of the Hue, Saturation, Brightness (HSB) model for the representation of the "color space" and from the representation of a color selector tool by means of a circle with the hues corresponding to different sectors, the saturation levels mapped along the rays and the brightness controlled by means of a separated slider. This tool is called color chooser. In order to build a similar tool, representing a "sound space", Barrass analyzes many different possibilities of correspondence between the dimensions of the color chooser and different sound parameters. In the ISS representation, the third dimension (the slider) becomes the height of a cylinder built on a circular pedestal similar to the color chooser. Thus, the ISS consists in a dimension with a categorical organisation of the information (the sectors of the circle), a dimension with a perceptual metric (ordered) along the radial spokes, and a vertical axle endowed with a perceptual metric too. Then, pitch, formants, static and dynamic timbers are alternatively mapped to the circle and the different mappings are tested with listening experiments. Finally, a "sound chooser" is designed, where the three dimensions of the ISS are related to Timbre, Brightness and

Pitch (TBP), the brightness corresponding to the radial dimension and the pitch to the height dimension.

In a more recent paper by Barrass and Kramer (1999), the authors discuss about sonification scenarios envisioning applications based on nano-guitars and garage-band-bacteria appearing to the stethoscope of a doctor. From these science fiction-like applications, a long series of considerations about sonification starts and an overview of the subject is drawn. The work analyzes a set of already existing applications ranging from auditory displays for visually impaired people to auditory feedback for collaborative work. The main points raised by the authors in terms of advantages of sonification are the possibility of perceiving cycles or temporal patterns in general as well as very short events and the possibility of perceiving multidimensional data sets (the ear is polyphonic, i.e. is able to separate multiple and simultaneous auditory streams one from the other). Problems of learnability, synthesis/design skills, nuisance and incomprehensibility are discussed with respect to different approaches to auditory display design such as earcons, auditory icons and parameter mappings.

A work related to that of Barrass for the representation of timbre spaces was done by Brazil and Fernström (2003). They presented a tool for accessing sounds or collections of sounds using sound spatialisation and context-overview visualisation techniques. Audio files are mapped to symbols and colors, and displayed in a 2D environment. There are multiple views: starfield view, TreeMap view, HyperTree view, TouchGraph view and a file information window is always available for detailed information. In addition to the visualisation components there are several interaction devices. Different sliders can be used to control various custom user classifications. Another tool is a simple text-based filter that exploits user classifications of the sounds. The system seems to improve performance in browsing audio files, but sometimes leads to ambiguity as users' classifications of sounds can differ. As a result of an evaluation test reported in the paper, the Hyperbolic layout (HyperTree) makes sound browsing both easy and enjoyable. Conversely, the TreeMap view provides poor visualisation of the data.

10.6.2 Interactive sonification

Hermann and Ritter (2004) propose a new approach to data sonification, starting with a deep investigation on the link between sound and meaning. The idea is to find a way to perform a data sonification without using musical listening and without any training. The Model Based Sonification (MBS) that the authors propose provides a natural mean for interacting with a sonification system and allows the development of auditory displays for arbitrary data sets. Both results are achieved using a virtual object in the interaction and a parameterised sound model as auditory display. The authors argue that the MBS has many advantages: few parameters to be tuned, a natural connection between sound and data, a soft learning slope, an intuitive interface, a continuous natural control. The MBS has been illustrated using an example of particle trajectories with a prototype of a tangible physical representation-interface. However, there is no real evaluation of the sonification method and of the interaction.

In a recent paper, Rath and Rocchesso (2005) explain how continuous sonic feedback produced by physical models can be fruitfully employed in HCI. The control metaphor which is used to demonstrate this statement is balancing a ball along a tiltable track. The idea of using a continuous feedback comes out by simply analyzing the natural behaviour, where we often refer to continuous sounds to get information about what is happening around us. Sonic feedback has the advantage that it can improve the effectiveness of interaction without distracting our focus of attention. The physical model of a rolling ball is used as a specific workbench, where the sound is particularly informative, conveying information about direction, velocity, shape and surface textures of the contacting objects.

The main characteristic of this model is its reactivity and dynamic behaviour: "the impact model used produces complex transients that depend on the parameters of the interaction and the instantaneous states of the contacting objects". The authors pointed out the importance of macroscopic characteristic renderings, such as the periodic patterns of timbre and intensity which are featured by a rolling sound: the rolling frequency is very important for the perception of size and speed. The physics-based algorithm developed in this work involves a degree of simplification and abstraction that implies efficient implementation and control. The principle is that of a cartoonification of sounds: a simplification of the models aimed at retaining perceptual invariants, useful for an efficient interaction (see also Section 10.7.2). The main idea here is to consider rolling objects that do not show perfect circular symmetry: the height of the center of mass will vary during movement, thus leading to periodic modulations of the effective gravity force. This approach has many advantages that can be summarised as follows:

- The synthesised sound is always new and repetition–free;
- There is no need to store large amounts of sound samples;
- All the ecological attributes can be varied on the fly allowing a continuous real time interaction with the model.

The rolling sound was evaluated in an interactive sonification task: subjects were asked to balance a virtual ball on a tiltable track, with and without a visual representation of the ball. It was found that

- The sound of the virtual rolling ball is easier to recognise than the sound of a real rolling ball; users can describe better the physical characteristics (e.g. the size) of the virtual ball than the real one;
- The cartoonification approach appears to be effective in such a control metaphor;
- Subjects intuitively understand the modelled metaphor without any learning phase;
- There is an improvement in performance when using sound feedback in a target-reaching task, from 9% for bigger displays to 60% for smaller displays;
- All subjects were able to solve the task using auditory feedback only.

This investigation suggests that continuous feedback of a carefully designed sound can be used for sensory substitution of haptic or visual feedback in embodied interfaces. Many multimodal contexts can benefit from cartoon sound models to improve the effectiveness of the interaction: video games and virtual environments are the more obvious cases.

Another example of interactive sonification is found in Avanzini et al. (2005). The paper starts with the description of a complete physical model of the complex mechanics of friction, taking into account numerical methods to make the model running in real time on low cost platforms. The main idea of the friction model is that of a "bristle-based" interpretation of the friction contact, which consists in a number of asperities (e.g. microscopic irregularities) of two facing surfaces. As an extension to prior art (de Wit et al., 1995), a parameter has been added in the definition of the friction force in order to simulate scraping and sliding effects other than the stick-slip phenomena. The model is divided in two main parts: the excitation and the resonators of the vibrating system. The resonating objects are modelled according to the modal synthesis approach as lumped mechanical systems. In order to use the model in interactive settings a numerical implementation is discussed. A decomposition of the system into a linear differential system coupled to a memoryless non-linear map is done in order to apply efficient numerical methods to the model. The final part of the article is the discussion of a number of applications of the model to the simulation of many everyday friction phenomena: all these examples explain how physical models can be used in a multimodal context, such as sound modelling for computer animation (see also Chapter 9). The model has been implemented as a plugin in pd[2] and then used in several examples of acoustic systems with induced frictional vibrations, such as wheel brakes, wineglass rubbing, and door squeaks. All these examples show a high degree of interactivity of the model: the user can control one of the virtual objects in the animation through a pointing device, controlling at the same time some of the physical parameters involved in the friction (for example, the force acting on the exciter). The positions and velocities which are returned by the synthesis engine can be used to drive both the graphic rendering and the

[2] `http://crca.ucsd.edu/~msp/software.html`

audio feedback.

10.7 Sound design

10.7.1 Sound objects

In the book entitled "Guide des objets sonores", Michel Chion (1983) tries to fix an omni-comprehensive yet synthetic review of the thinking of Pierre Schaeffer (1967). The work is a considerable effort aiming at making the intuitions and concepts developed by Schaeffer systematic. The main goals of the book are the definition of acousmatic music, the definition of reduced listening and of sound object, the definition of concrete (vs. abstract) music and the definition of a new solfège for the development of a new music.

The word "acousmatic" comes from the ancient Greek and means a sound that we hear, without seeing the source. Acousmatic is here meant as opposed to direct listening. The acousmatic situation corresponds to an inversion of the usual way of listening. It is not any more a question of studying how a subjective listening deforms the world. Conversely, listening itself becomes the phenomenon under study. Two new listening experiences due to the use of the tape recorder are mentioned as fundamental for the definition of the concept of acousmatic music: the looped tape and the cut of a sound of a bell (a sound without the attack). By listening to a sound repeated in a loop (identical to itself) and by the alteration of the identity of a sound by means of the suppression of its attack, one becomes aware of his own perceptual activity through a distortion of the natural listening conditions. This awareness is directly related to the second main concept defined by Schaeffer: reduced listening. This new way of listening is thought as opposed to what he calls trivial listening (a listening that goes directly to the causes, i.e. the events producing a sound), to pragmatic listening (a gallop can have the meaning of a possible danger or be just a rhythmical event) and cultural listening (that looks for a meaning). In other words, the reduced listening places out of our scope anything that is not the sound itself, even if related in a more or less direct way to it (sources, meanings, etc.). The sound itself becomes an object on its own, i.e. a sound object. In order to better specify the concept of sound object, Schaeffer adopts a negative approach. A sound object is nor the body of the sound source (sounding object), neither a physical signal. A sound object is nor a recorded sound neither a symbol on a score. Also, a sound object is not a state of our spirit. These negative specifications delimit what in a positive sense could be defined as "the sound itself".

By means of the manipulation of sound objects it becomes possible to build a new kind of music: the concrete music. Classical music starts from an abstract notation and the musical performance come afterwards. Conversely, the new music starts from the concrete phenomenon of sound and tries to extract musical values from it. In other words, mimicking the distinction between phonetics and phonology, the whole work of Schaeffer can be considered as a path from acoustics to what he defines as "acoulogy", i.e. from a gross sound, conceived as an acoustic object, to a sound that is analyzed and considered in a musical sense.

The ultimate goal of the development of a complete method for the analysis of sounds done by Schaeffer is the synthesis of new sounds as well as the production of a new music. The plan for this new methodology forms the "new solfège" and is articulated into 5 steps: typology, morphology, characterology, analysis and synthesis. Schaeffer developed only the first two steps, while the other three were only envisaged. More into detail, typology performs a first empirical sorting and defines sound classes. Morphology describes and qualifies sounds according to the results of typology. Characterology realises a sort of taxonomy and characterisation of the sound classes: a kind of new organology. Analysis looks for musical structures within the sound classes in terms of perceptual attributes. Finally, synthesis forms the innovative lutherie, i.e. a way for the creation of new sound objects according to the guidelines provided by the analysis. A noteworthy remark is that in a preliminary definition of what he calls analysis, Schaeffer defines the analytical tools in terms of the what he calls the natural perceptual fields, i.e. pitch, duration and intensity. These criteria are in a sense "natural", even if, for what concerns pitch, it seems to be quite a traditional choice. The tasks of typo-morphology are, thus, three: identifying (typology), classifying (typology) and describing (morphology) according to a reduced listening, i.e. independently from any reference to causes/origins of the sounds or to what they could evoke.

In this sense, the equivalent of a musical dictation in the context of the new solfège is the task of recognizing and defining the character and the declination of a sound object with the final goal of improving the listening skills. One of the main points in order to achieve an identification of the sound objects, is the definition of some segmentation criteria. This is not an evident task, since one does not want neither musical criteria nor natural systems of identification (source detection). The chosen units correspond to syllables. The distinction between articulation and prolongation, i.e. the identification of the breaking of the sonic-continuum in subsequent and distinct elements (consonants) and sounds with a structure that maintains its characteristics over time (vowels) is the way pointed out by the author, in order to identify the sound objects (syllables).

Finally, the author defines a set of descriptors that have some generality, even if they are possibly not as exhaustive as the author claims they are. The typology classification is based on the distinction between:

- Impulses;
- Iterations (sequences of impulses);
- Tonics (voiced sounds);
- Complex sounds (static mass with no defined pitch);
- Complex and variable sounds.

The morphological classification is based upon the subdivision of sounds according to:

- Matter criteria (mass, harmonic timbre, grain);
- Form criteria (allure - a kind of generalised vibrato- and dynamics);
- Variation criteria (melodic profile and mass profile).

In the context of the study on the soundscape, Murray Schafer (1994) talks as well about reduced listening by introducing the concepts of schizophonia and of sound looping, both related to the studies of Pierre Schaeffer. Schizophonia refers to the new listening scenario introduced by recording and reproduction of sounds, where the sources disappear from our visual feedback and are thus separated (schizo) from sounds (phonia). We do not see any source any more and the sounds become objects on their own. This is what he calls the effect on audio of the electric revolution (see Section 10.7.3).

A very good discussion about Schaeffer's and Chion's theories is available in the last chapter of a recent publication by Lombardo and Valle (2005). The authors provide also an organic and effective overview of the soundscape discipline and of the theoretical thinking concerning the subject of audiovision later discussed in this chapter (see Sections 10.7.3 and 10.7.5).

10.7.2 Sounding objects

In a paper by Rocchesso et al. (2003), three partners of the European Project "the Sounding Object"[3] draw their conclusions about three years of research . Their approach is complementary with respect to the concept of sound objects. The focus is not on the sound itself, but, on the contrary, the idea is to model the source in terms of its physical behaviour. Perception analysis, cartoonification/simplification and control of physically meaningful parameters were the three main guidelines of their work. The term cartoonification refers to the cartoons and to the technique of reducing complex audio and visual information to its essential elements. These reduced elements are then emphasised in order to make them easily and immediately decodable. One of the great advantages of a cartoonified version of reality is, thus, not only the easiness of the realisation but also an augmented intelligibility.

The paper presents some examples in terms both of sound design techniques and psychology of perception. First a psychology experiment was performed in order to explore the perception of sounds produced by filling/emptying bottles. The principle of cartoonification was illustrated by the

[3] http://www.soundobject.org

example of a cartoon mouse drinking with a straw. A gurgling sound was adopted with a varying timbre. The timbre variations are designed to convey in an effective (cartoonified) way the information about the filling level of the bottle. Another important aspect of the principle of simplification of the physical model, besides a higher intelligibility, is the possibility of using the complexity of human gesture in order to control the parameters of the model. This is crucial in order to obtain a natural sounding object.

As a general result of their experience, the authors propose a set of principles for sound design. The sound is designed according to a precise decomposition of the physical events occurring in the action. In terms of sound modelisation, each of these events is treated separately. According to these principles, the main identity of a large class of sounds can be defined and reproduced by means of a set of basic physical interaction models, i.e. those for the synthesis of impact sounds (Rath and Rocchesso, 2005) and friction sounds (Avanzini et al., 2005). In this way, with a further effort it is possible to enlarge the scope to the case of sounds due to breaking, crumpling, crushing, smashing and rolling. While the impact and friction sounds can be modelled as deterministic processes, these complex sounds can be simulated by means of stochastic processes controlling their temporal organisation. In other words, these phenomena can be conceived as an ensemble of many elemental phenomena, as impact and friction sounds, whose statistical behaviour presents some specific characteristics both for what concerns their temporal distribution and their energy. For example, the statistics of the surface impacts that take place, while an object is rolling, can be interpreted as Brownian motion (Corsini and Saletti, 1988). The crushing of a metallic box (for example a tin) can be thought about as the result of a temporal sequence of events of a Poisson process (Houle and Sethna, 1996). Also, breaking and crushing phenomena can find a representation in the sense of avalanche processes and of chaotic and self-organised systems (Sethna et al., 2001). As a following step, the quality of these sounds can be refined and enhanced by means of signal processing techniques. In conclusion, a higher level spatial and temporal organisation of the generated sounds is of extreme importance in order to enlarge the vocabulary and information contents that one wants to convey by means of sound.

The potentialities of cartoon sounds were deeply addressed by William Gaver (1993b,c) already in the early nineties. In his work, already discussed in Section 10.4, Gaver investigates a fundamental aspect of our way of perceiving the surrounding environment by means of our auditory system. As already mentioned, he considers the fact that our auditory system is first of all a tool for interacting with the outer world by means of what he defines as everyday listening. In this way, Gaver provides the starting and essential contribution to the perspectives of a sounding object design. In his works, he makes a fundamental distinction between three categories of everyday sounds: solid, liquid and aerodynamic. First he considers sounds produced by vibrating solids. Then he analyzes the behaviour of sounds produced by changes in the surface of a liquid. Finally he takes into consideration sounds produced by aerodynamic causes. Each of these classes is divided according to the type of interaction between materials. For example, sounds generated by vibrating solids are divided in rolling, scraping, impact and deformation sounds. These classes are denoted as "basic level sound-producing events". Each of them makes the physical properties of different sound sources evident. In a synaesthetical perspective, when listening to everyday sounds, one should be able to "hear" the size, the shape and the material of a sounding object.

At a higher level one has to consider three types of complex events: those defined by a "temporal patterning" of basic events (e.g. bouncing is given by a specific temporal pattern of impacts); "compound", given by the overlap of different basic level events; "hybrid events", given by the interaction between different types of basic materials (i.e. solids, liquids and gasses). According to Gaver, each of these complex events should potentially yield the same sound source properties, corresponding to the basic component events but also other properties (e.g. bouncing events can provide us information concerning the material but also the symmetry of the bouncing object). More in general, we can hear something that is not the size or the shape or the density of an object, but the effect of the combination of these attributes.

10.7.3 Soundscape

The word soundscape was born as a counterpart of landscape, denoting the discipline that studies sound in its environmental context, both naturalistic and within urban scenarios. This discipline grew up first in Canada, then propagating to other countries. A milestone publication on this subject was written by Murray Schafer (1994). Entitled Soundscape, Schafer's book is a long trip through a novel conception of sound. One of the main goals is the recovering of a clear hearing (claireaudience), and of a hi-fi (high fidelity) soundscape as opposed to the acoustically polluted an lo-fi (low fidelity) soundscape of our contemporary world. One of the main theses of the book is, in fact, that the soundscape is not an accidental by-product of a society, but, on the contrary, it is a construction, a more or less conscious "composition" of the acoustic environment we live in. Evaluation criteria are widely investigated by Schafer, leading to a basic classification of soundscapes into two categories, the already mentioned hi-fi and lo-fi scenarios that will be described more in detail later in this section. As a consequence of this analysis and classification, Schafer says that our subjective effort should be to pursue and progressively refine what he calls an "ear cleaning" process. An "ear cleaning" process means to become able to listen to, and evaluate the sounds of the surrounding environments in an analytical way, in order to become aware of the quality of a soundscape and be able to interact with it. It is a matter of taking care of ourselves. Hearing is an intimate sense similar to touch: the acoustic waves are a mechanical phenomenon and somehow they "touch" our hearing apparatus. Also, the ears do not have lids. It is thus a delicate and extremely important task to take care of the sounds that forms the soundscape of our daily life. On the contrary, the importance of the soundscape has been underestimated till nowadays and a process of cultural growth in this sense is extremely urgent. Schafer even writes that a lo-fi, confused and chaotic soundscape is a symptom of decadence of a society.

The main components of a soundscape are defined as keynotes, signals and sound prints. Keynotes are sounds related to the geography of a place, belonging somehow to the unconscious background of our perceptive world (the seagulls on the sea or the wind in a stormy mountain area). On the contrary a signal is anything that we listen to consciously and that conveys some information about the surrounding environment. Finally a sound print is something that belongs once more to the background, being a product of human and social life. Sound prints are the main concern of Schafer's investigation.

In the historical sections of the book, the analysis is always related to sociological aspects. The rural soundscape, the industrial "sound revolution" and the consequences of the electric revolution are analyzed in depth. In the pre-industrial society, the SPL in a rural village was never above 40 dB, except when the bells or the organ played or a mill was working. On the other side reports from the past tell us that the sound in the big towns of the pre-industrial era were unbearable. Nevertheless these sounds were variegated and their dynamics was spike-like and always changing. The main sound sources were people screaming (especially hawkers, street musicians and beggars), hand-worker activities, horses and other animals. As opposed to this kind of soundscape, the industrial revolution introduces continuous, not-evolving and repeating sounds. This is one of the main characteristic of a lo-fi soundscape. Schafer says that a spike-like and varying amplitude envelope was substituted by a continuous and steady amplitude envelope, which fixes a stable, persistent, and unnatural (psychologically disturbing) evolution of the sound dynamics.

Besides the industrial revolution, the electric revolution plays a particularly relevant role in the first part of the book: the electricity allows us to record, reproduce and amplify sounds. How this influenced the soundscape is also matter of discussion in the book. In particular, the concepts of schizophonia and of sound looping already mentioned in Section 10.7.2 are investigated. Another crucial point emerging from Schafer's analysis of the electric revolution is the outstanding use of bass frequencies in modern pop music, as compared to music of the past. All of these aspects deteriorate the quality of the "soundscape": the sound-to-noise ratio increases and we rapidly pass from a hi-fi soudscape to a lo-fi soundscape. The physical-symbolic meaning of low frequencies is quite clear: basses propagate farther and longer in time than high frequencies. Due to diffraction phenomena they bypass obstacles.

Also, it is difficult to localise a low-frequency sound source. The global effect is that of a sense of immersiveness that usually cannot be achieved in other musical traditions. As a remarkable exception, Schafer observes that a very similar immersive effect is a characteristic of a completely different scenario as that of the ancient Romanic and Gothic churches. When a choir is singing the reverberation creates the effect of prolongation, diffraction and delocalisation typical of bass sounds.

In the second part of his book, Schafer faces another main problem, that of a notation system for everyday sounds. His position is extremely critical towards ordinary representations, typically based on the spectrogram. He points out how this kind of representations misleads the attention from the auditory channel to the visual one. He suggests not to consider as serious an evaluation of sounds based only on some diagram: "if you don't hear it, don't trust it". Then, he defines a sort of taxonomy of sounds, a set of parameters relevant for the characterisation of a sound timbre, and a symbolic notation representing these parameters. This part ends with an extensive inquiry about how people in different cultural and geographical contexts judge more or less annoying different categories of sounds. An extremely interesting collection of data emerges from such an analysis. In general, both for the production of sounds as for the reaction that they generate, a sociological approach is always privileged.

In the last part of the book, Schafer moves towards the definition of an acoustic design practice. According to the author, the tasks of a sound designer should be the preservation of sound prints, especially those that are going to disappear, and the definition and development of strategies for improving a soundscape. In order to be successful in the latter task, the principles that a sound designer should follow are:

- To respect the human ear and voice. The SPL of the soundscape has to be such that human voices are clearly audible;
- To be aware of the symbolic contents of sounds;
- To know the rhythms and tempos of the natural soundscape;

- To understand the balancing mechanism by which an eccentric soundscape can be turned back to a balanced condition.

An interesting example of sound design for a urban context is given by the work of Karmen Franinovic and Yon Visell[4] with the sound installation Recycled Soundscape- Sonic Diversion in the City. The authors aim at stimulating people attention and awareness about the soundscape of their urban environment. What they promote is somehow an ear-cleaning process in the sense defined by Schafer. The installation is formed by a mobile recording cube (approximately one meter high), supporting a microphone placed in front of a parabolic surface in order to create a sharp directional capture pattern. By means of this microphone, users can select the sounds in the surrounding environment and record them. The installation is completed by two independent mobile cubes, which play back the recorded sound after a significant "recycling" processing performed by the computer. The idea is to get people involved in playful interactions in the urban setting and to make them sensitive about the cacophony that surrounds us in the city soundscape and let them try to put some order in the city. Quoting the words of the authors, "... these recordings of the context, made by previous and current users of the system ... are woven into the remixed soundscape".

10.7.4 Space and architecture

In his book, Schafer (1994) outlines the lack of awareness about sound in the architectural context. The main concern of architects is usually how to eliminate sound. No attention is devoted to active intervention in architectonic projects in terms of sonification of the environments. In this section we briefly discuss about sound and space from two different points of view: the relation between sound and architecture, meant as sonification of the environment, from one side, and the technical aspects related to the acoustic space rendering.

A very first and well-known example of an architectural project that gave extreme importance to sound was the Philips Pavilion at the Expo in

[4] `http://www.zero-th.org/`

Bruxelles in 1958. In that case Le Corbusier and Xenakis built a structure where sound played a fundamental role: the work was an organic project where space and sound were conceived together in order to be experienced as a unicum. By combining light, sound, and color, the Philips Pavilion was more than a building at the fair, it was a multimedia experience displaying the technological prowess of the Philips company. The music was composed by Edgar Varèse and entitled "Poème électronique". The radical concept behind this first experience is that sound modifies the perception of space. We could say that space is in sound or that (a different) space comes out, when sound rings. In this sense a new discipline as electroacoustic soundscape design, i.e. the electroacoustic sonification of the environments and buildings, takes a relevant place in the frame of sound design.

Recently, a remarkable philological work done as part of a European project called VEP (Virtual Electronic Poem) appeared, finalised at a virtual reconstruction of both the architectonic space of the Philips Pavilion and its acoustic space (Lombardo et al., 2005). The latter included the study of the acoustic features of the building and the reconstruction of the positioning of the loudspeakers. The work presented significant problems due to the lack of documentation and the necessity to recover the positions of the loudspeakers only by means of the incomplete and dark pictures of the interior of the Pavilion.

For what concerns the technical aspects, a good and concise overview on real time spatial processing techniques available up to 1999 for room simulation with application to multimedia and HCI can be found in a paper by Jot (1999). The author goes through the models of a) directional encoding and rendering over loudspeakers, including conventional recording and ambisonic B format, b) binaural processing and c) artificial reverberation, with an extension to the dynamic case for acoustic-source-distance rendering (Chowning's model) and to Moore's ray-tracing method. In this overview, the author points out the advantages and weak points of each approach, as for instance, the limit of the "sweet spot" for the methods of point a). He states the criteria for establishing perceptually-based spatial sound processing: tunability, configurability and, last but not least, computational efficiency and scalability. In particular the first criterion includes the definition of source direction (azimuth and elevation) and descriptor of the room. Configurability implies the possibility of changing output format (headphones vs. different loudspeaker configurations). Finally the author presents SPAT, the spatialisation tool realised by the IRCAM. The great novelty at the time for such a tool was the high-level interface. SPAT does not present to the user physical parameters but only perceptual parameters classified as: a) source perception (source presence, brilliance and warmth), b) source/room interaction (room presence), c) room perception (heaviness and liveliness). These parameters are chosen according to the studies of psychoacoustic done at IRCAM, specifically for the perceptual characterisation of room acoustic quality. A series of application scenarios is then analyzed, ranging from VR and multimedia to live performance and architectural acoustics.

10.7.5 Media

In the essay entitled "L'audio-vision. Son et image au cinéma" (Chion, 1990) two main theses appear from the very beginning of the book: 1) audiovision is a further dimension: different from bare vision and different from bare audio 2) between images and sound (music in particular) there is no necessary relationship. As an example of these theses, the author starts with the analysis of the movie "Persona" by Ingmar Bergman and of "Les Vacances de Monsieur Hulot" by Jacques Tati. Chion demonstrates that the bare vision of a sequence of mute images is something completely different from the audiovision of the same sequence (as an example, he describes the effect of the prologue of Persona played first without and then with the soundtrack). Also, contrasting audio and video situations are not only possible but sometimes extremely effective and rich of expressive contents. An example could be the scene on the beach in "Les Vacances de Monsieur Hulot", where a group of annoyed and silent people are "superimposed" on a joyful hilarious soundscape of children playing and screaming. The whole scene is of a great comic effect.

The author considers then the different roles of sound in a movie: sound has the function of temporal organisation and connection (overlapping effect) of isolated events. Also, it functions as a spatial connection: the acoustic connotation of the environment of a scene, for example its reverberation characteristics, gives unity to the scene itself. Another important aspect is the "ground" level of sound, i.e. the silence. Chion quotes Bresson: it was the synchronism between sound and image that introduced silence. Silence as pauses in between sound events. The mute cinema, on the contrary, is a continuous suggestion of sound. On the other side, there are sounds in the cinema used as a metaphor of silence: animals that cry in the far, clocks in the neighbor apartment, any very soft but present noise.

An interesting, possibly central matter of debate is given by off-field sounds. An articulated classification follows the analysis of all of the possible situations. Altogether with the "off sound" (the sound that does not belong to the time and space of the scene) and the "in-sounds" (the sounds whose sources appear in the scene), they form the so-called tri-circle. The off-field sounds can take different attributes: acousmatic, objective/subjective, past/present/future, giving raise to a wide scope of expressive possibilities. Also, off-field sounds can be of various kinds: trash (e.g. explosions, catastrophes noises), active (acousmatic sounds that provoke questions as "what is it?" or "where is it?") and passive (sounds that suggest an environment that embraces the images, giving a feeling of stability).

Another dimension of sound in the audiovisual context is its extension: null extension corresponds to an internal voice, while a wide extension could correspond, for example, to a situation where the traffic sounds from the near street are audible and amplified in an unrealistic way. Expressive effects can be obtained by playing with variations of the extension within a scene (e.g. alternatively from outdoor to indoor and to the mental dimension).

A further element is the definition of the listening point. One possibility is to refer to a spatial perspective: from which point in the space are we hearing? However, a subjective point of view is possible too: which character is listening? A particular case is given by weak sounds, which give the impression to be heard only by a single character, as if the sounds were near the ears. In all of the previous examples, the unrealistic element is essential. In general, there is no reason why the audiovision relationships (between images and sounds) should be the same as in real-life. A stylisation strategy of representation is possible and can open wide and various horizons for expressivity. This is also related to the discussion of Section 10.7.2.

In the followings of his book, Chion points out the difference between definition and fidelity: definition is a technical term denoting the range of reproduction/rendering possibilities of the system. On the contrary, fidelity is a dangerous term: fidelity evokes realistic reproduction that is a very arguable concept: cinema is a metaphoric representation, involving augmentation, unrealistic points of view, distortion of time and space, of soundscape and landscape. Sound should be veridical, not realistic. The goal of a veridical sound is to render the associated sensations, not to reproduce the sound realistically. A realistic sound, if detached from the image, is often not comprehensible, and could be a deception. Sound technicians are skeptic about the possibility of recognizing sources according to the produced sounds. Chion quotes a couple of examples: the same sound can be used in relationship with a smashed head in a war movie or a squeezed water-melon in a comic movie. The same gurgling sound can be used for a tortured Russian prince in the movie Andrej Rublev by Tarkowsky and for the gurgling of Peter Sellers in a comic movie. Reverberation also plays a role in this dialectics between realistic and unrealistic, contributing to a realistic rendering of the spatial dimension of sound. On the other side, unrealistic reverberation can give an impression of dematerialisation and symbolism. An interesting case of unrealistic rendering is given by the sounds and noises that children use to evoke the life of their puppets, dolls and little cars (especially their movement). Where do they come from? Or, once more, where does the cartoonification process draw his effectiveness from?

A significantly different case is given by the television. The main distinction between cinema and television lays in the different position occupied by sound. Television is rather an illustrated radio. Voices play always a principal role, in a sense, they are never off-field. Even in the news the images are rather a "decoration" of the verbal information. The radio-like attributes of television increase when the TV is constantly on, as in some public places. In this case the image is not any more the structural element, but only the exception, the "surprise". A particular case deserving some considerations is given by tennis matches. Tennis is the most "sonified" sport: the different impact sounds of the ball, the cries of the players, the audience exclamations. It is the only sport, where the speaker can stop talking even for 30 seconds and more.

Finally, it is necessary to spend a few words about video-clips. The structure of video-clips and their stroboscopic effect make them a special case. It is not any more a dramatic time, but rather the turning of the faces of a prism. The success of a video-clip relies mainly on a simple punctual synchronism between sound and images. Also, in the video-clip sound looses its linear character.

10.8 Perspectives

In this chapter we provided a wide overview of the multifaceted aspects and implications of sound design and auditory display. These research fields are relatively young and many are the open problems (more or less fundamental and more or less approachable and urgent considering the present state of the art). A first very general open question about sound design is if it is possible to claim that sound design is a discipline in the sense of what design is. In other words, is there anybody designing sounds with the same attitude Philippe Starck designs a lemon squeezer? Then, if the answer is negative, what is missing for a foundation of sound design as a discipline? As a particular case, product-sound design acquires every day a more and more relevant place in the loop process of product implementation and evaluation in the context of industrial design. Different definitions of Sound Quality have been proposed and different evaluation parameters have been defined for its quantification as loudness, sharpness, roughness and fluctuation strength. The main parameter, from which the others are more or less directly derived, is loudness, i.e. a psychoacoustic parameter used, somehow, to measure the aggressiveness of sound. However, more effective methods for defining and evaluating the aesthetics (in terms of "sensory pleasantness" and emotional contents) and the functionality of a sound have not been implemented yet and the development of appropriate methodologies of this kind is an urgent task for the growth of sound design as a mature discipline.

One of the possible approaches concerns the particular case of the design of ecological sounds by means of physical modelling techniques for sound generation. A set of flexible and effective algorithms for the synthesis of everyday sounds by means of perceptually-consistent parameters is still an open issue, even if a lot of work has been done in this sense (see Section 10.7.2). Intuitiveness and immediateness in defining the acoustic attributes of everyday sounds in correspondence to physical attributes in terms of materials, dimensions, shape and others would provide a powerful set of tools for sound designers.

From a more general point of view, one of the most important concepts drawn from the electroacoustic music experience and discussed in Section 10.7.1 is that of Sound Objects, introduced in a systematic way by Pierre Schaeffer. This remains a fundamental reference point for the whole practice of sound design. Various analytical methods have been proposed in this sense. However a systematic theory for the synthesis of new sound objects has not been developed yet. Effective techniques and guidelines for creating sounds according to any particular specification are still far from providing a general and well assessed methodology.

Sonification, intended as information representation by means of sound, is also an open research field as part of the more general context of auditory display. Even if a lot of work has been done, a systematic definition of sonification methodologies has not been assessed yet. Clear strategies and examples of how to design sound in order to convey information in an optimal way have not been developed so far. Sonification remains an open issue including communication theory, sound design, cognitive psychology, psychoacoustics and possibly other disciplines. By means of a cooperative action of these disciplines, a deeper analysis and understanding of the correlation existing within some given data sets and the correlation among the representing sounds as a result of a sonification process would be possible. However, one of the main problems of the sonification practice is that the result often tends to be repetitive and artificial, or irritating and fatiguing, or, on the contrary, distracting. A natural question emerges: could the contribution of a composer, accustomed to balance the sonic masses both in time and in synchronous situations, be decisive in order to avoid these kinds of problems? Is it meaningful to define a sonification practice that is informed by the practice of musical composition? Or, in a wider sense, is an art-technology collaboration a positive and, maybe, crucial element for a successful design of auditory displays and, more in general, for a successful interaction design?

Another interesting concept related to sonification and discussed in the chapter is the so called cartoonification process (see Section 10.7.2). Studies about the cartoonification of real-life representations are an open issue as well, especially from a psychological and communication theory point of view. The simplification introduced by a cartoonification process is extremely important in the context of sonification, but a deeper understanding of the criteria underlying the perception of cartoonified messages and representations of the real world is still necessary. On the other side, this approach seems extremely promising in terms of achieving successful results in the representation of information by means of sound.

Finally, active sonification of architectural or open-air spaces is an open issue too. No systematic study is available about how to design and modify a sound space (or a soundscape) in an active way, i.e. by means of the introduction of artificial/additional sounds (see Sections 10.7.3 and 10.7.4). This is an important area of sound design, dealing with the relationships between architectural spaces, environments, social contexts and the role and potential impact of sound on them.

Bibliography

F. Avanzini, D. Rocchesso, A. Belussi, A. Dal Palù, and A. Dovier. Designing an urban-scale auditory alert system. *IEEE Computer*, 37(9):55–61, Sept. 2004.

F. Avanzini, S. Serafin, and D. Rocchesso. Interactive simulation of rigid body interaction with friction-induced sound generation. *IEEE Trans. Speech and Audio Processing*, 13(5):1073–1081, 2005.

J.A. Ballas. Common factors in the identification of brief, miscellaneous, everyday sounds. *Journal of Experimental Psychology: Human perception and performance*, 19:250–267, 1993.

J.A. Ballas. The interpretation of natural sound in the cockpit. In N.A. Stanton and J.Edworthy, editors, *Human Factors in Auditory Warnings*, pages 91–112. Ashgate, Aldershot, UK, 1999.

M. Barra, T. Cillo, A. De Santis, U. Ferraro Petrillo, A. Negro, V. Scarano, T. Matlock, and P.P. Maglio. Personal webmelody: customized sonification of web servers. In *Proc.Int. Conf. on Auditory Display*, Espoo, Finland, July-Aug. 2001.

M. Barra, T. Cillo, A. De Santis, U.F. Petrillo, A. Negro, and V. Scarano. Multimodal monitoring of web servers. *Multimedia*, 9(3):32–41, July-Sep. 2002. URL `http://csdl.computer.org/comp/mags/mu/2002/03/u3toc.htm`. sound examples on the web.

S. Barrass. *Auditory Information Design*. Thesis submitted to the Australian National University, 1997.

S. Barrass and G. Kramer. Using sonification. *Multimedia Systems*, 7(1):23–31, Jan. 1999.

M.M. Blattner, D.A. Sumikawa, and R.M. Greenberg. Earcons and icons: their structure and common design principles. *Human-Computer Interaction*, 4: 11–44, 1989.

E. Brazil and M. Fernström. Audio information browsing with the sonic browser. In *Proc. Conf. on Coordinated and Multiple Views In Exploratory Visualization*, pages 26–33, London, Uk, July 2003.

A. S. Bregman. *Auditory scene analysis*. MIT Press, Harvard, MA, 1990.

S.A. Brewster. Using nonspeech sounds to provide navigation cues. *ACM Trans. on Computer-Human Interaction*, 5(3):224–259, Sept. 1998.

M. Chion. *Guide des objets sonores*. Éditions INA/Buchet Chastel, Paris, 1983.

M. Chion. *L'audio-vision. Son et image au cinéma*. Éditions Nathan, Paris, 1990.

F. Cifariello Ciardi. A multimodal toolkit for stock market data sonification. In *Proc. Int. Conf. on Auditory Display*, Sydney, Australia, July 2004.

G. Corsini and R. Saletti. A 1/f power spectrum noise sequence generator. *IEEE Trans. on Instrumentation and Measurement*, 37(4):615–619, 1988.

C. C. de Wit, H. Olsson, K. J. Åström, and P. Lischinsky. A new model for control of systems with friction. *IEEE Trans. Automat. Contr.*, 40(3):419–425, 1995.

W.T. Fitch and G. Kramer. Sonifying the body electric: Superiority of an auditory over a visual display in a complex, multivariate system. In G. Kramer, editor, *Auditory Display: Sonification, Audification and Auditory Interfaces*, pages 307–325. Addison-Wesley, Reading, Mass., 1994.

G.E. Frantti and L.A. Leverault. Auditory discrimination of seismic signals from earthquakes and explosions. *Bulletin of the Seismological Society of America*, 55(1):1–26, 1965.

S.P. Frysinger. Applied research in auditory data representation. In *Proceedings of the SPIE*, volume 1259, pages 130–139, 1990.

W. Gaver, T. Dunne, and E. Pacenti. Cultural probes. *Interactions*, pages 21–29, Jan. 1999.

W.W. Gaver. Synthesizing auditory icons. In *Proc. of the INTERACT and CHI Conf. on Human Factors in Computing Systems*, pages 228–235, Amsterdam, The Netherlands, Apr. 1993a.

W.W. Gaver. Using and creating auditory icons. In G. Kramer, editor, *Auditory Display: Sonification, Audification and Auditory Interfaces*, pages 417–446. Addison Wesley, 1994.

W.W. Gaver. What in the world do we hear? an ecological approach to auditory event perception. *Ecological Psychology*, 5(1):1–29, 1993b.

W.W. Gaver. How do we hear in the world? explorations of ecological acoustics. *Ecological Psychology*, 5(4):285–313, 1993c.

C. Hayward. Listening to the earth sing. In G. Kramer, editor, *Auditory Display: Sonification, Audification and Auditory Interfaces*, pages 369–404. Addison-Wesley, Reading, Mass., 1994.

J. Hereford and W. Winn. Non-speech sound in human-computer interaction: A review and design guidelines. *J. Educational Computing Research*, 11(3): 211–233, Mar. 1994.

T. Hermann and H. Ritter. Sound and meaning in auditory data display. *Proceedings of the IEEE*, 92(4):730–741, Apr. 2004.

P.A. Houle and J.P. Sethna. Acoustic emission from crumpling paper. *Physical Review*, 54(1):278–283, July 1996.

J.-M. Jot. Real-time spatial processing of sounds for music, multimedia and interactive human-computer interfaces. *Multimedia Systems*, 7(1):55–69, Jan. 1999.

G. Kramer. Some organizing principles for representing data with sound. In G. Kramer, editor, *Auditory Display: Sonification, Audification and Auditory Interfaces*, pages 185–222. Addison Wesley, Reading, Mass., 1994.

M. Kubovy and D. Van Alkenburg. Auditory and visual objects. *International Journal of Cognitive Science*, 80:97–126, June 2001.

V. Lombardo and A. Valle. *Audio e multimedia (in Italian)*. Apogeo, Milano, 2nd edition, 2005.

V. Lombardo, A. Arghinenti, F. Nunnari, A. Valle, H.H. Vogel, J. Fitch, R. Dobson, J. Padget, K. Tazelaar, S. Weinzierl, S. Benser, S. Kersten, R. Starosolski, W. Borczyk, W. Pytlik, and S. Niedbala. The virtual electronic poem (VEP) project. In *Proc. Int. Computer Music Conf.*, pages 451–454, Barcelona, Spain, Sep. 2005.

D.K. McGookin and S.A. Brewster. Understanding concurrent earcons: Applying auditory scene analysis principles to concurrent earcon recognition. *ACM Trans. Appl. Percept.*, 1(2):130–155, Oct. 2004.

K.V. Nesbitt and S. Barrass. Finding trading patterns in stock market data. *IEEE Computer Graphics and Applications*, 24(5):45–55, Oct. 2004.

R.D. Patterson. Auditory warning sounds in the work environment. *Philosophical Transactions of the Royal Society B*, 327:485–492, 1990.

M. Rath and D. Rocchesso. Continuous sonic feedback from a rolling ball. *IEEE Multimedia*, 12(2):60–69, 2005.

D. Rocchesso, R. Bresin, and Mikael Fernström. Sounding objects. *IEEE Multimedia*, pages 42–52, 2003.

C. Scaletti. Sound synthesis algorithms for auditory data representations. In G. Kramer, editor, *Auditory Display : Sonification, Audification and Auditory Interfaces*. Addison-Wesley, Reading, Mass., 1994.

P. Schaeffer. *La Musique Concrète*. Presses Universitaires de France, Paris, 1967.

M. Schafer. *Soundscape - Our Sonic Environment and the Tuning of the World*. Destiny Books, Rochester, Vermont, 1994.

J.P. Sethna, K.A. Dahmen, and C.R. Myers. Crackling noise. *Nature*, 410: 242–250, Mar. 2001.

S.D. Speeth. Seismometer sounds. *Journal of the Acoustical Society of America*, 33(7):909–916, 1961.

N.A. Stanton and J.Edworthy. Auditory warnings and displays: an overview. In N.A. Stanton and J.Edworthy, editors, *Human Factors in Auditory Warnings*, pages 3 – 30. Ashgate, Aldershot, UK, 1999a.

N.A. Stanton and J.Edworthy. Auditory warning affordances. In N.A. Stanton and J.Edworthy, editors, *Human Factors in Auditory Warnings*, pages 113–127. Ashgate, Aldershot, UK, 1999b.

C. Ware. *Information Visualization*. Morgan Kaufmann, Paris, 2004.

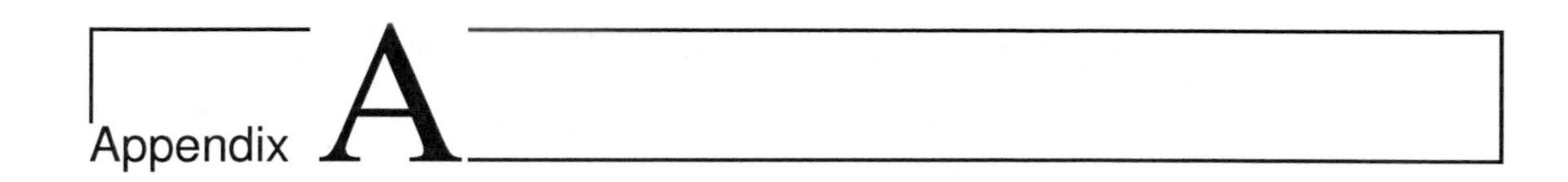

Controlling Sound Production

Roberto Bresin and Kjetil Falkenberg Hansen
Department of Speech, Music and Hearing, KTH, Stockholm

Matti Karjalainen, Teemu Mäki-Patola, Aki Kanerva, and Antti Huovilainen
Laboratory of Acoustics and Audio Signal Processing, Helsinki University of Technology

Sergi Jordà, Martin Kaltenbrunner, Günter Geiger, Ross Bencina
Music Technology Group, University Pompeu Fabra, Barcelona

Amalia de Götzen[1] and Davide Rocchesso[2]
[1]Department of Information Engineering, University of Padua
[2]Department of Art and Industrial Design, IUAV University of Venice

Introduction

This appendix is a collection of five contributions by different authors. Each contribution is self-contained and can be read separately. The scope ranges from new musical instruments to sonification of interactive books. The appendix starts with an introductory section, in which general concepts and problems related to sound control are presented. The authors outline the important issue of the choice of sound models that can provide suitable and responsive feedback in continuous control, as in the case of sound generated by body motion and sound as feedback in interaction. In the following sections, some recent and successful applications are illustrated. The first two, DJ Scratching and Virtual Air Guitar, focus on the control of virtual musical instruments, and in particular on control that generates sense through sound production. The last two sections, The reacTable and The Interactive Book, focus on the control of sounding objects, and are characterised by applications that control sounds that produce sense. The section sequence can also be seen as ordered according to an increasing level of sound model abstraction (from sampled sounds to cartoon sounds) and decreasing level of complexity of control gesture (from DJ scratching to simple sliding).

The authors of the different sections are:

A.1 Introduction
Roberto Bresin and Davide Rocchesso

A.2 DJ scratching with Skipproof
Kjetil Falkenberg Hansen and Roberto Bresin

A.3 Virtual air guitar
Matti Karjalainen, Teemu Mäki-Patola, Aki Kanerva, and Antti Huovilainen

A.4 The reacTable*
Sergi Jordà, Martin Kaltenbrunner, Günter Geiger, Ross Bencina

A.5 The interactive book
Amalia de Götzen and Davide Rocchesso

A.1 Background

[1] There are a few emerging facts that are conditioning our present approaches to the study of sound. Sensors of many different kinds are available at low cost and they can be organised into networks. Computing power is generously available even in tiny and low-power processors that can be easily embedded into artefacts of different nature and size. New design strategies that take advantage of these technological opportunities are emerging: physical computing, natural interaction, calm technologies are some of the many buzzwords that are being proposed as labels for these new trends in design. For the purpose of sound-based communication, the concept of embodied interaction (Dourish, 2001) is particularly significant. Embodiment is considered a property of how actions are performed with or through artefacts, thus embracing the position that treats meanings as inextricably present in the actions between people, objects, and the environment. A key observation that emerges from embodied interaction examples is that human interaction in the world is essentially continuous and it relies on a complex network of continuous feedback signals. This is significantly important if one considers that most interfaces to technological artefacts that are currently being produced are developed around switches, menus, buttons, and other discrete devices. The design of graphical user interfaces has been largely inspired by ecological psychology and concepts such as direct perception and affordances (Gibson, 1979). When designing embodied interfaces, we call for a reconciliation of ecological psychology and phenomenology that looks, with equal emphasis, at the objects and at the experiences. By means of physical modelling we can represent and understand the objects. By direct observation we can tell what are the relevant phenomena, which physical components are crucial for perception, what degree of simplification can be perceptually tolerated when modelling the physical reality. Specifically, sound designers are shifting their attention from sound objects to sounding objects, in some way getting back to the sources of acoustic vibrations, in a sort of ideal continuity with the experimenters of the early twentieth century, especially futurists such as Luigi

[1]Parts of this section are extracted and modified from a recent work by Rocchesso and Bresin (2007)

Russolo and his intonarumori. In the contemporary world, sounding objects should be defined as sounds in action, intimately attached to artefacts, and dynamically responsive to continuous manipulations. As opposed to this embodied notion of sound, consider an instrument that came shortly after the intonarumori, the theremin invented 1919 by Lev Termen. It is played by moving the hands in space, near two antennae controlling amplitude and frequency of an oscillator. Its sound is ethereal and seems to come from the outer space. This is probably why it has been chosen in the soundtracks of some science-fiction movies (see the documentary by Martin, 2001). Even though relying on continuous control and display, the lack of physical contact may still qualify the theremin as a schizophonic artefact, and it is not by coincidence that it is the only musical instrument invented in the twentieth century (the schizophonic age) that was used by several composers and virtuosi. Indeed, nephews of the theremin can be found in several recent works of art and technology making use of sophisticated sensors and displays, where physical causality is not mediated by physical objects, and the resulting interaction is pervaded by a sense of disembodiment.

A.1.1 Sound and motion

Sounds are intimately related to motion, as they are usually the result of actions, such as body gestures (e.g. the singing voice) or mechanical movements (e.g. the sound of train wheels on rails). In the same way as we are very accurate in recognizing the animate character of visual motion only from a few light points corresponding to the head and the major limb-joints of a moving person (Johansson, 1973), we are very sensitive to the fluctuations of auditory events in the time-frequency plane, so that we can easily discriminate walking from running (Bresin and Dahl, 2003) or even successfully guess gender of a person walking (Li et al., 1991). It is not a surprise that gestures are so tightly related with sound and music communication. A paradigmatic case is that of the singing voice, which is directly produced by body movements (see also Chapters 6.3 and 7 for overviews on gestures in music performance). In general, gestures allow expressive control in sound production. Another example is DJ scratching, where complex gestures on the vinyl and on the cross-fader are used for achieving expressive transformation of prerecorded sounds (Hansen and Bresin, 2004). In the context of embodied interfaces, where manipulation is mostly continuous, it is therefore important to build a gesture interpretation layer, capable to extract the expressive content of human continuous actions, such as those occurring as preparatory movements for strokes (see Dahl, 2004). Body movements preceding the sound production give information about the intentions of the user, smoother and slower movements produce softer sounds, while faster and sudden movements are associated to louder sounds. Gestures and their corresponding sounds usually occur in time sequences, and it is their particular time organisation that helps in classifying their nature. Indeed, if properly organised in time, sound events can communicate a particular meaning. Let us consider the case of walking sounds (Giordano and Bresin, 2006). The sound of a step in isolation is difficult to identify, while it gives the idea of walking if repeated a number of times. If the time sequence is organised according to equations resembling biological motion, then walking sounds can be perceived as more natural (Bresin and Dahl, 2003). In addition, if sound level and timing are varied, it is possible to communicate different emotional intentions with walking sounds. In fact, the organisation in time and sound level of structurally organised events, such as notes in music performance or phonemes in speech, can be controlled for communicating different emotional expressions. For instance in hyper- and hypoarticulated speech (Lindblom, 1990) and in enhanced performance of musical structure (Bresin and Friberg, 2000) the listener recognises the meaning being conveyed as well as the expressive intention on top of it. Research results show that not only we are able to recognise different emotional intentions used by musicians or speakers (Juslin and Laukka, 2003) but also we feel these emotions. It has been demonstrated by psychophysical experiments that people listening to music evoking emotions experience a change in biophysical cues (such as blood pressure, etc.) that correspond to the feeling of that specific emotion and not only to the recognition. Krumhansl (1997) observed that sad music produced largest changes in heart rate, blood pressure, skin conductance and temperature, while happy music produced largest changes in measures of respiration. Music and sound in general have therefore the power to effect the variation of many physiological parameters in our body. These results could be taken into account in the design of more engaging applications where sound plays an active role.

A.1.2 Sound and interaction

An important role in any controlling action is played by the feedback received by the user, which in our case is the sound resulting from the user's gestures on an object or a musical instrument. Therefore sound carries information about the user's actions. If we extend this concept and consider sounds produced by any object in the environment we can say that sound is a multidimensional information carrier and as such can be used by humans for controlling their actions and reactions relatively to the environmental situation. In particular, humans are able to extract size, shape, material, distance, speed, and emotional expression from sonic information. These capabilities can be exploited to use sound as a powerful channel of communication for displaying complex data. Interactive sonification[2] is a new emerging field where sound feedback is used in a variety of applications including sport, medicine, manufacturing, and computer games. There are many issues that have been raised in such applications, and answers are expected to come from interaction design, perception, aesthetics, and sound modelling. For instance, how do we achieve pleasant and effective navigation, browsing, or sorting of large amount of data with sounds? In the framework of the Sounding Object project , the concept of sound cartoonification has been embraced in its wider sense and applied to the construction of engaging everyday sound models. Simplified and exaggerated models have been proved to be efficient in communicating the properties of objects in actions, thus being excellent vehicles for informative feedback in human-artefact communication. For instance, it has been shown that temporal control of sound events helps in communicating the nature of the sound source (e.g. a footstep) and the action that is being performed (walking/running). The possibility of using continuous interaction with sounding objects allows for expressive control of the sound production and, as a result, to higher engagement, deeper sense of presence, and experiential satisfaction. Low-cost sensors and recent studies in artificial emotions enable new forms of interaction using

[2]See Hunt and Hermann (2005) for a recent overview of the field

previously under-exploited human abilities and sensibilities. For instance, a cheap webcam is sufficient to capture expressive gesture nuances that, if appropriately interpreted, can be converted into non-visual emotional cues. These new systems, albeit inexpensive and simple in their components, provide new challenges to the designer who is called to handle a palette of technologies spanning diverse interaction modalities. In the future, the field of interaction design is expected to provide some guidelines and evaluation methods that will be applicable to artefacts and experiences in all their facets. It is likely that the classic methods of human-computer interaction will be expanded with both fine-grained and coarse-grained analysis methods. On a small scale, it is often necessary to consider detailed trajectories of physical variables in order to make sense of different strategies used with different interaction modalities. On a large scale, it is necessary to measure and analyze the global aesthetic quality of experiences.

A.2 DJ scratching with Skipproof

A.2.1 Concept

Scratching has been established as a practice of treating the turntable and a sound mixer as a musical instrument, and is one of the usual DJ playing styles. Scratching and the related playing style *beat juggling* require much training for mastering the complex gesture control of the instrument.

Skipproof, a patch written for pd (Pure Data, Puckette, 1996), is both a virtual turntable and mixer, and an application for exploring the musical language of DJs. The name Skipproof is taken from a feature found on DJ-tools records called a *skip proof* section, where a sound (or set of sounds) are exactly one rotation long and repeated for a couple of minutes. If it happens that the needle jumps during a performance, the chances are quite good it will land on the same spot on the sound, but in a different groove. The main purpose is to "scratch" sound files using gesture controllers of different kinds.

Scratch performances are normally built up by the sequential executions of well-defined hand movements. Combinations of a gesture with the hand controlling the record and a gesture with the hand controlling the fader[3] on the mixer are called scratch techniques. These have become common language for DJs, and they refer to them by name (*baby*, *crab*, *flare*) and their characteristics (1-click *flare*, 2-click *flare*, reversed *tear*). About one hundred techniques were recognised and more than 20 analysed in previous studies (e.g. Hansen, 2002; Hansen and Bresin, 2004). The analysis focused on measuring the movement of the record and the fader. Based on the results, models of synthesised scratch techniques were constructed.

Originally, the software was intended to be a tool for reproducing and exploring the modelled scratch techniques with different characteristics. For instance, we could change speed and extent of the record movements while maintaining the fader gesture. The graphical user interface allows for easy experimenting with the models. We decided to develop the patch so it could

[3] The crossfader normally has a very short span from silent to full sound, often less than a millimeter. It is more realistic to consider this fader as a switch than a fader.

be used also for controlling sound files in a turntable-like manner. The user is not restricted to play sound files, but can control sound synthesis techniques such as physics-based models. In the end, Skipproof is combining features from turntable, mixer, vinyl records and modelled DJ performances.

A.2.2 Strategy-implementation

Skipproof most important feature is to allow execution and control of the scratch technique models, and to do this with customisable interfaces and controllers. The graphical user interface, made with GrIPD (Sarlo, 2003), serves as a collection of controllers a DJ normally would expect to find in addition to new controllers, see Fig. A.1.

All sound samples that are used with the patch are 1.6 seconds long and looped, reflecting the idea of a *skip proof* track. The sounds are taken from a DJ tool record and are all favored for scratching.

There are currently 12 typical scratch techniques ready to use in the patch. These are models based on analysis of recordings made by a professional DJ (Hansen, 2002). Each technique consists of a forward–backward movement of the record and synchronised actions with the fader. Changing the way a technique is performed is dependent on the controller gesture, and this connection between gesture and sound output can be customised. When using several controllers, different mapping schemes will be realised (Hunt and Kirk, 2000).

In the first version of Skipproof, the techniques can be altered in extent and speed of the movement, the two most important parameters in scratching (Hansen and Bresin, 2004). Both extent and speed can be exaggerated, so for example scratch techniques are performed faster and with larger movements than in a real situation.

All the elements in the GUI can be accessed by any controller. Until now, Skipproof has been played by a number of devices, including computer peripherals and tablets, MIDI devices, Max Mathews' *RadioBaton* (Boulanger and Mathews, 1997) and various sensors (e.g. *LaKitchen*'s Kroonde and Toaster sensor interfaces, Coduys et al., 2004). Among these controllers, the Radio-Baton in combination with sliders, buttons, magnetic field, flexion and light sensors has good potential to be a rewarding interface. Even a computer mouse in combination with other sensors is quite effective, in contrast to many new musical interfaces.

Recently, the concept of Skipproof has been successfully implemented and tested with the reacTable (see Section A.4). Scratch technique models were assigned to reacTable objects, where one object contained fader movement, one contained record movement and a third object contained the sound.

A.2.3 Results-expectations

A professional DJ has been using Skipproof in two live concert situations, see Fig. A.2. The first time, the RadioBaton was controlling the virtual turntable and technique models. The RadioBaton sticks were replaced with treated gloves, so the musician could control the turntable speed and the techniques easily, moving his hand in a 3-D space over the antennae.

In a second concert, the DJ used again the RadioBaton, but now he could trigger the models based on the hand gesture, as compared to the earlier version with button triggers. The approach toward a defined area of the antennae was measured in speed and distance, and this gesture determined scratch speed and extent. For this performance, a *light switch* was used to replace the fader. A light sensor was placed pointing directly toward a lamp, and the DJ could break the light beam by placing his hand close to the sensor, or by waving the hand after the light with outstretched fingers. In that way, controlled and rapid sound on–off events were possible, just like with a fader.

The DJ commented later that although there was a lack of mechanical feedback from the interface, it opened up to new possibilities. Controlling recorded techniques was considered to be hard, especially to get the correct tempo. A scratch DJ uses physical markings on the vinyl (stickers, label) to see where in a sound the pick-up is, and this feature is moved from the controller to the GUI in Skipproof. This takes time getting comfortable with and is not at all optimal.

Users without DJ experience have found the set-up with RadioBaton as turntable and light switch as fader to be both intuitive and exciting, and quite fast they could perform with it in a simple fashion. Using the reacTable (see Section A.4) as the controller interface is a very interesting approach that addresses the need from non-professional DJs to perform simple scratch sequences.

DJs seem overall to be interested and intrigued by new technology and possibilities. It is possible to build a system that enhances DJ performances of scratching, and it is desirable to experiment with the equipment currently preferred by DJs. Performing with models of scratch techniques is still a novel approach that needs to be tested more.

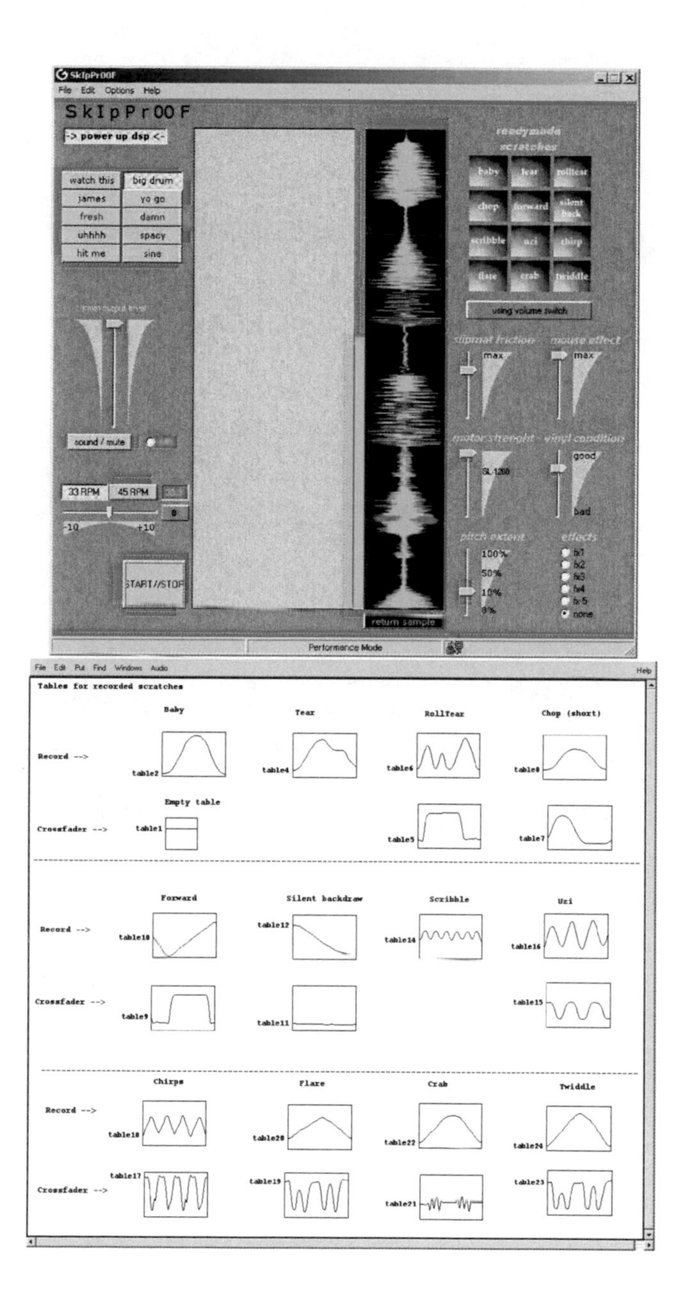

Figure A.1: Skipproof graphical user interface (left) and synthesised techniques (right). The GUI contains a combination of standard controllers found on turntables and mixers, and novel controllers such as technique triggers, visual representation of sound, sliders for changing control parameters, and more. The synthesised techniques are saved as sets of two tables; the upper is the record movement and the lower is the crossfader movement.

Figure A.2: DJ 1210 Jazz scratching with Skipproof in a concert. The Radio-Baton is operated with his right hand. Beside the computer screen, there is a lamp for a light sensor, on the floor, there is a rack of foot switches. The DJ's left hand is on the crossfader.

A.3 Virtual air guitar

A.3.1 Concept

A combination of hand-held controllers and a guitar synthesiser with audio effects is called here the Virtual Air Guitar (VAG). The name refers to playing an "air guitar", that is just acting the playing with music playback, while the term virtual refers to making a playable synthetic instrument. Sensing of the distance of hands is used for pitch control, the right hand movements for plucking, and the finger positions may in some cases be used for other features of sound production. The synthetic guitar algorithm supports electric guitar sounds, augmented with sound effects and intelligent mapping from playing gestures to synthesis parameters.

Electronic and computer-based musical instruments are typically developed to be played from keyboard, possibly augmented by foot, breath or other controllers. In VAG, we have explored the possibility to make an intuitive yet simple user interface for playing a particular virtual (synthetic) instrument, the electric guitar. In addition to the synthetic instrument and related audio effects (amplifier distortion and loudspeaker cabinet simulation) we have explored three different controllers for player interface: data gloves in a virtual room environment, special hand-held controller sticks, and webcam-based camera tracking of player's hands.

The first case (data glove control) is for flexible experimentation of possible control features, while the two others are intended for maximally simplified guitar playing, designed for wide audience visiting a science center exhibition. The two most important parameters needed in all cases are the pitch control (corresponding to fretting position) and the string plucking action. The pitch-related information is taken by measuring the distance of the two hands, which was found easier to use than the distance of left hand to a reference such as the players body. The string plucking action is most easily captured by the downward stroke of the right hand.

Here we present an overview of our strategy and implementation of the virtual air guitar, paying most attention to the user interface aspects from the players point of view. We then shortly present our results and expectations. A more detailed description[4] of the VAG can be found in Karjalainen et al. (2006).

A.3.2 Strategy-implementation

The VAG system is composed of (1) a guitar synthesiser with sound effects and audio reproduction, (2) a user interface consisting of handhold sensors, and (3) software to map user interface signals to expressive VAG playing.

The virtual instrument is a simulation of an electric guitar tuned to sound like the Fender Stratocaster, and it is realised using the Extended Karplus-Strong modelling technique described in Karjalainen et al. (1998). The preamplifier stages of a tube amplifier and loudspeaker cabinet were simulated digitally; the VAG also includes a delay unit and a reverberation unit integrated with the guitar synthesiser.

For the user interface, we have experimented with three approaches.

1. Data gloves and 3-D position-tracking in a cave-like virtual room (Fig. A.3). The finger flexure parameters can be mapped to sliding, string damping, selection of string or chords, or controlling different audio effects.
2. Small handhold control devices (Fig. A.4). The right-hand stick sends high-frequency pulses via a tiny loudspeaker, and the left-hand stick receives them via an electret microphone. The pitch is extracted from the propagation delay of the pulses, whereas the pluck is captured by an acceleration sensor microchip inside the right-hand stick.
3. Hand tracking by video image analysis (Fig. A.5). The user wears orange gloves, which correspond to blobs in video frames. A gesture extractor is informed by their location. A pluck is detected when the right hand passes through the imaginary guitar centerline. The guitar moves with the player, and the user's grip calibrates its size. Vibrato, slide, and string muting are extractable gestures. The user can choose among two playing modes: rhythm guitar and solo guitar, i.e. she can strum four different

[4]Web documents on the project, including videos of playing the VAG, are available at: `http://airguitar.tml.hut.fi` and `http://www.acoustics.hut.fi/demos/VAG`

power chords, or freely play a guitar solo on a pentatonic minor scale, with additional techniques such as fret sliding and vibrato.

The VAG's interface can use complex mapping rules and procedures from gesture to sound model. This mapping layer is called *musical intelligence* in the VAG system. The user's gestures are first interpreted into a meta language that describes guitar playing techniques on an abstract, musical level. For example, moving the right hand over the imaginary guitar strings in a strumming motion is interpreted as a pluck event. The musical intelligence thus contains both the rules by which gestures are converted into these musical events, and the implementations of the events for a certain sound model.

A.3.3 Results-expectations

The VAG is an entertainment device that is in line with the way of playing the air guitar - with showmanship, intensity and fun. The main effort has been to study how the hand positions and movements can be mapped to control typical playing of the electric guitar. This means a highly simplified user interface, which sets strict limits to what can be played by such a virtual guitar, certainly not satisfactory for a professional guitarist. It is, however, an interesting case of studying what can be done to demonstrate the basic features of playing a particular instrument with a given style, or to make inexpensive "toy" instruments for fun. This can then be augmented by extra functions in more complex controllers, for example using finger movements, foot pedals, etc. In such cases the challenge is to find expressive and intuitive forms of controls that may be useful also for professional musicians. Another way, to get rich sound from minimalistic controllers, is to use complex rule-based mappings from simple control signals to more advanced control signals for playing a virtual instrument.

Both the webcam and the control stick versions were placed on display at the Heureka Science Center in Finland in 2005 March. The webcam version became the most popular attraction of the music-related exhibition, being played over 60,000 times during the one year of the exhibition. It has also attracted international media attention, including numerous television shows, radio programs, popular magazine articles, and online articles. Currently, a commercial *Virtual Air Guitar game* is in development[5].

The VAG is a good example of a virtual instrument that requires special controllers and playing strategies, different from keyboard-oriented control. The simple versions described above are intended for toy-like applications, such as games, or instructional devices to characterise the most essential features of plucked string instrument playing.

Among future challenges, there are studies on more expressive control interfaces, which could be useful also for professional musicians. With respect to real guitar playing, there is much more freedom to apply different gesture parameters for virtual instrument control.

[5] `http://www.virtualairguitar.com`

Figure A.3: Soulful playing of a VAG in a virtual room using data gloves.

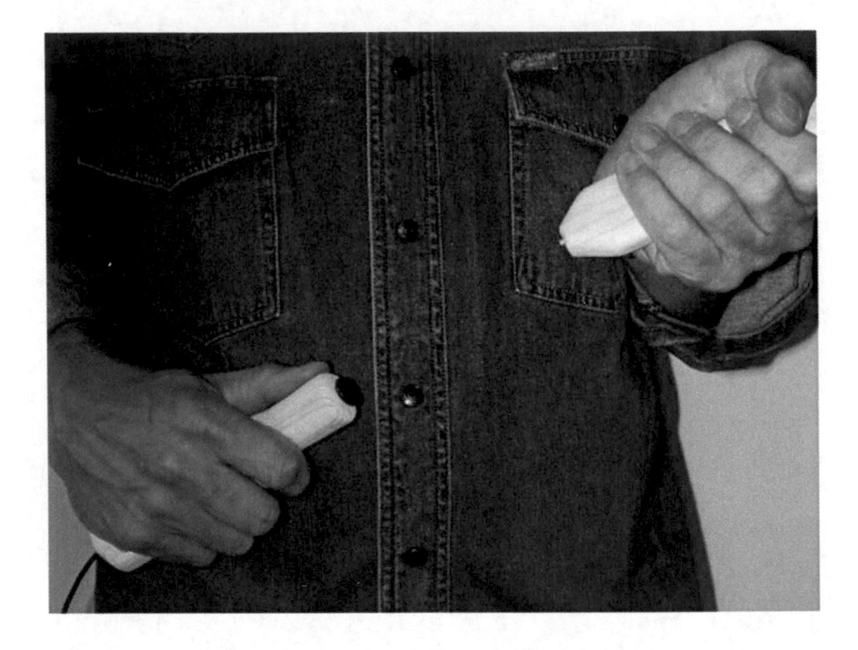

Figure A.4: Control sticks for playing the VAG. Right-hand stick (on the left) includes an acceleration sensor as well as a small loudspeaker to send the distance measurement pulse. Left-hand stick (on the right) receives the pulse by an electret microphone.

Figure A.5: Camera tracking of hand positions (orange gloves), as seen by the computer.

A.4 The reacTable

The reacTable is a multi-user electronic music instrument with a tabletop tangible user interface (Jordà et al., 2007; Kaltenbrunner et al., 2004). Several simultaneous performers share complete control over the instrument by moving and rotating physical objects on a luminous table surface. By moving and relating these objects, representing components of a classic modular synthesiser, users can create complex and dynamic sonic topologies, with generators, filters and modulators, in a kind of tangible modular synthesiser or graspable flow-controlled programming language.

In recent years there has been a proliferation of tabletop tangible musical interfaces. This trend started with the millennium with projects such as the Audiopad (Patten et al., 2002), Jam-o-drum (Blaine and Perkis, 2000) or SmallFish [6], and nowadays so many "musical tables" are being produced that it becomes difficult to keep track of every new proposal [7]. Is this just a coincidence or the result of a tabletop vogue? While arguably, not all the currently existing prototypes may present the same level of achievement or coherence, we believe that there are important reasons, perhaps often more intuited than stated, that turn musical instruments and tabletop tangible interfaces, into promising and exciting fields of crossed multidisciplinary research and experimentation.

A.4.1 Concept: Multithreaded musical instrument and shared control

New musical controllers and laptop performance Music controllers or new interfaces for musical expression (NIME) are experimenting an increasing attention from researchers and electronic luthiers. In parallel to this research bloom, the laptop is progressively reaching the point of feeling as much at home on stage as a saxophone or an electric guitar. However, the contemporary musical scene does not clearly reflect this potential convergence, and most

[6]SmallFish:http://hosting.zkm.de/wmuench/small_fish
[7]Kaltenbrunner,M.:http://www.iua.upf.es/mtg/reacTable/?related

laptop performers seem hesitant to switch towards the use of new hardware controllers, as if laptop performance and the exploration of post-digital sound spaces was a dialog conducted with mice, sliders, buttons and the metaphors of business computing. The reasons for this apparent rejection can be probably found in the inherent nature of computer-based music performance. In traditional instrumental playing, every nuance, every small control variation or modulation (e.g. a vibrato or a tremolo) has to be addressed physically by the performer. In digital instruments nevertheless, the performer no longer needs to control directly all these aspects of the production of sound, being able instead to direct and supervise the computer processes which control these details. These new type of instruments often shift the center of the performer's attention from the lower-level details to the higher-level processes that produce these details. The musician performs control strategies instead of performing data, and the instrument leans towards more intricate responses to performer stimuli, tending to surpass the note-to-note and the "one gesture-one acoustic event" playing paradigms present in all traditional instruments, thus allowing musicians to work at different musical levels and forcing them to take higher level and more compositional decisions on-the-fly (Jordà, 2005).

However, most of the music controllers currently being developed do not pursue this multithreaded and shared control approach, prolonging the traditional instrument paradigm instead. Many new musical interfaces still tend to conceive new musical instruments highly inspired by traditional ones, most often designed to be "worn" and played all the time, and offering continuous, synchronous and precise control over a few dimensions. An intimate, sensitive and not necessarily highly dimensional interface of this kind (i.e. more like a violin bow, a mouthpiece or a joystick, than like a piano) will be ideally suited for direct microcontrol (i.e. sound, timbre, articulation). However, for macrostructural, indirect or higher level control, a non-wearable interface distributed in space and allowing intermittent access (i.e. more like a piano or a drum), and in which control can be easily and quickly transferred to and recovered from the machine, should be undeniably preferred (Jordà, 2005).

TUIs: Making graphical user interfaces graspable Even if Graphical User Interface (GUI) conception and design may be central in most HCI related areas, not many new music instruments profit from the display capabilities of digital computers, whereas in the musical performance model we are discussing, in which performers tend to frequently delegate and shift control to the instrument, all affordable ways for monitoring ongoing processes and activities are especially welcome. Visual feedback potentially constitutes a significant asset for allowing this type of instruments to dynamically communicate the states and the behaviors of their musical processes (Jordà, 2002; Jordà and Wüst, 2003); it is the screen and not the mouse what laptop performers do not want to miss, and it is in this context where tabletop tangible interfaces may have a lot to bring.

Tangible User Interfaces (TUIs) combine control and representation within a physical artefact (Ullmer and Ishii, 2001). In table based tangible interfaces, digital information becomes graspable with the direct manipulation of simple objects which are available on a table surface. Combining augmented reality techniques that allow the tracking of control objects on the table surface, with visualisation techniques that convert the table into a flat screening surface, a system with these characteristics favors multi-parametric and shared control, interaction and exploration and even multi-user collaboration. Moreover, the seamless integration of visual feedback and physical control, which eliminates the indirection component present in a conventional screen + pointer system, allows a more natural, intuitive and rich interaction. With these considerations, it may not be a coincidence that in recent years, an increasing variety of tabletop tangible musical controllers or instruments, such as Audiopad (Patten et al., 2002) or the reacTable, is being developed.

A.4.2 The reacTable: Strategy and implementation

The reacTable consists of a translucent luminous round table - a surface with no head position or leading voice and with no privileged points-of-view or points-of-control - in which physical artefacts or pucks can be moved and rotated. Each puck represents a synthesiser module with a dedicated function

Figure A.6: Four hands at the reacTable

for the generation, modification or control of sound. Six functional groups exist each one associated with a different puck shape (audio generators, audio filters, controllers, control filters, mixers and global functions); different pucks within a same group show distinct symbols on their surfaces) (see Figures A.6 and A.7).

Modular synthesis goes back to the first sound synthesisers, in the digital and especially in the analogue domains, with Robert Moog's or Donald Buchla's Voltage controlled synthesisers (Chadabe, 1975). It also constitutes the essence of many visual programming environments for sound and music such as Max or Pd (Puckette, 1996). The reacTable outdoes these models by implementing what we call dynamic patching (Kaltenbrunner et al., 2004). Connections and disconnections between modules are not explicitly indicated by the performer, but automatically managed by means of a simple set of rules according to the objects' types, and their affinities and proximities with their neighbors. By moving pucks and bringing them into proximity with each other, performers on the reacTable construct and play the instrument at the same time. Since the move of any object around the table surface can alter existing connections, extremely variable synthesiser topologies can be attained resulting in a highly dynamic environment.

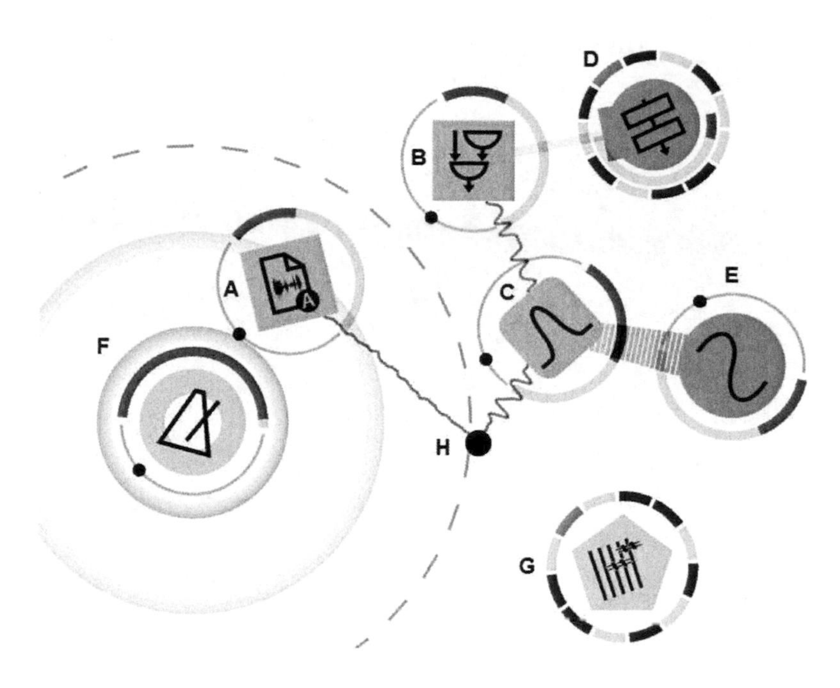

Figure A.7: A snapshot showing connections between several objects

Additionally, all reacTable objects can be spun as rotary knobs, which allow controlling one of their internal parameters, and a second parameter is controlled by dragging the finger around the objects' perimeter as shown in Figure A.8. Although the exact effects vary from one type of object to the other, rotation tends to be related with frequency or speed and finger dragging with amplitude. Finger interaction also allows to temporarily cutting (i.e. mute) audio connections, activating/deactivating discrete steps in objects such as step-sequencers or tonalisers (see D and G in Figure A.7) or even drawing envelopes or waveforms which will be "absorbed" by the nearest "wave or envelope-compatible" object.

In order to control a system of such complexity, with potentially dozens of continuous and discrete parameters, visual feedback becomes an essential component of the reacTable's interface. It was decided to avoid any type of alphanumerical and symbolic information and to banish any decorative display. All shapes, forms, lines or animations drawn by the visual synthesiser bring relevant information, and all the relevant information of the system (i.e. both the instantaneous results of the performers actions as well as the current

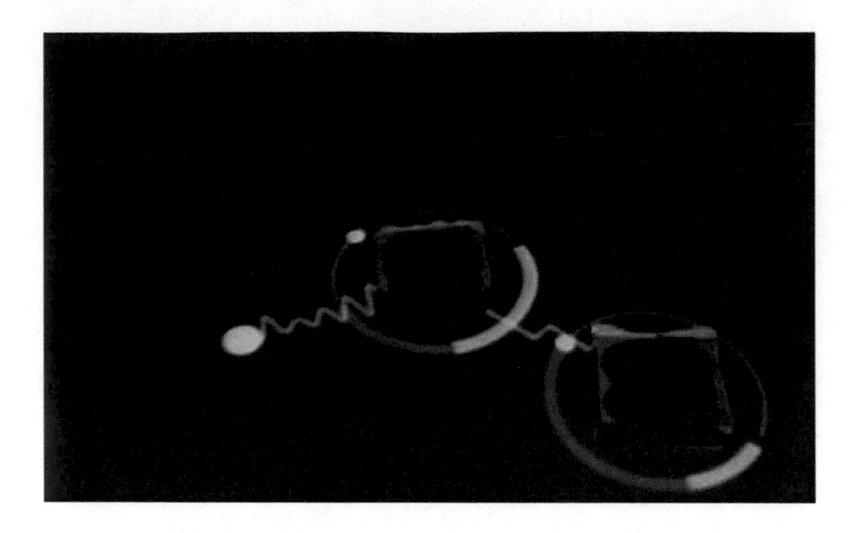

Figure A.8: Modifying a parameter with the finger

state of all the system controlled processes) is permanently displayed. The lines that show the connections between the objects convey the real resulting waveforms that are being produced or modified at each node (in the case of audio connections) or the density and intensity of the values they transport (in the case of control event-based connections). Low frequency oscillators, metronomes and other objects that vibrate at visible rates (i.e. below a few Hz) are animated with their precise heartbeat periods. All object states and internal parameters are also permanently monitored, such as in the white 180 degree circular fuel gauges that surround any object indicating their rotational values, in the dots that show the position of the second parameter slider (see all the objects in Figure A.7 except D and G, and the two objects in Figure A.8), or on the finger-activated discrete steps that surround objects such as the step-sequencers or the tonaliser (see respectively D and G in Figure A.7).

A.4.3 Results and expectations: Interacting with the reacTable

The reacTable was designed with both casual users and professional performers in mind, and it seeks to combine an immediate and intuitive access in a relaxed and immersive way, with the flexibility and the power of digital sound design algorithms and endless improvement possibilities and mastery. Since its first presentation in the summer of 2005, the reacTable has undergone a very active life outside of the laboratory. It has been exhibited in dozens of festivals, conferences or shows around the world, and has been played by several thousands users of all ages and different backgrounds (musicians, computer music and computer graphics experts; electronic music, digital art or computer game aficionados; teenagers, families with kids, etc.). The feedback has been very positive, often even passionate, showing that the reacTable can be very much enjoyable even without being fully understood.

In parallel to these public installations, many concerts have taken place during recent years, in festivals, clubs, art galleries or discotheques, which have turned the reacTable into an already mature musical instrument. Recently, it has even reached the rock stadiums as a regular instrument in Björk's last world tour, thus becoming one of the very few new digital instruments to have successfully passed the research prototype state.

The ideal combination of these two very different test-beds fulfills our initial goal, which was to build a musical instrument conceived for casual users at home or in installations, as well as for professionals in concert, and these two complementary test fronts keep bringing very relevant information into the continual reacTable design process refinement.

A.5 The interactive book

A.5.1 Concept

A book is a very well known object which everyone has used at least once in life. It plays an important role in children education: most of us learned colors, names of animals and numbers just "reading" or, better, interacting with some nice, colored pull-the-tab and lift-the-flap books. In the last decades children's books have been modified in order to use new interaction channels, inserting technology inside this old medium or using the book metaphor to develop new interfaces. It is quite clear that technology did not change too much for the book in thousand years: the history of books has seen new printing and composition techniques but the users are still basically dealing with the same artefact. Thus, the book as an object guarantees a high level of functionality.

Current commercial interactive books for children are very often similar to conventional colored stories with the addition of some pre-recorded sounds which can be triggered by the reader. The limitations of these books are evident: the sounds available are limited in number and diversity and they are played using a discrete control (typically a button). This means that sounds are irritating rather than being a stimulus to interact with the toy-book or allowing for learning by interaction.

Pull-the-tab and lift-the-flap books play a central role in the education and entertainment of most children all over the world. Most of these books are inherently cross-cultural and highly relevant in diverse social contexts. For instance, Lucy Cousins, the acclaimed creator of Maisy (Pina in Italy), has currently more than twelve million books in print in many different languages. Through these books, small children learn to name objects and characters, they understand the relations between objects, and develop a sense of causality by direct manipulation (Hutchins et al., 1986; Schneiderman, 2002) and feedback. The importance of sound as a powerful medium has been largely recognised and there are books on the market that reproduce prerecorded sounds upon pushing certain buttons or touching certain areas. However, such triggered sounds are extremely unnatural, repetitive, and annoying. The key for a successful exploitation of sounds in books is to have models that respond continuously to continuous action, just in the same way as the children do when manipulating rattles or other physical sounding objects. In other words, books have to become an embodied interface (Dourish, 2001) in all respects, including sound.

A.5.2 Strategy-implementation

In recent years, the European project "The Sounding Object"[8] was entirely devoted to the design, development, and evaluation of sound models based on a cartoon description of physical phenomena. In these models the salient features of sounding objects are represented by variables whose interpretation is straightforward because based on physical properties. As a result, the models can be easily embedded into artefacts and their variables coupled with sensors without the need of complex mapping strategies. Pop-up and lift-the-flap books for children were indicated as ideal applications for sounding objects (Rocchesso et al., 2003), as interaction with these books is direct, physical, and essentially continuous. Even though a few interactive plates were prototyped and demonstrated, in-depth exploitation of continuous interactive sounds in children books remains to be done.[9]

Everyday sounds can be very useful because of the familiar control metaphor: no explanation nor learning is necessary (Brewster, 2002). Moreover, it is clear that the continuous audio feedback affects the quality of the interaction and that the user makes continuous use of the information provided by sounds to adopt a more precise behavior. For example, the continuously varying sound of a car engine tells us when we have to shift gears. In this perspective sound is the key for paradigmatic shifts in consumer products. In the same way as spatial audio has become the characterizing ingredient

[8] `http://www.soundobject.org`

[9] The conception and realisation of an early prototype of a sound-augmented book were carried on by the second author as part of the Sounding Object project. Later on, students Damiano Battaglia (Univ. of Verona) and Josep Villadomat Arro (Univ. Pompeu Fabra, Barcelona) realised the sketches that are described in this paper as part of graduation projects, under the guidance of the authors.

for home theaters (as opposed to traditional TV-sets), continuous interactive sounds will become the skeleton of electronically-augmented children books of the future. The book-prototype is designed as a set of scenarios where narration develops through sonic narratives, and where exploration is stimulated through continuous interaction and auditory feedback. Through the development of the book, the class of models of sounding objects has been deeply used and verified. The physical models of impacts and friction have been used to synthesise a variety of sounds: the steps of a walking character, the noise of a fly, the engine of a motor bike, and the sound of an inflatable ball.

The integration and combination of the sound models available from the Sounding Object project in an engaging tale has been studied. The first step was to create demonstration examples of interaction using different kinds of sensors and algorithms. During this phase the most effective interactions (i.e. easier to learn and most natural) have been chosen, and several different scenarios were prepared with the goal of integrating them in a common story. The scenarios use embedded sensors, which are connected to a central control unit. Data is sent to the main computer using UDP messages through a local network from sensors and the sound part is synthesised using custom designed pd[10] patches. These pd patches implement a set of physical models of everyday sounds such as friction, impacts, bubbles, etc. and the data coming from sensors are used to control the sound object model in real time. In the following paragraph an investigation scenario will be described.

The *steps* scenario shows a rural landscape with a road; an embedded slider allows the user to move the main character along the road, and all movement data are sent to the computer, where the velocity of the character is calculated and a sound of footsteps is synthesised in real-time. The timing, distance, and force of the sound of each step is modified as a function of the control velocity. Fig. A.9 shows a preliminary sketch, while fig. A.10 shows the final prototype with the embodied sensor.

[10] http://www.pure-data.info

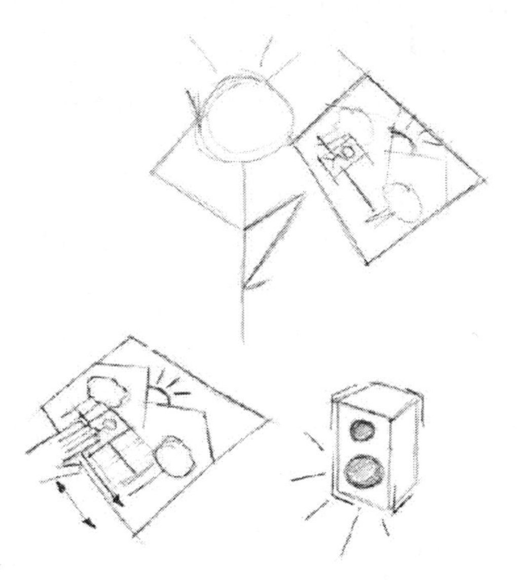

Figure A.9: The user is looking at the scene, identifies the moving part and tries to move the character generating sound

A.5.3 Results and expectations

Our investigation shows that in the near future lift-the-flap books for children will be augmented by sounds that respond continuously and consistently to control gestures. The sample scenario shown in the previous paragraph demonstrates the effectiveness of sound as an engaging form of feedback and the possibility of embedding real-time physics-based models of everyday sounds in small inexpensive stand-alone systems. A relevant part of future work will concentrate on real-world tests with children that will enhance the playability/usability of prototype books. Another aspect which will be further developed is the embedding and the sophistication of the technologies used.

Figure A.10: Interaction through slider: the footsteps scenario prototype

Bibliography

T. Blaine and T. Perkis. Jam-o-drum, a study in interaction design. In *Proc. Symposium on Designing Interactive Systems*, pages 165–173, NY, 2000. ACM Press.

R. Boulanger and M. Mathews. The 1997 Mathews radio-baton and improvisation modes. In *Proc. Int. Computer Music Conference*, Thessaloniki, Greece, 1997.

R. Bresin and S. Dahl. Experiments on gestures: walking, running, and hitting. In D. Rocchesso and F. Fontana, editors, *The Sounding Object*, pages 111–136. Mondo Estremo, Florence, Italy, 2003.

R. Bresin and A. Friberg. Emotional coloring of computer-controlled music performances. *Computer Music Journal*, 24(4):44–63, 2000.

S. Brewster. Non-speech auditory output. In Jacko J. and Sears A., editors, *The Human-Computer Interaction Handbook*, pages 220–239. Lawrence Erlbaum, 2002.

J. Chadabe. The voltage-controlled synthesizer. In John Appleton, editor, *The development and practice of electronic music*. Prentice-Hall, New Jersey, 1975.

T. Coduys, C. Henry, and A. Cont. Toaster and kroonde: High-resolution and high-speed real-time sensor interfaces. In *Proc. Conf. on New Interfaces for Musical Expression*, pages 205–206, Hamamatsu, Shizuoka, Japan, 2004.

S. Dahl. Playing the accent - comparing striking velocity and timing in an ostinato rhythm performed by four drummers. *Acta Acustica united with Acustica*, 90(4):762–776, 2004.

P. Dourish. *Where the Action Is: the foundations of embodied interaction*. MIT Press, Cambridge, MA, 2001.

J.J. Gibson. *The Ecological Approach to Visual Perception*. Lawrence Erlbaum Ass., Cambridge, MA, 1979.

B. Giordano and R. Bresin. Walking and playing: whatâ€™s the origin of emotional expressiveness in music? In *Proc. of Int. Conf. on Music Perception and Cognition*, page 149, Bologna, Italy, 2006.

K. Falkenberg Hansen. The basics of scratching. *Journal of New Music Research*, 31(4):357–365, 2002.

K. Falkenberg Hansen and R. Bresin. Analysis of a genuine scratch performance. In *Gesture-Based Communication in Human-Computer Interaction; Proc. Int. Gesture Workshop*, pages 519–528, Genova, Italy, 2004. Springer Verlag.

A. Hunt and T. Hermann. Special issue on Interactive Sonification. *IEEE Multimedia*, 12(2), 2005.

A. Hunt and R Kirk. Multiparametric control of real-time systems. In M. Battier, J. Rovan, and M. Wanderley, editors, *Trends in Gestural Control of Music*. FSU - Florida, April 2000.

E. L. Hutchins, J. D. Hollan, and D. A. Norman. Direct manipulation interfaces. In D. Norman and S. W Draper, editors, *User-Centred System Design*, pages 87–124. Lawrence Erlbaum Associates, Hillsdale, New Jersey, 1986.

G. Johansson. Visual perception of biological motion and a model for its analysis. *Perception and Psychophysics*, 14:201–211, 1973.

S. Jordà. FMOL: Toward user-friendly, sophisticated new musical instruments. *Computer Music Journal*, 26(3):23–39, 2002.

S. Jordà. *Digital Lutherie: Crafting musical computers for new musics performance and improvisation*. PhD thesis, Universitat Pompeu Fabra, Barcelona, 2005.

S. Jordà and O. Wüst. Sonigraphical instruments: From FMOL to the reacTable*. In *Proceedings of the 2003 International Conference on New Interfaces for Musical Expression (NIME-03)*, pages 70–76, Montreal, 2003.

S. Jordà, G. Geiger, M. Alonso, and M. Kaltenbrunner. The reacTable: Exploring the synergy between live music performance and tabletop tangible interfaces. In *Proceedings of the first international conference on "Tangible and Embedded Interaction" (TEI07)*, pages 139–146, Baton Rouge, Louisiana, 2007.

P.N. Juslin and J. Laukka. Communication of emotions in vocal expression and music performance: Different channels, same code? *Psychological Bulletin*, 129(5):770–814, 2003.

M. Kaltenbrunner, G. Geiger, and S. Jordà. Dynamic patches for live musical performance. In *Proc. International Conference on New Interfaces for Musical Expression (NIME-04)*, pages 19–22, 2004.

M. Karjalainen, V. Välimäki, and T. Tolonen. Plucked-string models: From the Karplus-Strong algorithm to digital waveguides and beyond. *Computer Music Journal*, 22(3):17–32, 1998.

M. Karjalainen, T. Mäki-Patola, A. Kanerva, and A. Huovilainen. Virtual air guitar. *J. Audio Eng. Soc.*, 54(10):964–980, October 2006.

C. L. Krumhansl. An exploratory study of musical emotions and psychophysiology. *Canadian Journal pf Experimental Psychology*, 51(4):336–352, 1997.

X. Li, R. J. Logan, and R. E. Pastore. Perception of acoustic source characteristics: Walking sounds. *The Journal of the Acoustical Society of America*, 90(6): 3036–3049, 1991.

B. Lindblom. Explaining phonetic variation: a sketch of the H&H theory. In Hardcastle and Marchal, editors, *Speech production and speech modeling*, pages 403–439. Kluwer, Dordrecht, 1990.

S. Martin. *Theremin, an electronic odissey*. MGM, 2001. Documentary on DVD.

J. Patten, B. Recht, and H. Ishii. Audiopad: A tag-based interface for musical performance. In *Proceedings of the 2003 International Conference on New Interfaces for Musical Expression (NIME-02)*, pages 11–16, Dublin, 2002.

M. Puckette. Pure data. In *Proc. Int. Computer Music Conference*, pages 269–272, Hong Kong, August 1996.

D. Rocchesso and R. Bresin. Emerging sounds for disappearing computers. In N. Streitz, A. Kameas, and I. Mavrommati, editors, *The Disappearing Computer*, pages 233–254. Springer, 2007.

D. Rocchesso, R. Bresin, and M. Fernström. Sounding objects. *IEEE Multimedia*, 10(2):42–52, April 2003.

J. A. Sarlo. Gripd: A graphical interface editing tool and run-time environment for pure data. In *Proc. Int. Computer Music Conference*, pages 305–307, Singapore, August 2003.

B. Schneiderman. *Leonardo's laptop: human needs and the new computing technologies*. MIT press, Cambridge, 2002.

B. Ullmer and H. Ishii. Emerging frameworks for tangible user interfaces. In John M. Carnoll, editor, *Human Computer Interaction in the New Millenium*, pages 579–601. Addison-Wesley, Reading, MA, 2001.